Green Materials and Advanced Manufacturing Technology

Green Engineering and Technology: Concepts and Applications

Series Editors

Brujo Kishore Mishra

GIET University, India and Raghvendra Kumar, LNCT College, India

The environment is an important issue these days for the whole world. Different strategies and technologies are used to save the environment. Technology is the application of knowledge to practical requirements. Green technologies encompass various aspects of technology that help us to reduce the human impact on the environment and create ways of sustainable development. This book series seeks to enlighten readers regarding green technology in different ways, aspects, and methods. This technology helps people to understand the use of different resources to fulfil needs and demands. Some points that will be discussed include the combination of involuntary approaches, government incentives, and a comprehensive regulatory framework meant to encourage the diffusion of green technology in the underdeveloped countries and developing states of small islands, which require unique support and measures to promote the green technologies.

Green Automation for Sustainable Environment
Edited by Sherin Zafar, Mohd Abdul Ahad, M. Afshar Alam, Kashish Ara Shakeel

AI in Manufacturing and Green Technology
Methods and Applications

Edited by Sambit Kumar Mishra, Zdzislaw Polkowski, Samarjeet Borah, and Ritesh Dash

Green Information and Communication Systems for a Sustainable Future
Edited by Rajshree Srivastava, Sandeep Kautish, and Rajeev Tiwari

Handbook of Research for Green Engineering in Smart Cities
Edited by Kanta Prasad Sharma, Abdel-Rahman Alzoubaidi, Shashank Awasthi, Ved Prakash Mishra

Green Internet of Things for Smart Cities
Concepts, Implications, and Challenges

Edited by Surjeet Dalal, Vivek Jaglan, and Dac-Nhuong Le

Green Materials and Advanced Manufacturing Technology
Concepts and Applications

Edited by Samson Jerold Samuel Chelladurai, Suresh Mayilswamy, Arun Seeralan Balakrishnan, S. Gnansekaran

For more information about this series, please visit: https://www.crcpress.com/Green-Engineering-and-Technology-Concepts-and-Applications/book-series/CRCGETCA

Green Materials and Advanced Manufacturing Technology

Concepts and Applications

Edited by
Samson Jerold Samuel Chelladurai,
Suresh Mayilswamy, Arun Seeralan Balakrishnan
and S. Gnanasekaran

CRC Press is an imprint of the
Taylor & Francis Group, an **informa** business

First edition published 2021
by CRC Press
6000 Broken Sound Parkway NW, Suite 300, Boca Raton, FL 33487-2742

and by CRC Press
2 Park Square, Milton Park, Abingdon, Oxon, OX14 4RN

Library of Congress Cataloging-in-Publication Data

Names: Chelladurai, Samson Jerold Samuel, editor.
Title: Green materials and advanced manufacturing technology : concepts and applications / edited by Samson Jerold Samuel Chelladurai, Suresh Mayilswamy, ArunSeeralan Balakrishnan, and S. Gnanasekaran.
Description: First edition. | Boca Raton, FL : CRC Press, 2021. | Series: Green engineering and technology: Concepts and applications | Includes bibliographical references and index.
Identifiers: LCCN 2020033580 (print) | LCCN 2020033581 (ebook) | ISBN 9780367521066 (hardback) | ISBN 9781003056546 (ebook)
Subjects: LCSH: Manufacturing processes. | Materials. | Green products.
Classification: LCC TS183 .G737 2021 (print) | LCC TS183 (ebook) | DDC 670--dc23
LC record available at https://lccn.loc.gov/2020033580
LC ebook record available at https://lccn.loc.gov/2020033581

ISBN: 978-0-367-52106-6 (hbk)
ISBN: 978-1-003-05654-6 (ebk)

Typeset in Times
by Deanta Global Publishing Services, Chennai, India

Contents

Preface

The main focus of this collection of scientific book chapters is on recent theoretical and practical advancements in green materials – their blends, composites related to mechanical properties, microstructural characterization, tribology and advanced manufacturing technologies used in aerospace and automobile industries. This volume provides important original and theoretical experimental results that use non-routine technologies often unfamiliar to readers. Chapters in this book present novel applications of more familiar experimental techniques and analyses of composite problems, such as welding techniques that indicate the need for new experimental approaches.

This book highlights the recent trends in the fields of green composites, metal matrix composites, ceramic matrix composites, machinability studies of metals and composites using surface grinding, drilling, electrical discharge machining, joining of metals using friction stir welding, shielded metal arc welding, linear friction welding, surface modification using laser cladding, types of dust collectors in waste management and recycling in industries. We anticipate that this book will be of significant interest to scientists working on basic issues surrounding composites and advanced manufacturing technology, as well as those who work in industry in applied problems such as processing. Because of the multidisciplinary nature of the chapters, this book will attract a broad audience, including chemists, materials scientists, physicists, manufacturing and chemical engineers and processing specialists who are involved and interested in the future frontiers of blends. This book gives insight into and a better understanding of the development of green materials and advanced manufacturing technology used in various manufacturing sectors.

The editors are grateful to their management, institutions and family members as well as for the effort of our contributors who have authored the book chapters. Additionally, without the tireless effort of the publication staff of CRC Press, this book would have not been possible.

Editors

Samson Jerold Samuel Chelladurai, PhD, is an Associate Professor in the Department of Mechanical Engineering at Sri Krishna College of Engineering and Technology (SKCET), Coimbatore, Tamil Nadu, India. He completed his bachelor's degree in mechanical engineering in 2008 and his master's degree in manufacturing engineering in 2010. He secured university first rank with a gold medal for his outstanding academic performance in his post-graduate studies. He completed his Doctor of Philosophy in the area of Composite Materials from Anna University, Chennai, India in 2018. Chelladurai specializes in manufacturing of metal matrix composites, characterization of composites, tribology and optimization techniques. He has published many research articles in SCI and Scopus indexed journals, five books, of which *Engineering Graphics* has reached a large audience and three patents. He has received grants worth of $89,000. Dr Chelladurai has delivered many guest lectures to faculty members and students and received several awards from SKCET and other organizations, including Grants Received Award, Best Research Paper Award, Best Faculty Award, Star Performer Award, Shri. P.K. Das Memorial Best Faculty Award in Mechanical Engineering, Best Academic Non-Circuit Faculty, and Young Researcher Award in Mechanical Engineering. He has served on editorial/advisory boards and as a reviewer for reputed journals.

Suresh Mayilswamy, PhD, is Associate Professor in the Department of Robotics and Automation Engineering, PSG (Poolaimedu Samanaidu Govindasamy) College of Technology, India. He received his bachelor's degree in mechatronics engineering and his master's in computer integrated manufacturing. He earned his PhD in mechanical engineering from Anna University, Chennai, India, in 2014. His PhD research focused on theoretical and experimental investigations of orientations of brake pad in vibratory part feeders. His current research focuses on advanced manufacturing technology, process automation and industrial robotics. To his credit, he has published nearly 73 technical papers in refereed and impact factored international and national journals as well as at international conferences and national conferences. He has authored three books in the field of industrial automation, assembly automation and contemporary research in engineering and technology. He is keenly interested in the development of an advanced curriculum for the future engineers of other institutions and has participated actively as a board of studies member to frame syllabi for mechatronics engineering. He has received awards, including outstanding faculty of the institution and best faculty of the year from SKCET and PSG Institute of Advanced Studies for his academic and research works in the year 2013 and 2018, respectively. He is a recognized reviewer of Science Direct, Springer, and ACTA Press among others.

Arun Seeralan Balakrishnan, PhD, is an academician in mechatronics engineering at the school of engineering at Asia Pacific University (APU) of Technology and Innovation, Kuala Lumpur, Malaysia. He received his diploma in electrical and electronics engineering in 1995 and continued his studies, graduating as a mechanical engineer in 2000. He moved up to industrial engineer, contributing his knowledge in the field of automation engineering to a specific industry Coimbatore, India. After 7 years of experience in the industry, he graduated with his Master of Engineering degree in 2010. He has acted as an academician for the past 9 years at APU, Malaysia. Since 2013, he has been a chartered Engineer of IMechE (UK) and holds a position as an Academic Liaison Officer to support upcoming young engineers to enhance their knowledge. He has published a book on engineering drawing using Auto-CAD. He earned his Doctor of Philosophy in Engineering degree from the University of Tenaga National in Kuala Lumpur, Malaysia. He has published many technical papers in reputed international journals that extend his contribution to the field of engineering. He is also a research member of the APcorE research centre in APU. Since 2018, he has been a member of the Board of Engineers, Malaysia and actively takes part in research activity.

S. Gnanasekaran, PhD, is an Assistant Professor in the Department of Mechanical Engineering at Sri Sakthi Institute of Engineering and Technology, Tamil Nadu, India. He completed his Bachelor of Engineering in Mechanical Engineering in 2008 and his master's degree in CAD/CAM in 2013. He completed his Doctor of Philosophy in Welding and Surface Engineering, and his research was funded by the University Grants Commission – Department of Atomic Energy. He has published many research articles in SCI and Scopus indexed journals and published five patents in IPR journal. He has published five book chapters in various publications. He organized a faculty development programme, which was funded by Anna University, Chennai, India. He received several awards from international conferences for his technical work and also received best faculty award for his immeasurable contribution toward student welfare. He has served on editorial/advisory boards and as a reviewer for reputed journals.

Contributors

Vivek Aggarwal
Department of Mechanical Engineering
I.K. Gujral Punjab Technical University
Kapurthala, Punjab, India

Sarfraj Ahmed
Boiler Maintenance Department
National Thermal Power Corporation Limited (NTPC) - Steel Authority of India Limited (SAIL) under Singh Engineering Works (SEW)
Bhilai, India

A. Umesh Bala
Department of Manufacturing Engineering
Annamalai University
Chidambaram, India

N. Balaji
Mechanical Engineering
Sri Krishna College of Engineering and Technology
Coimbatore, India

S. Balasubramani
Mechanical Engineering
Sri Eshwar College of Engineering
Coimbatore, India

V. Balasubramanian
Centre for Materials Joining and Research

and

Department of Manufacturing Engineering
Annamalai University
Chidambaram, Tamil Nadu, India

Amit Bansal
Department of Mechanical Engineering
I.K. Gujral Punjab Technical University
Kapurthala, Punjab, India

K. P. Boopathiraja
Mechanical Engineering
Bannari Amman Institute of Technology
Sathyamangalam, India

B. Deepanraj
Jyothi Engineering College
Cheruthuruthy, Thrissur, India

Vikas Dhawan
Shri Guru Govind Singh Tricentenary University
Gurugram, India

Amoljit Singh Gill
Department of Mechanical Engineering
I.K. Gujral Punjab Technical University
Kapurthala, Punjab, India

Piyush Gulati
Lovely Professional University
Punjab, India

B. Guruprasad
Alagappa Chettiar Government College of Engineering and Technology
Karaikudi, Tamilnadu, India

R. Jeyakumar
Mechanical Engineering
Sri Krishna College of Engineering and Technology
Coimbatore, India

R. Suresh Kumar
Mechanical Engineering
Sri Eshwar College of Engineering
Coimbatore, India

Rajeev Kumar
Lovely Professional University
Punjab, India

Arjun Kundu
Department of Metallurgical and Materials
National Institute of Technology
Raipur, India

M. Madhan
Mechanical Engineering
Velammal Engineering College
Chennai, India

S. Magibalan
Mechanical Engineering
K.S.R College of Engineering
Tiruchengode, India

M. Mahalingam
Government College of Engineering
Bodinayakkanur, Theni District
Tamil Nadu, India

D. Mala
University College of Engineering
Panruti, Tamil Nadu, India

D. Manikandan
Centre for Materials Joining and Research

and

Department of Manufacturing Engineering
Annamalai University
Chidambaram, Tamil Nadu, India

Suresh Mayilswamy
Department of Robotics and Automation Engineering
PSG College of Technology
Coimbatore, India

Munish Mehta
School of Mechanical Engineering
Lovely Professional University
Phagwara, Punjab, India

C. Mukundhan
Centre for Materials Joining and Research

and

Department of Manufacturing Engineering
Annamalai University
Chidambaram, Tamil Nadu, India

K. Murugan
Mechanical Engineering
Government Polytechnic College Valangaiman
Thiruvarur, Tamilnadu, India

S. Navaneethakrishnan
Department of Mechanical Engineering
Erode Sengunthar Engineering College
Perundurai, Erode, India

A. Palanisamy
Department of Mechanical Engineering
Surya Engineering College
Erode, Tamilnadu, India

Vijay Petley
Materials Group
Gas Turbine Research Establishment (GTRE)
Bengaluru, Karnataka, India

G. Prabhakaran
Mechanical Engineering
KCG College of Technology
Chennai, India

C. Rajarajan
Centre for Materials Joining and Research

and

Department of Manufacturing Engineering
Annamalai University
Chidambaram, Tamil Nadu, India

C. Rajendran
Sri Krishna College of Engineering and Technology
Coimbatore, India

T. Ramakrishnan
Mechanical Engineering
Sri Eshwar College of Engineering
Coimbatore, India

R. Ramamoorthi
Mechanical Engineering
Sri Krishna College of Engineering and Technology
Coimbatore, India

M. Sakthivel
Mechanical Engineering
Anna University Regional Campus
Coimbatore, India

Samson Jerold Samuel Chelladurai
Mechanical Engineering
Sri Krishna College of Engineering and Technology
Coimbatore, India

R. Sathiyamoorthy
Mechanical Engineering
Institute of Road Transport
Bargur, India

N. Senthilkumar
Adhiparasakthi Engineering College
Melmaruvathur, Tamil Nadu, India

K. Shanmugam
Department of Manufacturing
Annamalai University
Chidambaram, India

Shubham Sharma
Department of Mechanical Engineering
I.K. Gujral Punjab Technical University
Kapurthala, Punjab, India

Gursharan Singh
Department of Mechanical Engineering
I.K. Gujral Punjab Technical University
Kapurthala, Punjab, India

Jai Inder Preet Singh
Lovely Professional University
Punjab, India

Jujhar Singh
Department of Mechanical Engineering
I.K. Gujral Punjab Technical University
Kapurthala, Punjab, India

Manpreet Singh
Lovely Professional University
Punjab, India

Sehijpal Singh
Guru Nanak Dev Engineering College
Ludhiana, Punjab, India

V. Sivabharathi
Department of Mechanical Engineering
St. John College of Engineering and Management
Palghar, Maharashtra, India

S. Sivamani
Engineering Department
Salalah College of Technology
Salalah, Oman

S. P. Sivapirakasam
Mechanical Engineering
National Institute of Technology
Tiruchirappalli, India

P. Sivaraj
Centre for Materials Joining and Research

and

Department of Manufacturing Engineering
Annamalai University
Chidambaram, Tamil Nadu, India

S. Sivasankaran
Department of Mechanical Engineering

and

College of Engineering
Qassim University
Buraidah, Saudi Arabia

R. Sureshkumar
Mechanical Engineering
Sri Eshwar College of Engineering
Coimbatore, Tamilnadu

T. Tamizharasan
Green Pearl India (Pvt.) Ltd
Kattangulathur, Tamil Nadu, India

R. Varahamoorthi
Department of Manufacturing Engineering
Annamalai University
Chidambaram, India

P. Ashoka Varthanan
Mechanical Engineering
Sri Krishna College of Engineering and Technology
Coimbatore, India

S. Venkatesh
Mechanical Engineering
National Institute of Technology
Tiruchirappalli, India

Shweta Verma
Materials Group
Gas Turbine Research Establishment (GTRE)
Bengaluru, Karnataka, India

M. Vijayanand
Engineering Department
Salalah College of Technology
Salalah, Oman

K. P. Yuvaraj
Mechanical Engineering
Sri Krishna College of Engineering and Technology
Coimbatore, India

1 Parametric Optimization of Surface Roughness and Surface Temperature during Minimum Quantity Lubrication (MQL) and Conventional Flood Lubrication Techniques in Surface Grinding of Mild Steel

A Performance Comparison and Analysis

Gursharan Singh, Jujhar Singh, Shubham Sharma, Amoljit Singh Gill, Munish Mehta and Suresh Mayilswamy

CONTENTS

1.1 INTRODUCTION

Because of economic and ecological stresses, industry is looking for methods of reducing lubricant use during metal cutting operations (Liew and Hsien 2015). Environmental preservation and the protection of human health exist at the forefront of industry operations on a global scale (Goldberg 2012). The overall objective of international standards is to protect the environment in balance with socioeconomic needs (Inasaki 2012). Many organizations that signed up for these requirements would need to measure and demonstrate the reduction of the following five elements in consumption/damage: toxic emissions to air; the release of hazardous effluents into water; focussing on waste management; land pollution and the use of natural resources and raw materials (Astakhov 2010). Any machining process aims at reducing the machining costs by increasing performance and productivity (Morgan et al. 2012). The cutting efficiency is influenced by several parameters and their interaction with one another; for example, the interaction between cutting fluids (e.g., power, type of tool holder, tool clamping), workpiece (e.g., material, type, application method, application quantity, flow rate, pressure); cutting tool (e.g., tool material, hardness, tool coating, tool size, tool length, number of edges, angles, nose radius); machine tool (e.g., rigidity, power, workpiece: material type of tool holder, tool clamping); hardness, size and machining parameters (e.g., speed, feed, depth of cut, type of operation) (Kuram et al. 2013). Many parameters influence the cutting performance and their interaction with each other, for example, the interactions between cutting fluids (e.g., form, application process, quantity of application, flow rate, pressure), cutting tool (e.g., tool material, hardness, tool coating, tool size, tool length, number of edges, angles, radius of nose) and machine tool (rigidity, strength, form of tool holder, tool clamping). Cutting fluid also accounts for around 20% of the total manufacturing cost, so optimum use of machining oil becomes a top priority to maximize the profit in any manufacturing industry (Roy et al. 2018). Cutting fluid can be divided into four categories: machining oil, soluble oil (emulsion added in base oil), synthetic fluid and semisynthetic oil. Machining oils are derived from vegetables and petroleum. It is used in low machining parameters (low speed, feed and depth of cut). Soluble oils are normally droplets of oil in water. Synthetic and semisynthetic oils are mainly derived from chemicals. They are used for machining operations with high machining speed and feed rate (Rao and Srikant 2006). The major drawback of machining fluids is their negative effects on humans and the environment. The disposal of machining fluid is also a major challenge, as most of the machining fluid causes harm to the environment (Astakhov 2010). Modern industries are exploring new methods of reducing lubricant/coolant consumption during machining operations. Apart from increasing the production cost, excessive use of coolant also causes problems related to workers' health and environmental contamination (Weinert et al. 2005). To keep tool wear under control, proper heat removal and lubrication are very important. The heat-removal capacity of flooded cooling was found to be very effective at lower speeds, but its thermal conductivity worsens with an increase in machining speed. Thus, minimum quantity lubrication (MQL) came into existence and became the best alternative for flooded cooling (Barczak et al. 2010). In the MQL process, a much smaller amount of coolant/lubricant is mixed with air to form an aerosol, which is sprayed with a nozzle at high pressure in the cutting zone (Varadharajan et al. 2002). The system is composed of a discharge nozzle, atomizer and cutting fluid sump, etc. The atomizer serves as an ejector in which the coolant is atomized by high pressure air. The atomized coolant is then supplied to the machining site by air in a low-pressure distribution system. Because of the Venturi effect in the mixing chamber, where it is held at a steady hydraulic rate, partial vacuum draws the cutting fluid from the oil sump. The air passing through the mixing chamber atomizes the coolant flux into the particulate aerosol of microns. Sprinkled in the cutting zone as mist, this aerosol serves as both a coolant and as a lubricant, penetrating deep into the workpiece of the tool's interface. Surface grinding is a precise machining process; it is characteristic for producing high surface quality and components with the least tolerance (Dogra et al. 2018). In the grinding process, as material is removed, the abrasives plunge and slide against the workpiece, causing an increase in the temperature of the cutting zone. If a coolant is not used, heat accumulation in the grinding zone can cause serious defects in the finished parts, such as metallurgical changes, dimensional wear rate and residual stress (Kaplonek et al. 2019). Thus, lubrication becomes an important necessity for the grinding fluids, along with chip removal and grinding zone cooling to improve process efficiency. Such fluids tend to affect the surroundings and health of the user. The cost goes beyond viability, making it a negative factor as well. MQL can thus act as a good alternative to conventional approaches for these processes (Hadad and Hadi 2013).

The efficiency of surface grinding depends on maintaining efficient wheel speed, cut depth, movement of the workpiece and an effective supply of coolant during the process. Thus, the selection of the correct grinding parameters and the cutting fluids parameters improve the surface quality and reduce the cutting zone temperature as well as the cutting force (Balan et al. 2013). There have been many studies on the machining characteristics using conventional fluid, dry and MQL techniques. Tawakoli et al. (2010) found the effect of various cooling fluids on surface quality in MQL grinding. Hadad and Hadi (2013) compared the grindability of two types of steel with MQL methodology; the surface roughness was found to be slightly lower than that of fluid cooling in the case of hardened stainless steel but higher in the case of aluminium alloy. In MQL, some lubrication level is obtained to fulfil the primary functions of the grinding fluid. Compared to fluid grinding, traditional MQL grinding with oil does not offer a better cooling efficiency. Tangjitsitcharoen (2010) proposed that machining under an MQL environment resulted in better surface quality, lower machining temperature, longer tool life and lower tool wear as compared to an environment of wet and dry machining. Alves et al. (2011) established fluid methodology of MQL behaviour to establish fluid methodology through the assembly of special nozzles. MQL steel grinding was performed in a compressed air flow using pulverized vegetable oil. They measured the quality of the surface and ground diametric wear on the wheel. The promising results were achieved with MQL for tool wear and surface integrity. They got the best MQL results in air time: 26.4 m/s and lubricant: 40 ml/h. Lee et al. (2012) investigated the process of microgrinding mixed nanoparticles with lubricating oils and found that nanofluid MQL is efficient in reducing grinding force and improving surface quality. Vazquez et al. (2015) and Sarıkaya et al. (2016) used MQL technology together with compressed air to atomize and spray the cutting fluid into the cutting area. They found that this not only significantly reduced the use of the cutting fluid but also allowed the cutting fluid to enter in the area of cutting. Wang et al. (2016) investigated the lubrication properties of the nickel-based alloy GH4169 with seven standard vegetable and paraffin oils during flood grinding and MQL grinding at the wheel/workpiece interface. The vegetable oils had better lubrication properties than both mineral oil and flood cutting fluid because of the high binding strength and low friction coefficient.

Yang et al. (2017) explored the critical maximum undeformed equivalent chip thickness by different lubrication conditions for the brittle-ductile transition in zirconia ceramic grinding. The accuracy of the predictive grinding force model and the predictive surface topography model for zirconia grinding under various lubrication conditions is substantially improved. Literature reports suggest that because of enhanced lubrication and decreased friction of the grain-workpiece interface, MQL improves surface strength, tool wear and cutting forces (Lukasz and Andre 2012; Mao et al. 2013; Chetan et al. 2016). Roy and Ghosh (2014) observed that, in the case of MQL, the tool-tip temperature was 10–30% lower compared with flooded cooling. Better surface quality and superior capacity to dissipate heat were also achieved during machining under MQL environment. Li et al. (2013) suggested that the machining performance of the MQL environment, suspended by nanoparticles, would be better compared with dry grinding, flooded cooling grinding and traditional MQL grinding. An enormous reduction in surface roughness and temperature was achieved by the inclusion of nanoparticles in the MQL environment. Sharma et al. (2016) stated that the use of MQL with different oils and nanofluids produce better surface quality when compared with dry matching. Chakule et al. (2017) studied the effect of parameters of machining on the phase of grinding under MQL conditions. It was observed that the surface roughness of the work specimen using soluble oil under MQL was much lower than those using wet and dry grinding areas. This was accomplished because of the enhanced level of cooling and lubrication under MQL. Guo et al. (2017) mixed castor oil with other vegetable oils and measured their lubrication efficiency using MQL grinding method on the nickel-based alloy GH4169. It was concluded that the mixed oils' comprehensive lubrication efficiency was higher than castor oil's, and soybean/castor oil obtained the highest efficiency. Mia et al. (2018) reviewed the cleaner manufacturing process for machining of hardened steel from the environmental friendliness viewpoint.

To build the model for sustainability evaluation, the environmental approach to the Pugh matrix was implemented, and the research results showed that MQL is an environmentally friendly, low-consumption and efficient method of lubrication and cooling. Roya et al. (2019) concluded that the use of MQL environment during machining operation allowed considerable reduction in machining temperature and favoured removal of chips between the interface of the tool samples. This helps to reduce the frictional force that allows the finished product to have superior surface quality. Roya et al. (2019) observed that the MQL environment has excellent heat-dissipation capability of the machining fluid, which helps to reduce machining temperature and decrease wear on the tool surface for building up edge formation and adhesion. Thus, the MQL environment produces superior surface quality, lower machining temperature and lower tool wear with the least coolant/lubricant consumption. Seyedzavvara et al. (2019) investigated the cooling and lubrication properties of graphite nanofluids during MQL grinding. It was concluded that, at extreme cutting conditions, the grinding temperature and grinding force are lower than other lubrication methods. This is due to the existence of graphite nanoparticles at the wheel-workpiece interface that form tribofilm on the outer surface of the workpiece. The conclusions of studies carried out by researchers on grinding using MQL indicate that, owing to better lubrication and reduced friction of the grain-workpiece interface, surface quality, tool wear and cutting forces are improved with MQL process. However, the relationship between the surface quality of dry, flooded and MQL conditions is still not clear.

In this paper, a number of experiments are carried out under dry, flooded and MQL conditions to investigate the grinding performance such as surface roughness. A proper selection determines the range of the cutting parameters which are helpful for economical machining as compared with flooding or dry cutting techniques. The aim of the present paper is to provide a brief analysis of the various effects of MQL on cooling capability and surface roughness during grinding operations.

1.2 MATERIALS AND EXPERIMENTAL SETUP

The experiment was performed on the mild steel (HR1) with the main motive to optimize the best possible cooling technique and thus improve the surface property, such as surface roughness, and reduce the residual stresses developed during the surface grinding operation. First, with the help of a power hexa tool, the workpiece was cut into a 200× 100 × 25 mm (lbt) size from a mild steel plate. Partial roughening of the surface was done in the shaper machine in which the thickness of the workpiece was reduced to 1 mm. When the surface roughening process was completed, four slots opposite to the roughing surface were cut at equal distance with the help of the shaper machine. The main propose of the slots is to protect the thermocouple wire during the temperature measurement of the interface of the grinding wheel and workpiece. After completion of the slotting process, four holes, 2 mm in diameter, were drilled in each slot at equal distance. In these blind holes, thermocouples were affixed to measure the temperature produced during the grinding operation. The MQL system basically used cutting fluid as the carrier for the lubrication. In the MQL system, a mixture of cutting fluid and air was sprayed on the grinding zone with the help of nozzles, into which the cutting fluid enters from one inlet while the lubricant enters from another, mixed in the mixing chamber and thus sprayed at the interface of grinding wheel and workpiece during the operation. The schematic view of the surface grinding is shown in Figure 1.1.

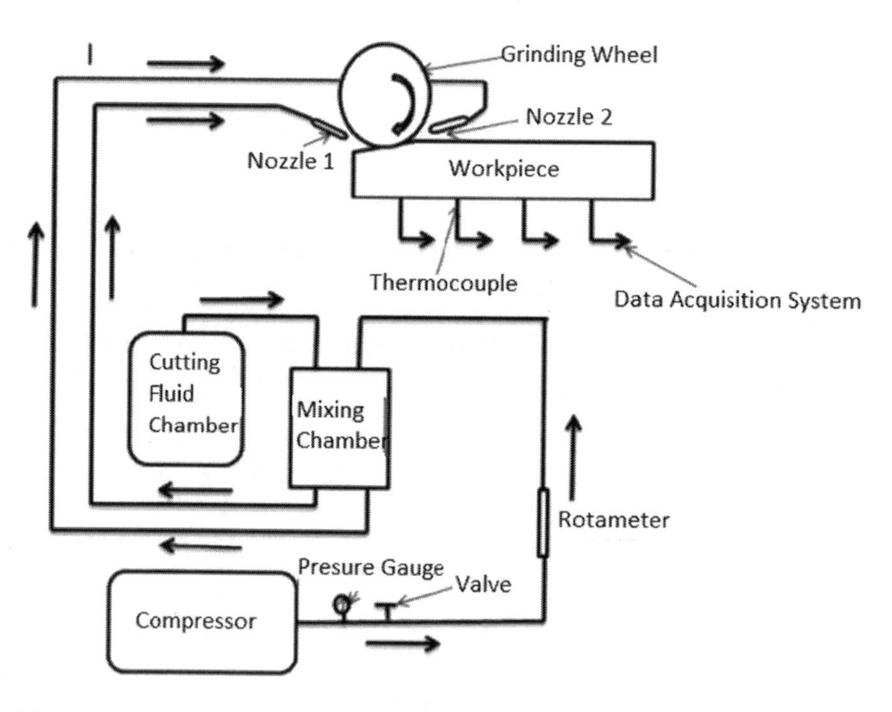

FIGURE 1.1 Schematic diagram of surface grinding.

Grinding a mild steel workpiece with aluminium oxide abrasives could be better for surface integrity and grinding efficiency. The detailed parameters of grinding wheels are listed in Table 1.1. Before every test, the grinding wheel was dressed with the help of 0.5 carat diamond dresser to obtain consistency in all grinding. The constant working pressure of the air was 5.5 bar and maintained through the pressure regulator. A "Jaspin" acrylic rotameter, with a range of 0–150 litres per minute, was used to count the flow rate of compressed air coming from the compressor. The cutting fluid, at a constant flow rate of 65 ml/hr, was continuously fed to nozzles under the action of gravitational force. The lubricant and water ratio used to make the cutting fluid was 6% by volume. For flood and MQL grinding operation, Procut soluble cutting oil at 6% by volume was used. The flood cutting fluid rate was 1100 L/hr, and MQL cutting fluid rate was 65 ml/hr. The machined surface quality and the grinding energy consumption are directly determined by the grinding temperature, so it is a very critical measurement parameter during the grinding process. The grinding parameters during this experimentation mention are shown in Table 1.1.

During the cutting field, the grinding heat is produced mainly by elastic deformation, plastic deformation, chip formation and friction actions. Higher grinding temperature may easily cause thermal damage to the surface and can deteriorate the integrity of the surface. The grinding temperature was measured via a multichannel temperature measurement device in the seven working conditions. The K-type 2 mm diameter thermocouple was placed in the 2 mm diameter thermocouple mounting slot to test the surface temperature of the grinding. The quality of the surface has a strong effect on workpiece durability and serviceability. Hence, the quality of the surface is a crucial parameter for determining the grinding path. In the work, the experiments were carried out under the seven working conditions to evaluate the impact of the

TABLE 1.1
Grinding Parameters

Equipment Parameter	Parameter Value
Grinding wheel	Al2O3
Wheel diameter	Φ200 × 13 mm
Work-table speed	0.52 m/min
Work material	Mild steel HR1
Grinding depth	15 μm
Wheel speed	2800 rpm
Grinding method	Up and down grinding
Environment	Dry, Flood, MQL
MQL coolant flow rate	65 ml/hr
Air pressure	5.5 bar
MQL coolant	Procut soluble cutting oil content 6%
Main motor	2 HP, 3 Phase
Wheel balancing stand recommended	One
Flood flow rate	1100 l/hr

dry, flooded and MQL conditions on surface integrity regarding surface quality and surface roughness. The Surftest SJ-310, a portable surface-roughness measurement tool was used to analyse the machined surface of the workpiece and measure the roughness of the surface. A total of eight experiments were performed, as shown in Table 1.2, in which two experiments are on the dry grinding operation, two experiments are on the flood grinding operation, two experiments are on the MQL grinding operation in which both of the nozzles are on the same side of the grinding wheel [MQL(BNSS)], one experiment is on the MQL grinding operation in which both of the nozzles are on opposite sides of the grinding wheel [MQL(BNOS)], and one experiment for MQL in which only one nozzle is used to supply the mixture of air and cutting fluid [MQL(SN)] during the operation. An aluminium oxide grinding wheel of a 200 mm diameter and 13 mm width are used in the grinding operation. During the experimentation, speed of grinding wheel and depth of cut were kept constant as 2800 rpm and 15 μm respectively. Also, the cutting fluid flow rate and air pressure are kept constant throughout the experiment. There are three cases in the grinding operation to measure the temperature and surface roughness.

1.2.1 Dry Grinding Operation

In dry grinding operation, the grinding operation is performed without cutting fluid. Dry grinding operation is performed only when surface finish is not so important. Sharma et al. (2015) found that, during dry grinding operation, the surface temperature of the workpiece is at its maximum, which causes poor surface finish, and some residual stress is also developed in the grinding surface, which causes dimension inaccuracy in the grinding surface.

1.2.2 Flood Grinding Operation

In flood grinding operation, only cutting fluids are supplied during the operation. The temperature produced during the operation is measured by the thermocouples

TABLE 1.2
Design of Experiments under Grinding Operation

Experiment No.	Wheel Speed (rpm)	Depth of Cut (μm)	Work-Table Speed (m/min.)	Cutting Condition
1	2800	15	0.52	Dry
2	2800	15	0.52	Dry
3	2800	15	0.52	Flood
4	2800	15	0.52	Flood
5	2800	15	0.52	MQL(SN)
6	2800	15	0.52	MQL(BNSS)
7	2800	15	0.52	MQL(BNOS)

FIGURE 1.2 MQL grinding with both nozzles on the same side.

placed in the blind holes, and the surface finish of the workpiece is observed by the surface-roughness tester.

1.2.3 MQL Grinding Operation

In this MQL process, the mixture of air and lubricant is supplied at the interface of the workpiece and grinding wheel by the help of a nozzle during the grinding operation. There are two cases of MQL surface grinding operation, as mentioned in Table 1.2.

1.2.3.1 MQL Grinding with Both Nozzles Same Side [MQL(BNSS)]

In this condition, two nozzles are used during the surface grinding operation to supply the mixture of compressed air and cutting fluid. Both of the nozzles carry the mixture of coolant and air from the same side of the grinding. The grinding wheel is rotated in a clockwise direction, as shown in Figure 1.2.

1.2.3.2 MQL Grinding with Both Nozzles on Opposite Sides [MQL(BNOS)]

In this condition, one nozzle is positioned in the direction of grinding-wheel rotation and the second nozzle is positioned in the opposite direction of grinding-wheel rotation. MQL supply rate and air pressure in both of the nozzles are constant throughout the process. The positioning of the nozzles is shown in Figure 1.3.

All the experimental results were plotted graphically and comparatively studied to recognize the best technique and also to determine the effect of various grinding operations i.e. dry, flood and MQL grinding operations on surface temperature and surface roughness.

1.3 RESULTS AND DISCUSSION

1.3.1 Surface Temperature in Dry Grinding Operation

Figure 1.4(a) shows the surface temperature versus time (seconds) for the first experiment run, The surface temperature increases with respect to time and the maximum

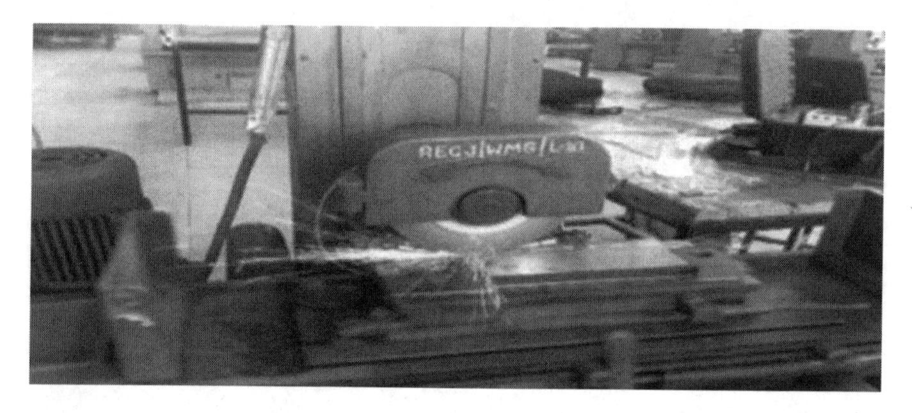

FIGURE 1.3 MQL grinding operation with both nozzles on opposite sides.

temperature of 125°C is obtained after 260 seconds. As there is no lubrication or coolant, surface temperature increases continuously. Figure 1.4(b) shows the surface temperature versus time (seconds) for the second experiment run with identical parameters to the first one. The maximum temperature of 135°C is obtained after 290 seconds. Both the first and second runs of the experiment give similar surface temperature during dry grinding.

1.3.1.1 Dry Grinding Operations with Maximum Temperature per Pass

Figure 1.5 shows the first and second run experiments during dry grinding with maximum temperature per pass. During the grinding operation, the time it takes for the grinding wheel to make a complete pass in the forward direction (up grinding) is 3 seconds; a complete pass in the backward direction (down grinding) is also 3 seconds.

In both up grinding and down grinding, the grinding operation is performed, and maximum temperature in each up- and down-grinding operation is taken into consideration by comparing the temperature of all the sensors. The maximum temperature per pass during the dry grinding experiments is almost the same because all the grinding parameters are the same for both of the experiments.

1.3.2 Surface Temperature during Flood Grinding Operation

The third and fourth experiments of the series were run during flood grinding operation in which the cutting fluid is supplied at the interface of the grinding wheel and the workpiece. Figure 1.6(a) shows the variation of surface temperature with time during flood grinding. The maximum surface temperature of 48°C is obtained during flood grinding. The variation of surface temperature with time during flood grinding for the fourth experiment run is shown in Figure 1.6(b).

1.3.2.1 Flood Grinding Operations with Maximum Temperature per Pass

Figure 1.7 shows the third and fourth experiments of the series, where were run during flood grinding with maximum temperature per pass. In this flood grinding

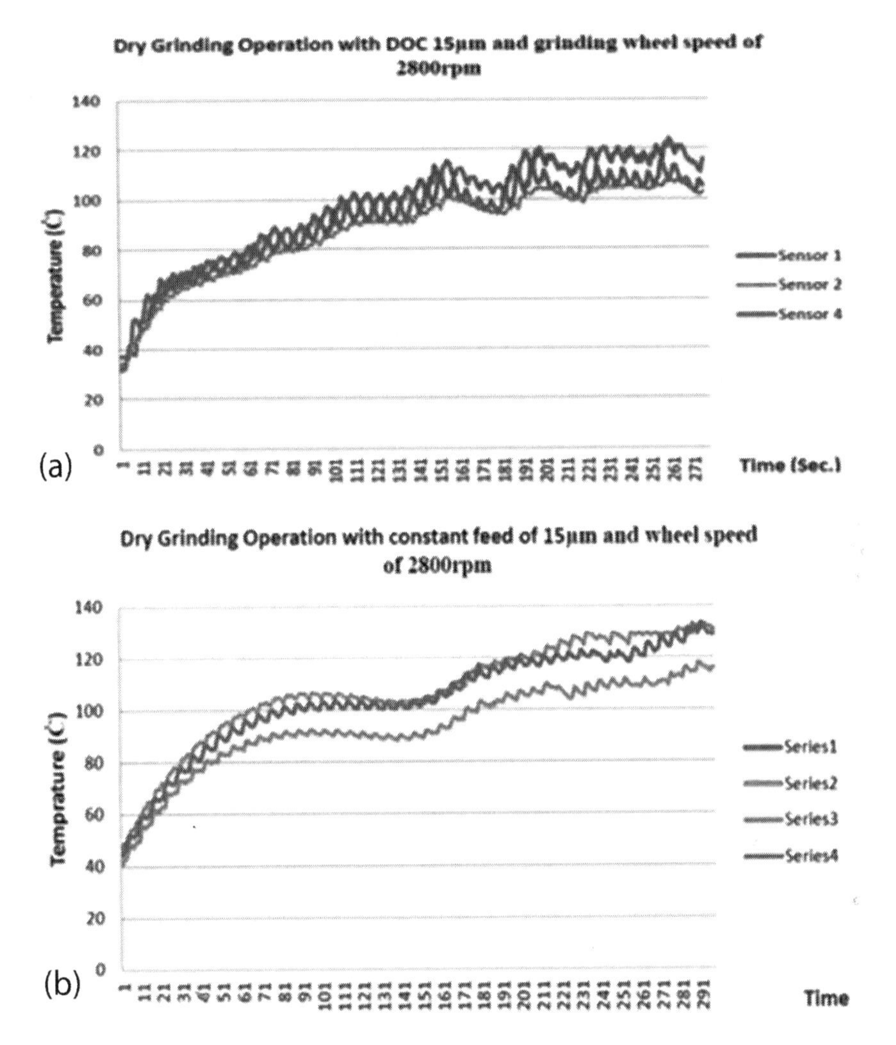

FIGURE 1.4 (a) Dry grinding operation with DOC 15 μm and grinding-wheel speed of 2800 rpm for the first experiment run. (b) Dry grinding operation with DOC 15 μm and grinding-wheel speed of 2800 rpm for the second experiment run.

operation, there are up flood grinding operation and down flood grinding operation. Every up and down flood grinding operation take 3 seconds, and the maximum temperature during each pass of flood grinding operation is taken. The maximum temperature per pass is almost the same because all the grinding conditions are the same, and the MQL flow rate is also the same during the operation.

1.3.3 Surface Temperature in MQL Grinding Operation

In the MQL grinding operation, a mixture of cutting fluid and air was sprayed at the interface of the grinding wheel through a nozzle. In the fifth experiment of the series,

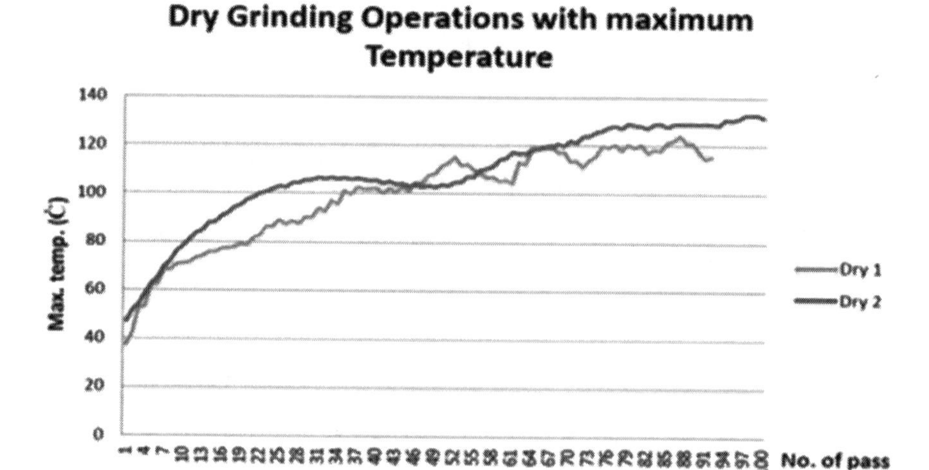

FIGURE 1.5 Dry grinding experiment 1 and 2 with maximum temperature per pass.

a single nozzle (SN) is placed toward the grinding contact zone. The depth of cut was 15 μm. Figure 1.8 shows the variation of surface temperature with time during MQL grinding with a single nozzle.

During the sixth experiment of the series, both of the nozzles are placed on the same side (BNSS) toward the grinding contact zone of the grinding wheel. Figure 1.9 shows the variation of surface temperature with time during MQL grinding when both of the nozzles are on the same side (BNSS).

During the seventh experiment of the series, both of the nozzles are placed on opposite sides (BNOS), one toward the grinding contact zone and another toward the grinding wheel directly. Figure 1.10 shows the variation of surface temperature with time during MQL grinding when both of the nozzles are on opposite sides (BNOS).

1.3.4 Temperature Comparison in Various Grinding Operations

In dry conditions, the temperature of the grinding surface has the highest values, as shown in Figure 1.11. This is due to the lack of cooling and lubrication. Such high temperatures may result in severe damage to the ground surface integrity, including burning of material and plastic flow, causing cracks and pitting marks on the workpiece surface (Tao et al. 2017). In contrast, grinding with flood cooling generates the lowest temperatures as compared with other cooling conditions because of the enhanced heat-transfer capability of this cooling method. High flow rate of fluid in flood cooling facilitates the dissipation of heat from the grinding zone and reduces energy partition of the workpiece during grinding.

In the flood grinding operation, the surface temperature of the workpiece is minimal, but the cost of flood grinding operation is high because of the large amount of cutting fluid needed, and it is difficult and costly to recycle and dispose of the cutting fluid. Thus, we can say the temperature of the surface of the workpiece during the MQL and flood grinding operations is almost the same, and far less cutting fluid is

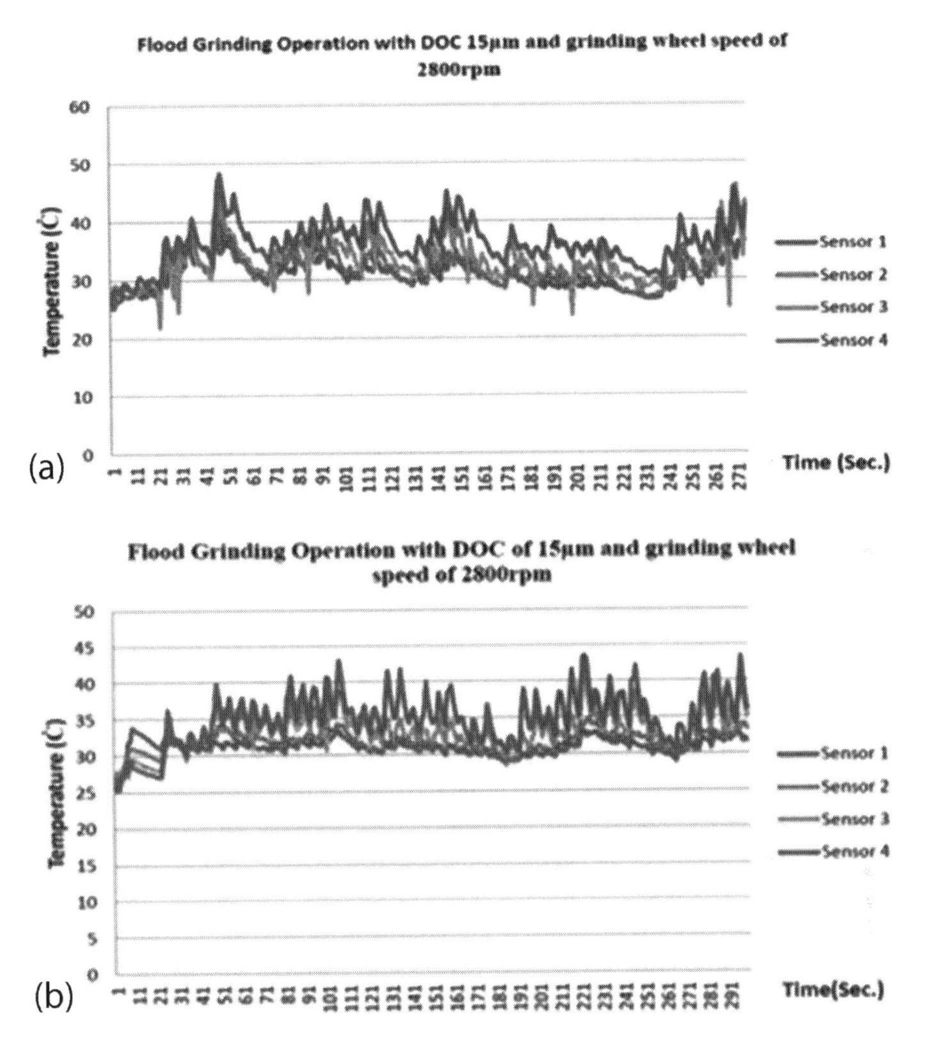

FIGURE 1.6 (a) Flood grinding operation with DOC 15 µm and grinding-wheel speed of 2800 rpm in the third experiment. (b) Flood grinding operation with DOC of 15 µm and grinding-wheel speed of 2800 rpm in the fourth experiment.

consumed during MQL than flood grinding operation. Thus, we can say that MQL grinding operation is better than dry and flood grinding operation to reduce the temperature of the grinding surface during the operation.

1.3.5 Temperature Comparison in Various MQL Grinding Operations

In all MQL grinding experiments, the grinding parameters, such as grinding-wheel speed, cut depth and workpiece size, and MQL parameters, such as the MQL flow rate and air pressure, are constant. The grinding surface temperature is at minimal as shown in Figure 1.12. When a single nozzle is positioned in the direction of the

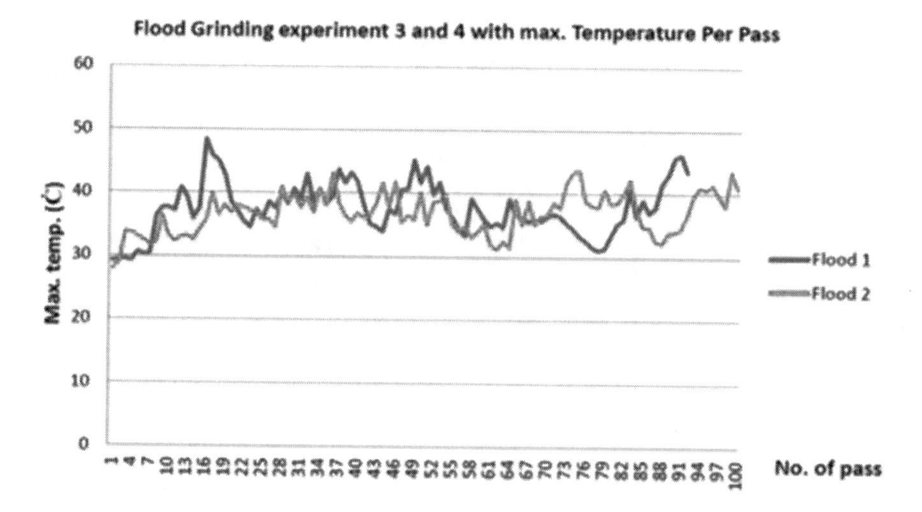

FIGURE 1.7 Flood grinding experiment 3 and 4 with maximum temperature per pass.

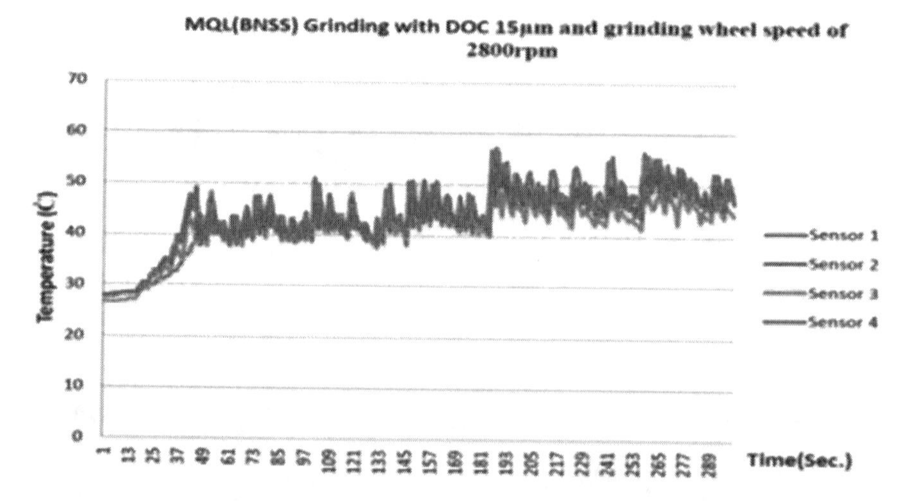

FIGURE 1.8 MQL(SN) grinding operation with DOC 15 μm and grinding-wheel speed of 2800 rpm.

grinding contact zone and the surface temperature is optimum when both nozzles are mounted in the opposite direction (BNOS) to the grinding contact zone then the surface temperature is optimum.

The surface temperature of grinding surface reduces either when both the nozzles are on the same side or when a single nozzle is working with MQL; however, placing both of the nozzles on opposite sides (BNOS), one toward the grinding contact zone and another toward the grinding wheel, both with the same MQL flow rate and air pressure, is not as effective at reducing the surface temperature of the grinding surface.

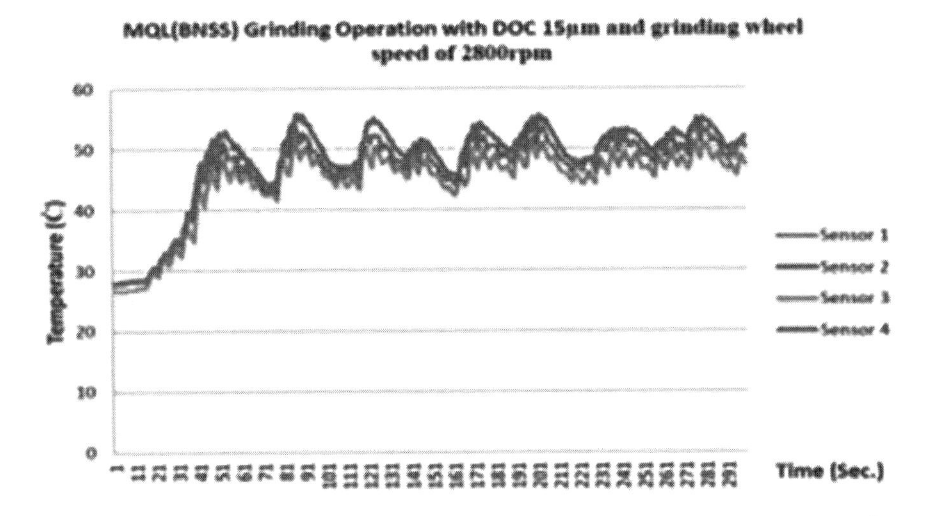

FIGURE 1.9 MQL(BNSS) grinding with DOC 15 μm and grinding-wheel speed of 2800 rpm.

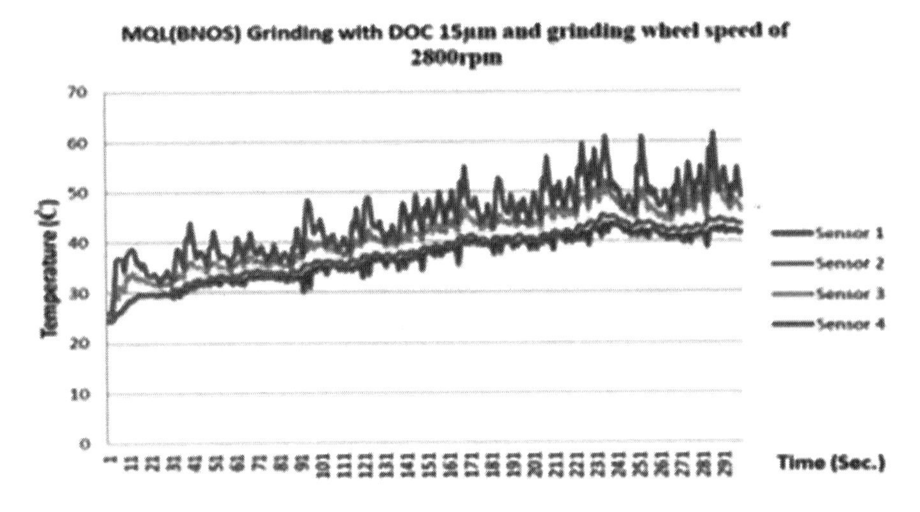

FIGURE 1.10 MQL(BNOS) grinding with DOC 15 μm and grinding-wheel speed of 2800 rpm.

1.3.6 Measurement of Surface Roughness

1.3.6.1 Surface Roughness in Dry Grinding Operation

A Surftest SJ-310 portable surface-roughness measurement tool was used to observe the machined workpiece surface roughness. During the whole process, the grinding-wheel speed of 2800 rpm, depth of cut 15 μm and workpiece movement 0.52 m/min

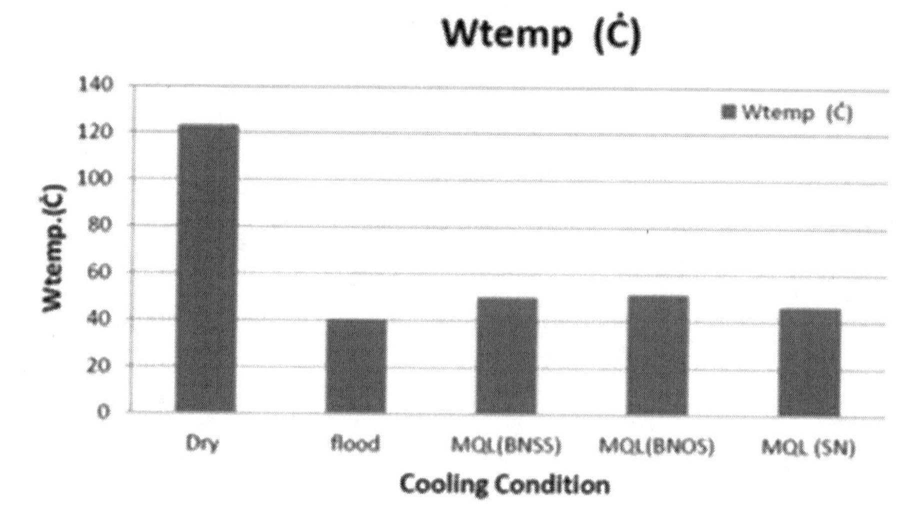

FIGURE 1.11 Temperature comparison in various grinding operations.

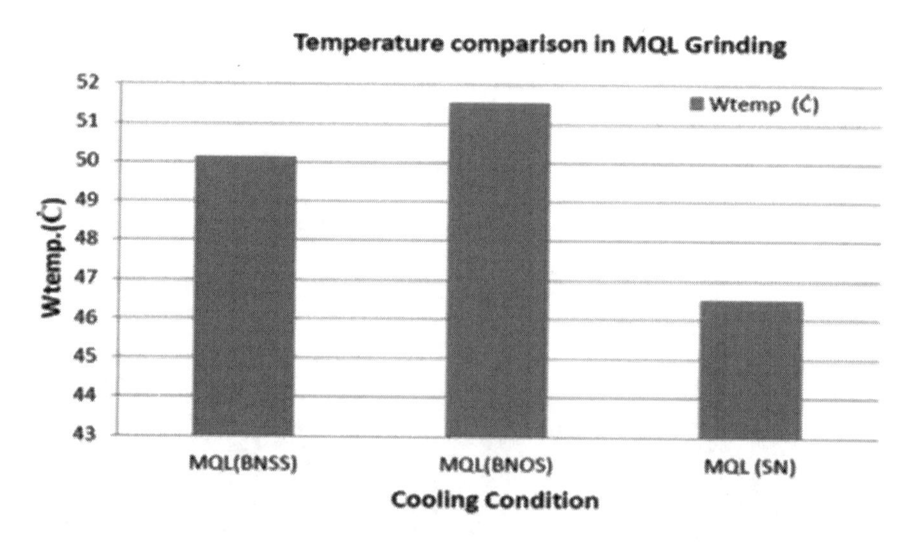

FIGURE 1.12 Temperature comparison in MQL grinding.

is kept constant. Figure 1.13 shows the variation in the surface roughness during dry grinding operation. The maximum surface roughness is seen by 9 μm.

1.3.6.2 Surface Roughness in Flood Grinding Operation

Figure 1.14 shows the up and down variation in the surface roughness during flood grinding operation. The maximum surface roughness is seen by 6.5 μm. Lower surface roughness is noted for this operation than during dry grinding operation. The lowest surface roughness found in flood machining may be due to the comparatively

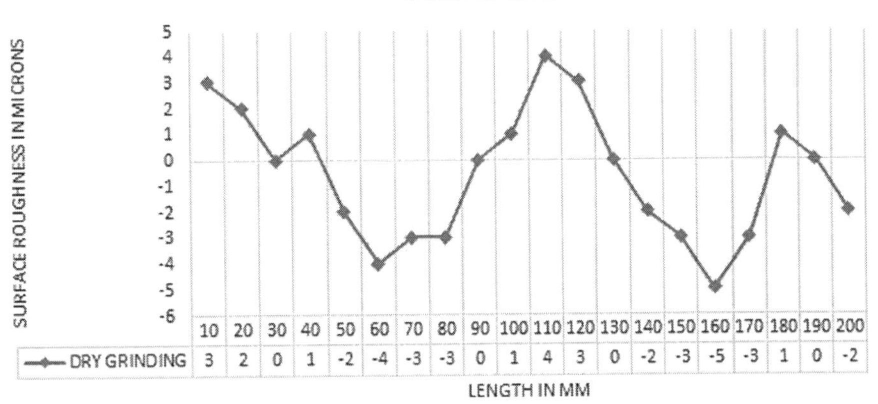

FIGURE 1.13 Dry grinding operation with DOC 20 μm and grinding-wheel speed of 2800.

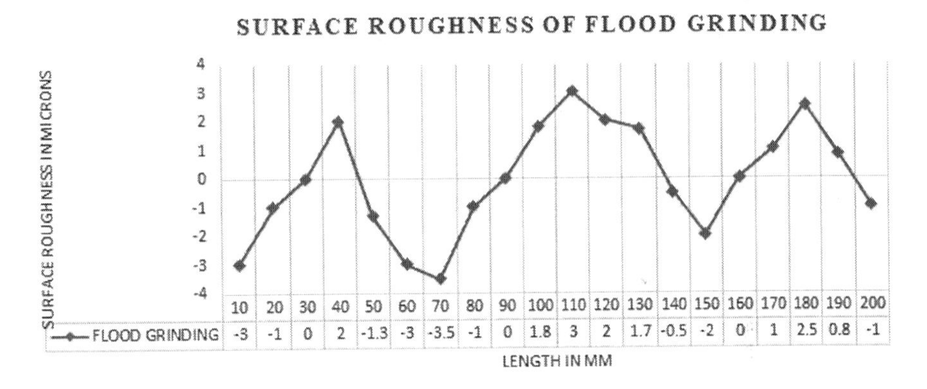

FIGURE 1.14 Flood grinding operation with 15 μm and grinding-wheel speed of 2800 rpm.

low wear of the tool along with continuous emulsion surface contact at the tool chip and interfaces of the tool workpiece. The low tool wear was an influencing factor for the lower surface roughness found in the cryogenic machining (Hong et al. 2002). Therefore, flood grinding operation gives better surface roughness than dry grinding operation.

1.3.6.3 Surface Roughness in Various MQL Grinding Operations

In the MQL grinding operation, experiments use three lubrication methods: SN, BNSS and BNOS. The combined graph of surface roughness from starting point to end point is illustrated in Figure 1.15. The surface-roughness variation is seen by 5 μm during the single nozzle cooling method. Moreover, there is 6 μm surface roughness variation seen in BNSS, which is greater than the SN method by 1 μm. During the third experiment of MQL grinding operation (BNOS) a large surface

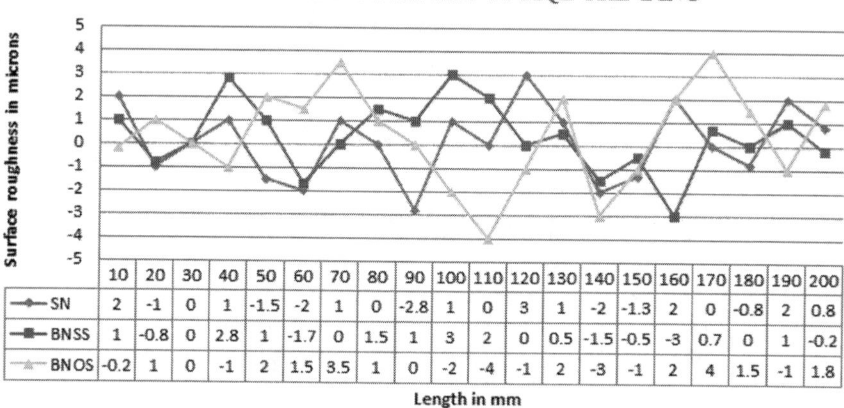

	10	20	30	40	50	60	70	80	90	100	110	120	130	140	150	160	170	180	190	200
SN	2	-1	0	1	-1.5	-2	1	0	-2.8	1	0	3	1	-2	-1.3	2	0	-0.8	2	0.8
BNSS	1	-0.8	0	2.8	1	-1.7	0	1.5	1	3	2	0	0.5	-1.5	-0.5	-3	0.7	0	1	-0.2
BNOS	-0.2	1	0	-1	2	1.5	3.5	1	0	-2	-4	-1	2	-3	-1	2	4	1.5	-1	1.8

FIGURE 1.15 MQL grinding operations with 15 μm and grinding-wheel speed of 2800 rpm.

roughness variation is noted by 8 μm. These graphical values show that the SN method provides better finishing than other operations.

1.3.7 Surface Roughness Comparison in Various Grinding Operations

In the flood grinding operation, the surface roughness of the workpiece is also the same as MQL(SN), but the cost of flood grinding operation is high because of the large amount of cutting fluid needed, and it is difficult and costly to recycle and dispose of the cutting fluid. The temperature in the MQL grinding operation is slightly higher than that of flood grinding operation but less than that of dry grinding operation. The comparison of surface roughness during different grinding conditions is shown in Figure 1.16.

As it is clear from the Figure 1.16, we can say that the surface roughness of the workpiece from the MQL(SN) and flood grinding operations is noted as the same, but the consumption of cutting fluid is far less in MQL(SN) than flood grinding operation. Thus, we can say that MQL(SN) grinding operation is better than dry and flood grinding operation to reduce the temperature and surface roughness of the grinding surface during the operation.

1.4 CONCLUSION

Mild steel workpiece grinding was made using the three conditions of cooling viz dry, flood and MQL. The MQL grinding had three conditions: both of the nozzles were on the same side of the grinding wheel, both of the nozzles are on opposite sides of the grinding wheel, and only one nozzle is working during the operation. The grinding parameters such as grinding-wheel speed, depth of cut and the workpiece movements are also kept constant during the dry, flood and MQL. The conclusions drawn from the study are as follows.

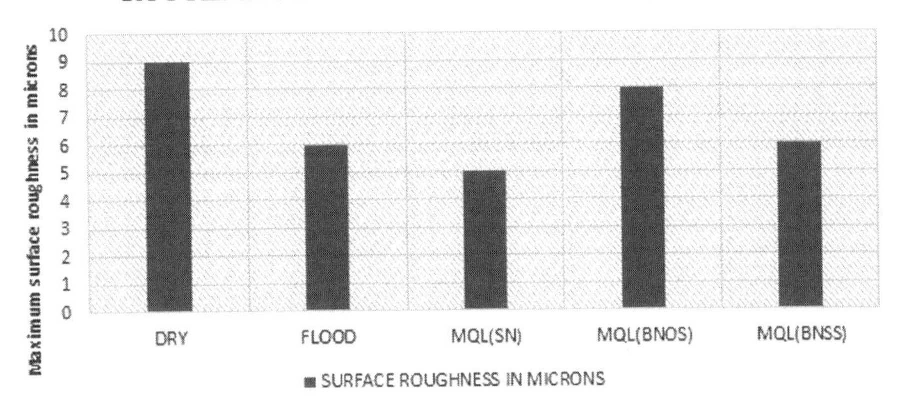

FIGURE 1.16 Surface roughness in various grinding operations.

- Minimum surface roughness noted during MQL(SN) grinding operation is 5 μm; however, maximum surface roughness was 9 μm during dry surface grinding operation and 6 μm during flooded grinding operation.
- Minimum surface temperature noted was 40°C during flooded grinding operation. However, 45°C is noted during MQL(SN) grinding operation when 60–100 ml/hr. coolant is used.
- MQL(SN) is deemed to be the best cooling method because the minimum surface roughness value was noted at 5 μm and surface temperature value was noted at 45°C.

DECLARATION OF CONFLICT OF INTEREST

The authors declare that they have no conflict of interest with respect to the research, authorship and/or publication of this article.

FUNDING

The authors did not receive any research funding or grants from any organization.

ETHICAL APPROVAL

This article does not contain any studies with human participants or animals performed by the author.

ACKNOWLEDGEMENTS

The authors wish to acknowledge the Department of RIC-RESEARCH, INNOVATION AND CUNSULTANCY, IK-Inder Kumar (RIC, IK) Gujral Punjab

Technical University, Kapurthala, Punjab, India for providing the opportunity to conduct this research.

REFERENCES

Alves M.C.S., Bianchi, E.C., Aguiar, P.R. and Canarim, R.C. 2011. Influence of optimized lubrication cooling and minimum quantity lubrication on the cutting forces, on the geometric quality of the surfaces and on the micro-structural integrity of hardened steel parts. *Revista Materia*, 16(3): 754–766.

Astakhov, V.P. 2010. Machining and machinability of materials. *International Journal of Machining and Machinability of Materials*, 7: 1–16.

Balan, A.S.S., Vijayaraghavan, L. and Krishnamurthy, R. 2013. Minimum quantity lubricated grinding of Inconel 751 alloy. *International Journal of Materials and Manufacturing Processes*, 28(4): 430–435.

Barczak L.M., Batako A.D.L. and Morgan M.N. 2010. A study of plane surface grinding under minimum quantity lubrication (MQL) conditions. *International Journal of Machine Tools and Manufacture*, 50: 977–985.

Chakule, R.R., Chaudhari, S.S. and Talmale, P.S. 2017. Evaluation of the effects of machining parameters on MQL based surface grinding process using response surface methodology. *Journal of Mechanical Science and Technology*, 31(8): 3907–3916.

Chetan, Ghosh, S. and Rao, P.V. 2016. Environment friendly machining of Ni-Cr-Co based super alloy using different sustainable techniques. *International Journal of Materials and Manufacturing Processes*, 31(7): 852–859.

Dogra, M., Sharma, V.S., Dureja, J.S. and Gill, S.S. 2018. Environment-friendly technological advancements to enhance the sustainability in surface grinding – A review. *Journal of Cleaner Production*, 197: 218–231.

Goldberg, M. 2012. Improving productivity by using innovative metal cutting solutions with an emphasize on green machining. *International Journal of Machining and Machinability of Materials*, 12: 117–125.

Guo, S., Li, C., Zhang, Y., Wang, Y., Li, B., Yang, M., Zhang, X. and Liu, G. 2017. Experimental evaluation of the lubrication performance of mixtures of castor oil with other vegetable oils in MQL grinding of nickel-based alloy. *Journal of Cleaner Production*, 140: 1060–1076.

Hadad, M. and Hadi, M. 2013. An investigation on surface grinding of hardened stainless steel S34700 and aluminum alloy AA6061 using minimum quantity of lubrication (MQL) technique. *The International Journal of Advanced Manufacturing Technology*, 68(9–12): 2145–2158.

Hong, S.Y., Ding, Y. and Jeong, J. 2002. Experimental evaluation of friction coefficient and liquid nitrogen lubrication effect in cryogenic machining. *Machining Science and Technology*, 6: 235–250.

Inasaki, I. 2012. Towards symbiotic machining processes. *International Journal of Precision Engineering and Manufacturing*, 13(7): 1053–1057.

Kapłonek, W., Nadolny, K., Sutowska, M., Mia, M., Pimenov, D.Y. and Gupta, M. K. 2019. Experimental studies on MoS2-treated grinding wheel active surface condition after high-efficiency internal cylindrical grinding process of INCONEL® alloy 718. *Micromachines*, 10(4): 1–19.

Kuram, E., Ozcelik, B. and Demirbas, E. 2013. Environmentally friendly machining: Vegetable based cutting fluids. In: Davim, J.P., (ed.), *Green Manufacturing Processes and Systems, Materials Forming, Machining and Tribology*. Berlin Heidelberg: Springer-Verlag, pp. 23–47.

Lee, P.H., Nam, J.S., Li, C. and Lee, S.W. 2012. An experimental study on micro-grinding process with nanofluid minimum quantity lubrication (MQL). *International Journal of Precision Engineering and Manufacturing*, 13(3): 331–338.

Li, C.H., Li, J.Y., Wang, S. and Zhang, Q. 2013. Modeling and numerical simulation of the grinding temperature field with nanoparticle jet of MQL. *Advances in Mechanical Engineering*, 5: 1–9.

Liew, W. and Hsien, Y. 2015. *Towards Green Lubrication in Machining*. Springer Briefs in Green Chemistry for Sustainability. Springer.

Lukasz, M.B. and Andre, D.B. 2012. Application of minimum quantity lubrication in grinding. *International Journal of Materials and Manufacturing Processes*, 27: 406–411.

Mao, C., Zhang, J. and Huang, Y. 2013. Investigation on the effect of nanofluid parameters on MQL grinding. *International Journal of Materials and Manufacturing Processes*, 28(4): 436–442.

Mia, M., Gupta, M.K., Singh, G., Krolczyk, G. and Pimenov, D.Y. 2018. An approach to cleaner production for machining hardened steel using different cooling-lubrication conditions. *Journal of Cleaner Production*, 187: 1069–1081.

Morgan, M.N, Barczak, L. and Batako, A. 2012 Temperatures in fine grinding with minimum quantity lubrication (MQL). *International Journal of Advanced Manufacturing Technology*, 60: 951–958.

Rao, D.N. and Srikant, R.R. 2006. Influence of emulsifier content on cutting fluid properties. *Proceedings of the Institution of Mechanical Engineers, Part B: Journal of Engineering Manufacture*, 220: 1803–1806.

Roy, S. and Ghosh, A. 2014. High-speed turning of AISI 4140 steel by multi-layered TiN top-coated insert with minimum quantity lubrication technology and assessment of near tool-tip temperature using infrared thermography. *Journal of Engineering Manufacturing*, 228 (9): 1058–1067.

Roy, S., Kumar, R., Das, R. K. and Sahoo A.K. 2018. A comprehensive review on machinability aspects in hard turning of AISI 4340 steel. *Materials Science and Engineering*, 390: 1–9.

Roya, S., Chand, S., Kumar, R. and Das, R.K. 2019. A brief review on machining operations conducted using different machining inserts under minimum quantity lubrication environment. *Materials Today: Proceedings*, 18: 3134–3143.

Roya, S., Kumar, R., Sahoo A.K. and Das, R.K. 2019. A brief review on effects of conventional and nano particle based machining fluid on machining performance of minimum quantity lubrication machining. *Materials Today: Proceedings*, 18: 5421–5331.

Sarıkaya, M., Yılmaz, V. and Gullu, A. 2016. Analysis of cutting parameters and cooling/lubrication methods for sustainable machining in turning of Haynes 25 superalloy. *Journal of Cleaner Production*, 133: 172–181.

Seyedzavvara, M., Shabgard, M. and Mohammadpourfard, M. 2019. Investigation into the performance of eco-friendly graphite nanofluid as lubricant in MQL grinding. *Machining Science and Technology*, Taylor & Francis Group.23(4): 569–594

Sharma, A.K., Tiwari, A.K. and Dixit, A.R. 2016. Effects of minimum quantity lubrication (MQL) in machining processes using conventional and nanofluid based cutting fluids: A comprehensive review. *Journal of Cleaner Production*, 127: 1–18.

Sharma, V.S., Singh, G. and Sorby, K. 2015. A review on minimum quantity lubrication for machining processes. *International Journal of Materials and Manufacturing Processes*, 30(8): 935–953.

Tangjitsitcharoen, S. 2010. Monitoring of dry cutting and applications of cutting fluid. *Proceedings of the Institution of Mechanical Engineers: Journal of Engineering Tribology*, 224(16): 209–219.

Tao, Z., Yaoyao, S., Laakso, S. and Jinming, Z. 2017. Investigation of the effect of grinding parameters on surface quality in grinding of TC4 titanium alloy. *Procedia Manufacturing*, 11: 2131–2138.

Tawakoli, T., Hadad, M.J. and Sadeghi, M.H. 2010. Investigation on minimum quantity lubricant-MQL grinding of 100Cr6 hardened steel using different abrasive and coolant-lubricant types. *International Journal of Machine Tools and Manufacture*, 50(8): 698–708.

Varadharajan, A.S., Philip, P.K. and Ramamoorthy, B. 2002. Investigations on hard turning with minimal cutting fluid application (HTMF) and its comparison with dry and wet turning. *International Journal of Machine Tools and Manufacture*, 42:193–200.

Vazquez, E., Gomar, J., Ciurana, J. and Rodríguez, C.A. 2015. Analyzing effects of cooling and lubrication conditions in micromilling of Ti6Al4V. *Journal of Cleaner Production*, 87: 906–913.

Wang, Y., Li, C., Zhang, Y., Yang, M., Li, B., Jia, D., Hou, Y. and Mao, C. 2016. Experimental evaluation of the lubrication properties of the wheel/workpiece interface in MQL grinding using different types of vegetable oils. *Journal of Cleaner Production*, 127: 487–499.

Weinert, K., Inasaki, I., Sutherland, J.W. and Wakabayashi, T. 2005. Dry machining and minimum quantity lubrication. *CIRP Annals – Manufacturing Technology*, 53(2): 511–537.

Yang, M., Li, C., Zhang, Y., Jia, D., Zhang, X., Hou, Y., Li, R. and Wang, J. 2017. Maximum undeformed equivalent chip thickness for ductile-brittle transition of zirconia ceramics under different lubrication conditions. *International Journal of Machine Tools and Manufacture*, 122: 55–65.

2 Importance of Dust Collectors in Waste Management and Recycling in Industries

S. Venkatesh, S. P. Sivapirakasam and M. Sakthivel

CONTENTS

2.1 INTRODUCTION TO DUST COLLECTORS

Today, industries and society are anticipating a waste-free environment. Industries such as steel, manufacturing, thermal power and chemical engineering plants, as well as foundries and petroleum, cement and gold manufacturing industries are encountering issues as they attempt to recycle scraps, powder particles and excess materials created during manufacturing (Boysan et al. 1982). For example, collecting 1 gram of gold scrap from the grinding or machining of gold saves approximately ₹3000. Therefore, collecting the scraps/emission from every manufacturing process is vital, as it increases the efficiency of the plant, reduces raw material waste, reduces energy consumption and reduces the environmental impact. Moreover, these scraps affect the health of humans and contaminate natural resources. According to the Prevention and Pollution Control Act (1981), the particles less than 2.5 μm are expressed as $PM_{2.5}$ (fine particulates). The particles less than 10 μm are expressed as PM_{10} (coarse particulates). These particulate matters are also referred to as flying dust particles (Licht 1988). These small particulates are easily inhaled by humans, which, subsequently, penetrate deep into the lungs, creating health issues such as DNA mutation, heart disease and lung cancer (Cudahy and Helsel 2000).

To collect these scraps, industries are utilizing different kinds of dust collectors. These dust collectors separate the scraps/particles from manufacturing processes to shelter the environment (Peukert and Wadenpohl 2001). Generally, there are five types of dust collectors that are used to separate the particulate matter from manufacturing processes, namely wet scrubbers, fabric filters, electrostatic precipitators, unit collectors and inertial separators. There are three types of inertial separators: baffle chambers, settling chambers and cyclone separators (Peukert and Wadenpohl 2001).

2.2 ELECTROSTATIC PRECIPITATOR

The electrostatic precipitator separates the flying scraps from the air/gas medium by electrostatic forces. This equipment consists of discharge wires or electrodes and gathering plates. The electrical pitch is generated between the electrodes and gathering plates by applying the elevated voltage to the electrodes. Afterward, the air/gas is ionized to supply the ions around the discharge wire. The scrap particles surrounded in the air/gas are excited by ions when it passes between the electrodes and gathering plates, generating a strong Coulomb force. This force causes the scrap particles to settle on the collecting plates (Dixkens and Fissan 1999). Afterward, these particles are collected by rubbing with a wire brush, knocking the collecting plates and washing with water. The collected scrap particles are then recycled. While electrostatic precipitators have high purification efficiency and can be used for collecting scrap particles of less than 1 μm, they have a high installation and maintenance cost and require a larger area for installation. The schematic view of an electrostatic precipitator is shown in Figure 2.1.

2.3 FABRIC FILTERS

The schematic view of fabric filters, also called baghouses, is shown in Figure 2.2. The series of baghouses are arranged in a cylindrical container for collecting the scrap particles. When the scrap particles pass through the filter, they accumulate on the filter surface. After reaching sufficient low pressure drop within the filter, the scraps are cleaned by the shakers or pulse jet. It has high collection efficiency and it is easy to operate, but it is not suitable for high-temperature places and requires a large space for installation.

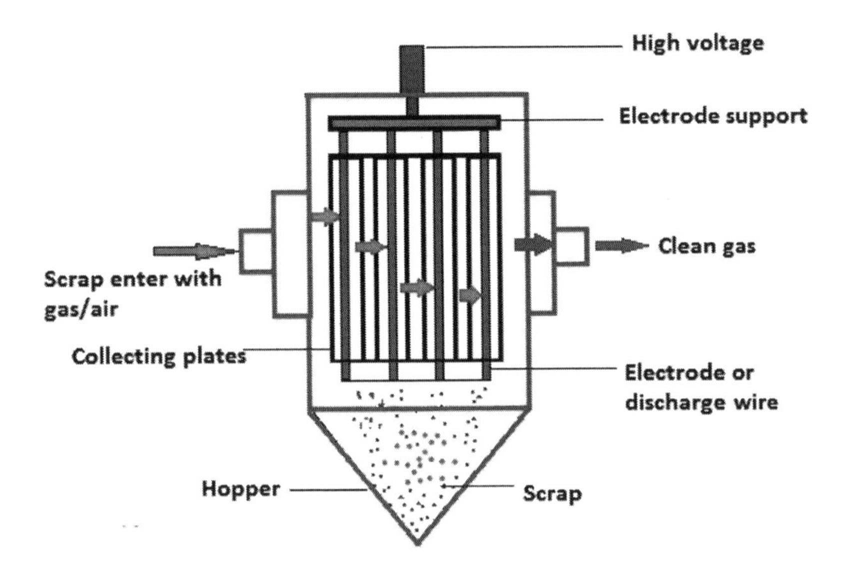

FIGURE 2.1 Electrostatic precipitator.

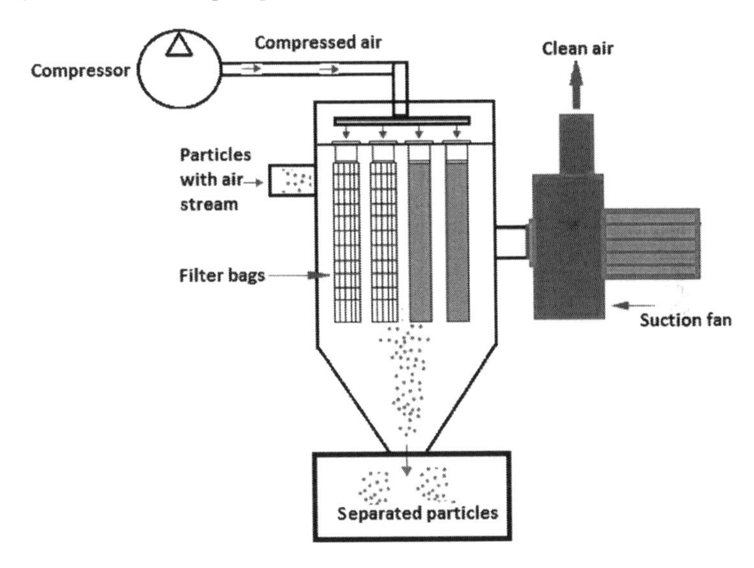

FIGURE 2.2 Fabric filters.

2.4 UNIT COLLECTORS

Unit collectors are small in size and contain a suction blower and various types of particle separators. There are two forms available: fabric collectors and cyclone collectors. This type of dust collector is most suitable for portable and isolated dust-producing operations. This type of dust collector is used at points along moving

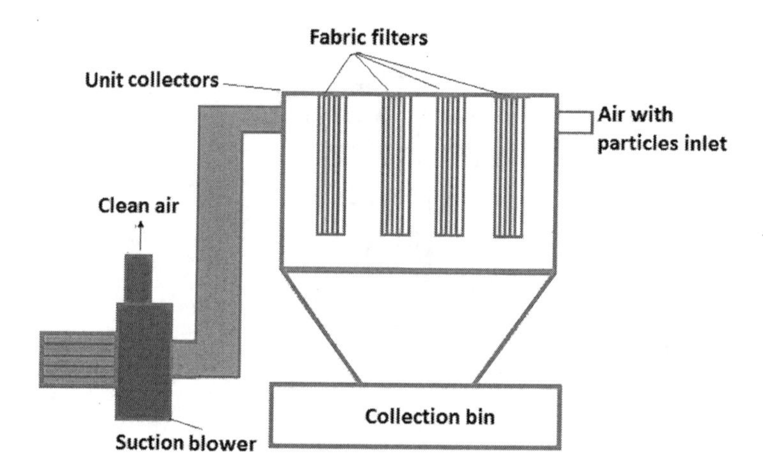

FIGURE 2.3 Unit collectors.

conveyor belts. It requires less space and low initial cost; however, maintenance and service cost are high and requires separate storage space for collecting the dust particles. In this equipment, a suction blower is used to suck the micron-size scrap particles from the manufacturing area. Afterward, these particles pass through the fabric filters where they are collected, allowing clean air to be sent into the atmosphere. The unit collector is shown in Figure 2.3.

2.5 WET SCRUBBERS

Generally, liquid is used in wet scrubbers to divide the waste particles from the fluid medium. Most industries utilizing wet scrubbers usually use water the scrubbing liquid. The water is injected into the air stream containing waste particles. It is noted that higher particle separation efficiency has been achieved by increasing the contact of liquid and gas stream. Generally, three configurations are available in wet scrubber: gas-liquid contact, gas-humidification and gas-liquid separation (Azzopardi 1992). Further, the wet scrubbers are classified such as low capacity, low to intermediate capacity, intermediate to high capacity and high capacity scrubbers. The Venturi scrubbers are categorized as high-energy scrubbers.

Wet scrubbers have a higher collection efficiency compared to other dust-collecting equipment (Pak and Chang 2006). Hence, these types of scrubbers are recommended in foundry and steel plants. The Venturi scrubber has three sections: throat, convergent and divergent or diffuser sections. Usually, the particulates are injected at the convergent section with the gas or air stream. The liquid droplets are sprayed against the air stream in the throat section. Afterward, the particulates confined by the liquid in the throat section are transferred to the diffuser. Then, the captured scraps are gathered by the cyclone attached to the Venturi scrubber. The Venturi scrubbers consume more energy. In addition, operating and maintenance costs are high, and recycling water requires additional power usage and equipment. A diagram of the Venturi scrubber is shown in Figure 2.4.

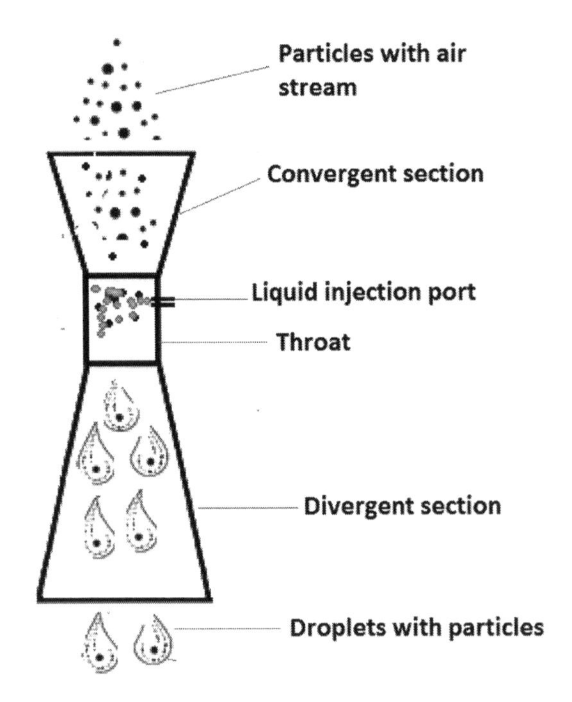

FIGURE 2.4 Venturi scrubber.

2.6 INERTIAL SEPARATORS

Inertial separators remove the particulate matter from the fluid or gas medium by centrifugal, inertial and gravitational forces. Generally, industries use three types of inertial separators: baffle chambers, settling chambers and centrifugal collectors. The baffle chambers and settling chambers are not suitable for mineral processing industries. The cyclone separators, Venturi scrubbers and electrostatic precipitators are mostly used in mineral processing industries.

2.6.1 Baffle Chambers

The baffle chambers utilize baffle plates to divide the scraps from the air stream. They are used primarily as a dust cleaner. The waste scraps mixed with the air enter the baffle chamber through the inlet. The direction of the gas stream is diverted by the baffle plates located inside the chamber, causing the larger particles to settle in the collection bin. Afterward, these scraps are transferred for recycling. The layout of the baffle chamber is shown in Figure 2.5.

2.6.2 Settling Chamber and Centrifugal Collectors

The settling chamber has a simple duct-type cross-section design, which is shown in Figure 2.6. Initially, the high-velocity gas stream enters at the inlet of the settling

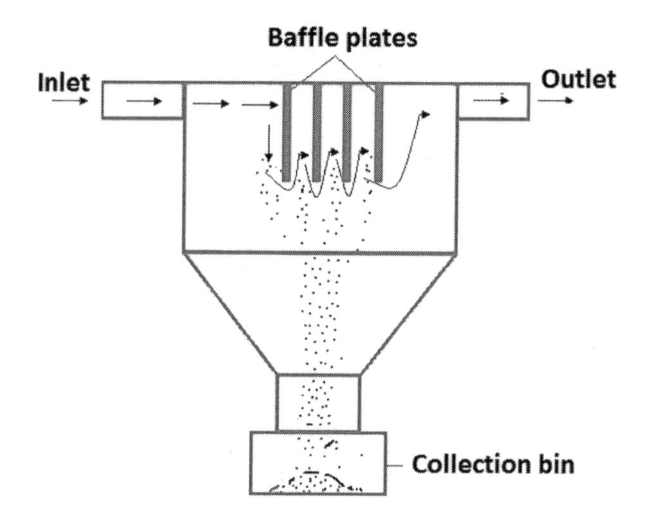

FIGURE 2.5 Baffle chamber.

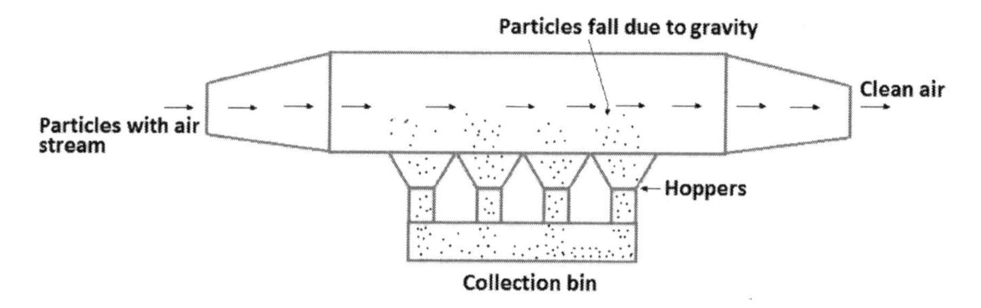

FIGURE 2.6 Settling chamber.

chamber. Once inside, the velocity of the gas is decreased because of the duct-type cross sections, causing the heavier particles to settle in the collection bin. These collected particles are then sent to be recycled.

Centrifugal collectors, another type of inertial separator, use centrifugal force for scrap separation. Cyclone separators, also called centrifugal collectors, are briefly discussed in the next section.

2.7 CYCLONE SEPARATOR

2.7.1 Cyclone Separator Working Principle

Cyclone separators use a centrifugal force to separate the scraps or powder from the fluid medium without using any additional filters. Generally, the hydro cyclones are used to separate particulates from a liquid medium, whereas gas cyclones are used to separate particulates from an air or gas medium. Also referred to as centrifugal separators, they consist of four important parts: the upper cylindrical part known as

TABLE 2.1
Design Factors for the Standard Model Cyclones

Design/Terms	D/D	b/D	a/D	h/D	De/D	H/D	S/D	B/D
Lapple (1939)	1	0.25	0.5	2	0.5	4	0.625	0.25
Stairmand (1951)	1	0.2	0.5	1.5	0.5	4	0.5	0.375
Swift (1969)	1	0.21	0.44	1.4	0.4	3.9	0.5	0.4
Dirgo and Leith (1985)	1	0.3	0.5	3.5	0.333	6	0.558	0.375

a barrel, lower conical part, inlet and outlet. In this type of dust collector, the vortex is generated by converting the inertia force of the gas-solid stream to the centrifugal force. In cyclone separators, the gas-solid stream enters at the top of the unit through the inlet. The gas-solid particles then travel toward the conical part, creating an external vortex. Because of the increasing air velocity in the outer vortex, centrifugal force is developed that affects the scraps. The larger particles have high inertia force on the outer vortex, which forces the scraps to strike the wall of the cyclone and fall to the bottom of the unit where it is collected. Afterward, the gas reaches the lower conical section and starts to move inward toward the centre of the vortex where it flows up, escaping through the outlet as pure gas or air. The separated particles are settled at the collection chamber, which is connected to the base of the cyclone separator (Alexander 1949). There are three types of cyclone separators: tangential inlet, axial-flow and bottom inlet cyclone separators. The design factors of the standard cyclone separators are given in Table 2.1.

2.7.2 Tangential Inlet Cyclone Separator

In a tangential inlet cyclone, the waste particles mixed with air enter the cyclone tangentially with an elevated velocity. Hence, the inertia force of the gas-solid particles is changed into centrifugal force. Therefore, the particles are forced to strike the wall of the cyclone where the scraps are isolated from the gas. The particles spin around the vortex finder region until they are transferred to the conical section and then settle in the collection chamber. The pure gas exits to the atmosphere through the outlet tube. In tangential inlet type cyclone separators, three types of inlets are available including tangential inlet, helical inlet and involutes inlet (Richards 2000). Tangential inlet cyclone separators are suitable for all industries but are typically used in cement plants, powder metallurgical industries and foundries, etc. A schematic view of a tangential inlet cyclone is shown in Figure 2.7.

2.7.3 Axial-Flow Cyclone Separator

The axial-flow cyclone separator has coaxial inlet and outlet ports. The axial-flow cyclone separator receives the gas-laden particles through the axial inlet then it

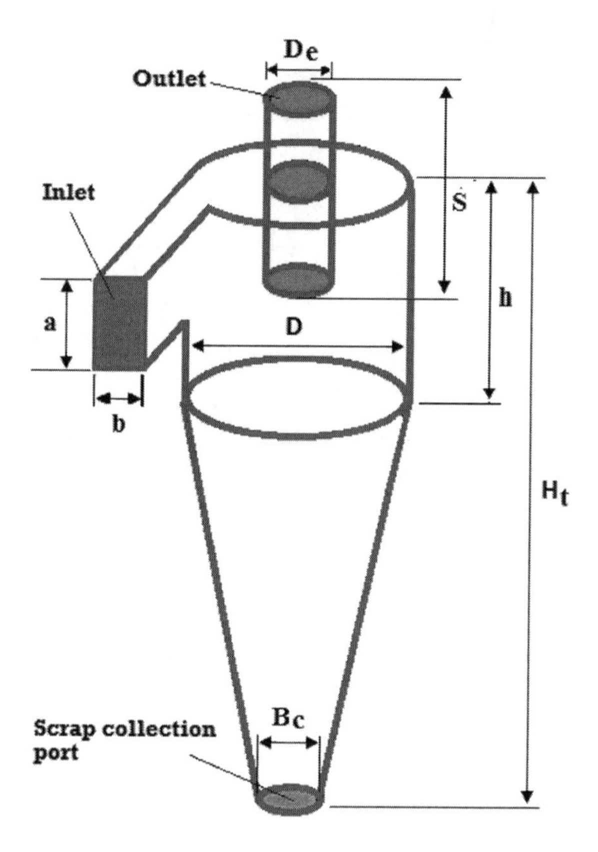

FIGURE 2.7 Schematic diagram of tangential inlet cyclone separator.

spins by the guide vanes arranged inside the cylindrical body. The spinning action separates the particles from the gas medium. Subsequently, the particles are confined at the bottom of the cyclone, and the pure gas exits through the coaxial outlet (Richards 2000). A schematic diagram of an axial-flow cyclone separator is shown in Figure 2.8. These types of separators are used in power plants and the chemical engineering industry.

2.7.4 Bottom Inlet Cyclone Separator

The bottom inlet cyclone has a tangential inlet just above the conical section. The outlet of the cyclone separator is tangential to the cylindrical body which is located at the top of the barrel. In this type of cyclone separator, the gas-laden particles enter through the tangential bottom inlet. After that, the particles spin inside the cyclone by centrifugal force. Then, the scraps gradually settle in the collection chamber. Subsequently, the clean gas exits through the tangential outlet (Richards 2000). A schematic diagram of a bottom inlet cyclone separator is shown in Figure 2.9. Most of the large-scale industries are using bottom inlet and tangential inlet cyclones separators to isolate particulates from the gas stream (Venkatesh et al. 2019).

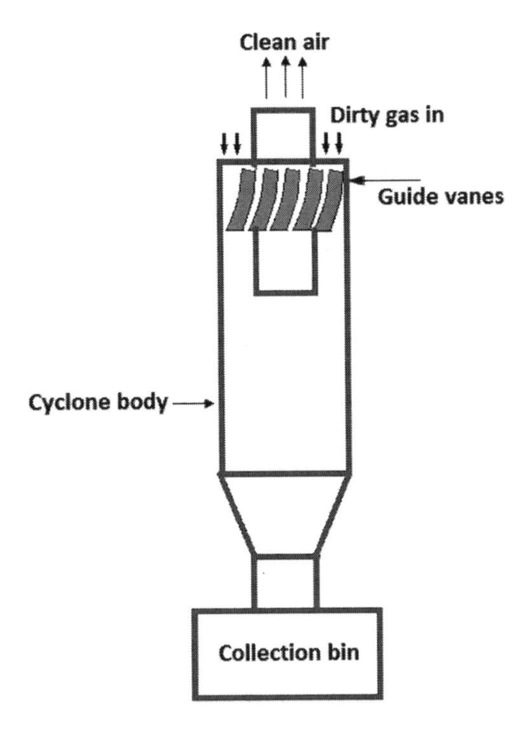

FIGURE 2.8 Schematic diagram of axial-flow cyclone separator.

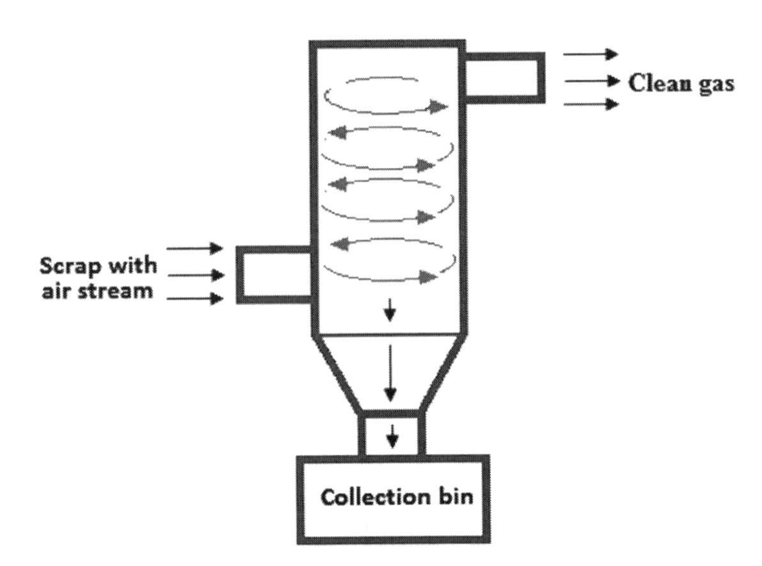

FIGURE 2.9 Schematic diagram of bottom inlet cyclone separator.

2.7.5 Cut-Off Diameter and Pressure Drop

The cut diameter and pressure drop are vital parameters. By these factors, one can analyze the competence of the cyclone. The head (H_d), cut diameter (X_{50}), number of turns (N_e), pressure drop (Δp), Euler number (E_u) and Stokes number (Stk_{50}) can be determined by Eq. 2.1–2.9 (Shepherd and Lapple 1939; Barth 1956; Sinnott 2005). Eq. 2.1, Eq. 2.2 and Eq. 2.3 can be applied only if the Venturi is attached to the tangential inlet and bottom inlet cyclone separators. In another case, Eq. 2.4, Eq. 2.5 and Eq. 2.6 can be used for computing the H_d, X_{50} and N_e if the Venturi is not attached to the cyclone separator (normal cyclone without Venturi). In these equations, the inlet length of the Venturi (a_v) and inlet width of the Venturi (b_v) is replaced by cyclone inlet length (a) and inlet width (b). Eq. 2.7, Eq. 2.8 and Eq. 2.9 can be applied for calculating the pressure drop, Euler number and Stokes number for both cases of cyclone separator (with and without Venturi Venturi).

If the Venturi is connected to the cyclone separator, then H_d, X_{50} and N_e equations can be written as

$$H_d = \frac{Ka_v b_v}{D_e^2} \tag{2.1}$$

Where K = 17 to 18 is a constant.

$$X_{50} = \left[\frac{9\mu b_v}{2\pi N_e V_i (\rho_p - \rho_g)} \right]^{0.5} \tag{2.2}$$

$$N_e = \frac{1}{a_v}\left[h + \frac{H_t - \mathrm{h}}{2} \right] \tag{2.3}$$

If the Venturi is not connected to the cyclone separator (normal cyclone without Venturi), then H_d, X_{50} and N_e equations can be written as

$$H_d = \frac{Kab}{D_e^2} \tag{2.4}$$

Where $K = 17$ is a constant.

$$X_{50} = \left[\frac{9\mu b}{2\pi N_e V_i (\rho_p - \rho_g)} \right]^{0.5} \tag{2.5}$$

$$N_e = \frac{1}{a}\left[h + \frac{H_t - h}{2} \right] \tag{2.6}$$

$$\Delta p = 0.5 \rho_g V_i^2 H_d \tag{2.7}$$

$$E_u = \frac{\Delta p}{0.5\rho_g V_i^2} \tag{2.8}$$

$$Stk_{50} = \frac{\rho_p X_{50}^2 V_i}{18\mu D} \tag{2.9}$$

2.7.6 Applications

- Cyclone separators are largely utilized in aerosol sampling, mineral processing, casting, chemical engineering, agricultural and pharmaceuticals industries, etc.
- They are also used in food processing industries.
- They are used in spray dryers, fluidized bed reactors and dust sampling equipment.
- Cyclone separators are connected with Venturi scrubbers and electrostatic precipitators in casting industries and powder metallurgical industries.
- They are used in centralized vacuum cleaning systems to collect scrap particles produced from grinding fibre materials.

2.7.7 Advantages of the Cyclone Separator

The advantages of cyclone separators compared with other dust separation equipment include:

- They have a simple design and flexibility. The separated dust particles are dry and recyclable.
- Fabrication of the cyclone separators is very easy compared to other types of dust collectors because there are no moving parts. They can be easily fabricated using sheet metals.
- They are suitable for high temperature and pressure applications. They are robust and operating and maintenance cost is less. Additional filter equipment is not required at the outlet.

2.7.8 Disadvantages of the Cyclone Separator

Cyclone separators have some disadvantages compared to other devices, including

- If the scrap sizes fall below the cut-off diameter, then its efficacy in gathering these particles is reduced. This equipment is less effective when the scrap size falls under 5 μm.
- It may produce high pressure drops when the geometric design is poor. It gives less collection efficiency when compared with the wet scrubbers, electrostatic precipitators and fabric filters. Erosion inside the cyclone is possible when separating abrasive scrap.

REFERENCES

Alexander, R.M. 1949. Fundamentals of cyclone design and operation. *Proceedings of the Australian Institute of Mineral and Metallurgy NS*, 152: 203–228.

Azzopardi, B.J. 1992. Gas-liquid flows in cylindrical Venturi scrubbers: Boundary layer separation in the diffuser section. *Chemical Engineering Journal*, 49: 55–64.

Barth, W. 1956. Design and layout of the cyclone separator on the basis of new investigations. *Brennstoff-Warme-Krafat (BWK)*, 8: 1–9.

Boysan, F., Ayers, W.H. and Swithenbank, J. 1982. A fundamental mathematical modeling approach to cyclone design. *Transactions of the Institution of Chemical Engineers*, 60: 222–230.

Cudahy, J. and Helsel, W. 2000. Removal of products of incomplete combustion with carbon. *Waste Management*, 20: 339–345.

Dirgo, J. and Leith, L. 1985. Cyclone collection efficiency: comparison of experimental results with theoretical predictions. *Aerosol Science and Technology*, 4: 401–415.

Dixkens, J. and Fissan, H. 1999. Development of an electrostatic precipitator for off-line particle analysis. *Aerosol Science and Technology*, 30: 438–453.

Licht, W. 1988. *Air pollution control engineering*. New York, NY: Marcel Dekker, 1–10.

Peukert, W. and Wadenpohl, C. 2001. Industrial separation of fine particles with difficult dust properties. *Powder Technology*, 118: 136–148.

Pak, S. and Chang, K. 2006. Performance estimation of a Venturi scrubber using a computational model for capturing dust particles with liquid spray. *Journal of Hazardous Materials*, 138(Part-B): 560–573.

Richards, R.J. 2000. *Control of particular matter emissions student manual*. Durham, North Carolina, United States: Air Pollution Training Institute.

Shepherd, C.B. and Lapple, C.E. 1939. Flow pattern and pressure drop in cyclone dust collectors. *Industrial and Engineering Chemistry*, 31: 972–984.

Sinnott, R.K. 2005. *Chemical engineering design*. Oxford, UK: Elsevier Science.

Stairmand, C.J. 1951. The design and performance of cyclone separators. *Transactions of the Institution of Chemical Engineers*, 29: 356–383.

Swift, P. 1969. Dust control in industry. *Steam Heat Engineer*, 38: 453–456.

Venkatesh, S., Sakthivel, M., Avinasilingam, M., Gopalsamy, S., Arulkumar, E. and Hariprasanth, D. 2019. Optimization and experimental investigation in bottom inlet cyclone separator for performance analysis. *Korean Journal of Chemical Engineering*, 36: 929–941.

3 Influence of Nickel-Based Cladding on the Hardness and Wear Behaviour of Hard-Faced Mild Steel Using E-7014 Electrode Using Shielded Metal Arc Welding

Gursharan Singh, Shubham Sharma, Jujhar Singh, Vivek Aggarwal, Amit Bansal and Suresh Mayilswamy

CONTENTS

3.1 INTRODUCTION

For industries in any country, particularly those in developing countries such as India, the durability and longevity of materials are highly valued. Industrial organizations, whether they are in the manufacturing/assembly or service field, derive their reputation and prestige from their products' longevity and reliability. Deterioration of products from wear and corrosion can cause losses in economic and reputational terms. Although researchers have already paid considerable attention to developing new techniques and methods for arresting and managing wear- and corrosion-related problems, more work is still needed. It is estimated that developing and using improved wear- and corrosion-management methods, more than 30 percent of wear- and corrosion-related costs can be minimized (D'Oliveira et al. 2006).

Using higher quality, higher cost wear- and corrosion-resistant metals and enhancing the wear and corrosion resistance of current metals and alloys by adding some modifications to the surface can reduce problems related to wear and corrosion. Recently, thermal spraying, chemical vapour deposition and physical vapour deposition methods have been frequently considered among the numerous commercially feasible surface-coating techniques, in which a metal is poured over another surface in order to increase surface hardness and make it resistant to abrasion, impact, corrosion, galling and cavitation. This process is called hard facing because the surfaces deposited are harder than the base metal (Sunil and Kumar 2010).

Hard facing can be applied by a number of welding processes. Selection of the most appropriate welding method for a given job depends on different factors, such as the type of hard facing, component feature, base metal structure, welding equipment functionality and weld component repair status. Hard facing has emerged as a significant process to resolve the issue of wearing in various types of steel, improving the surface properties of components in terms of hardness and wear and corrosion resistance. Some welding techniques used to perform hard facing are mainly used to prolong or boost the service life of the engineering components and to reduce their expense. For hard facing of components, several welding processes can be used, such as oxy-acetylene, shielded metal arc welding (SMAW), gas tungsten arc welding (GTAW), gas metal arc welding (GMAW), flux-cored arc welding (FCAW), plasma arc welding (PAW) and electro-slag welding (ESW) (Okechukwu et al. 2017).

SMAW is the most frequently used hard-facing method because of its low cost, high flexibility and easy access to a wide range of sizes and types of hard-facing electrodes. Welding can be done on jobs of any shape or size and in any position and at any location. The welder can control the heat input rate, dilution by base metal and can easily cover irregular areas. Welders can use the stringer bead technique to minimize heat input or the weave technique to maximize heat input. The welder can direct the arc of the molten puddle and control the size of the puddle to minimize dilution of the base metal. SMAW is basically a fusion welding process. First, contact is made between the electrode and the workpiece to create an electrical circuit, and then an arc is created by separating the conductors. A diagram of SMAW is shown in Figure 3.1.

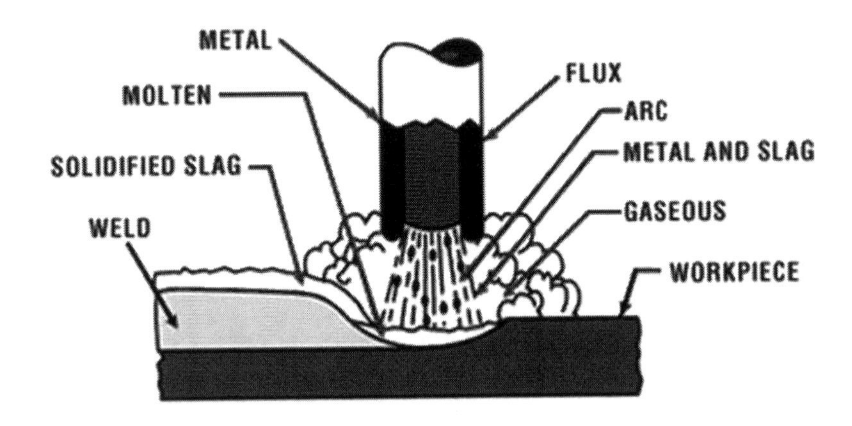

FIGURE 3.1 Hard facing through shield metal arc welding.

The arc is sustained by an electrical discharge from an ionized particle path called plasma. The electrical energy in the arc is converted into intense heat that reaches a temperature from 3000°C to 4000°C. By melting the electrode and edges of parent metal below the flame, the heat creates a pool of molten metal. This pool of molten metal is the weld upon solidification. This method is flexible and applies in all welding positions and in all metals for which electrodes were produced (Buchanan et al. 2007).

Fernandez et al. (2003) evaluated the effect of abrasive wear on nickel-based alloy coatings with a focus on factors such as the load applied, tungsten carbide (WC) reinforcement particles, abrasive grain size and environment. The results showed that the grain size of the abrasive and the reinforcing particles of the WC both have a major abrasive effect. It compares strongly with the effects of the applied load and the environment, having a much smaller effect and a very low effect respectively. Chatterjee and Pal (2003) studied the action of solid particle erosion (SPE) of various hard-facing electrodes deposited on grey cast iron (ASTM 2500) using quartz sand and iron ore as erodent particles. They concluded that the hardness is not a valid indicator of hard-facing soil erosion resistance. With rising volume fraction of carbides, the erosion rate decreases when soft iron ore particles are used as erodents. Buchanan et al. (2007) studied hypereutectic and hypoeutectic abrasive wear actions of Fe-Cr-C hard facings in microstructure terms. Using two industrial hard-facing electrodes, SMAW deposited the coatings onto a grey cast iron substratum. The hardness of the hypereutectic coating was found to be substantially greater than that of the hypoeutectic coating. In both cases maximum hardness inside the first layer deposited was achieved. Kashani et al. (2007) investigated the use of super alloy nickel and cobalt base as wear-resistant hard-facing materials on H11 tool steel substrates using tungsten inert gas (TIG) welding technique. Wear measurements were performed at room temperature and 550°C using a pin-on-disk wear tester. The findings showed wear at high temperatures was much lower. Inconel 625 demonstrated the desired reaction to wear. Coronado et al. (2009) studied the effect of welding processes on abrasive resistance to wear, using four kinds of hard-facing alloys deposited in a single- and triple-layer pattern with two separate welding processes i.e. FCAW and SMAW. Results showed that the FCAW in single and triple layers offers better wear resistance. Gualco et al. (2010) studied the effect of gas shielding, heat input and post-weld heat treatment (PWHT) on microstructural assessment and wear resistance on modified martensitic tool steel AISI H1E3. Throughout the semiautomatic gas-shielded arc welding process, the tubular metal-coated wire was used for hard facing. They found shielding gas showed little effect on the chemistry of weld deposits. High heat production resulted in greater precipitation of carbide, and thus lower hardness, whereas low heat input resulted in higher hardness. PWHT resulted in a reduction of 30 percent wear resistance as compared to welded specimens. Rajeev et al. (2017) investigated the effect of combining base metal and filler metal during the construction of hard-facing welds on the resistance to abrasive wear. In making three layers of hard-facing metals, two types of hard-facing electrodes with a different chemical composition were used. They concluded that the wear strength is highly dependent on the wear parameters as well as the morphological structure of hard-facing metals. Kumar and Singh (2019) investigated the effect of various formulations of chromium powder, deposited by paste coating process, on mild steel. Three different metal powder compositions were selected for hard-facing material. They observed that cold rolled (CR) 90 percent is the best composition among the three compositions under study in terms of deposit quality, hardness achieved and wear rate. It is clear from the literature that hard facing though welding coating helps to improve the wear resistance of metal and alloy. In this paper, we investigate the effect of hard facing using SMAW coating (welding of nickel and sodium silicate) on the wear and microhardness characteristics of mild steel plate. During this study, we also evaluated the effect of process parameters such as weight percentage of nickel in the form of paste and current on wear resistance and microhardness.

3.2 MATERIAL AND METHODOLOGY

The four mild steel specimens (100 × 30 × 10 mm) were taken as the base metal or substrate material and were deposited by SMAW after paste application on three hard-facing substrate materials. In order to prevent experimental errors, the specimens were thoroughly prepared mechanically and chemically cleaned (i.e. emery paper, acetone, grinding and so on) before depositing the hard-faced material. The general chemical composition of mild steel is given in Table 3.1.

TABLE 3.1
Chemical Composition of Base Metal

Carbon	Silicon	Manganese	Phosphorus	Sulfur	Iron
0.29	0.284	1.52	0.012	0.009	97.885

TABLE 3.2
Chemical Composition of AWS E7014 Electrode

Element	C	Mn	Si
Percentage	0.07	0.57	0.24

3.2.1 Selection of Electrode

An AWS E-7014 electrode was used in the form of iron powder for welding the mild steel. This electrode has features of a smooth arc, good arc stability, low spatter and medium to low penetration. It also provides outstanding slag removal and bead appearance. The general chemical composition of the electrode is given in Table 3.2.

3.2.2 Selection of Metal Powders

In order to prepare pastes, the hard-facing powder (nickel powder) was mixed in different proportions in sodium silicate (binder), which was applied to the surface in the form of a coating to be hard faced.

3.2.3 Selection of Welding Process

SMAW is mostly used for hard facing because of its easy operation and low cost. This type of welding can be used for jobs of any size and shape, in any position and at any location. The welder can control heat input rate. The welder will direct the arc on the molten puddle and monitor the puddle size to reduce dilution through the base metal (for example, in the welding of austenitic manganese steel). The setup for SMAW is as shown in Figure 3.2.

3.2.3.1 Welding Equipment

SMAW was used for metal deposition. SMAW is, in essence, a process of fusion welding. First, contact is made between the electrode and the workpiece to create an electrical circuit, and then an arc is created by separating the conductors. The technical specification of the welding machine utilized for this experiment is shown in Table 3.3.

To assess the impact of two independent parameters, a three-level factorial design of (3 ^ 2 = 9) nine trials was selected. The use of three-level factorial architecture helps to reduce experimental runs. A factor's upper limit is represented as (H), lower as (L) and intermediate as (I). The direct and indirect parameters were kept unchanged, even under consideration. The defined values of process parameters are given in Table 3.4 with their units and notations.

Table 3.5 presents the design matrix developed to conduct the nine factorial design trials runs of 3 ^ 2. The response parameter (wear) is reported, as per the

FIGURE 3.2 Submerged arc welding setup.

TABLE 3.3
Technical Specifications of Welding Machine

Model	Manufacturer	Voltage (V)/Phase, Frequency (Hz)	Setting Range (AC) A	Open Circuit Voltage (V)	Dimensions L*W*H (mm)	Weight (kg)
Origo Arc 200	Panasonic limited, Kolkata	230/1 50	5–180	60–75	380×180×300	25

TABLE 3.4
Process Parameters and Their Levels for Experimentation

			Levels		
Sr.no.	Parameters	Units	Upper (H)	Intermediate (I)	Lower (L)
1.	Nickel Percentage	--------	90	80	70
2.	Current	Ampere	150	130	110

design matrix, by conducting experiments or further investigations. The sample was categorized into three lots (HH, HI, HL; II, IH, IL; LL, LI, LH) and three samples of each load were selected. A total of nine samples were examined with three groups.

The model was developed using the regression method, and the analysis of variance method was used to check the significance of the model. Design-Expert (DX8) software was used to implement a factorial design consisting of nine experiments and to develop a model showing the relationships between the response wear and process parameters (welding current and percentage of Ni) for coded values of (H), (I), (L) for each parameter. The pin-on-disc wear testing is shown in Figure 3.3.

TABLE 3.5
Design Matrix for Experimentation

Sample No.	Sample Name	Nickel %	Current
1	HH	H	H
2.	HI	H	I
3.	HL	H	L
4.	IH	I	H
5.	II	I	I
6.	IL	I	L
7.	LH	L	H
8.	LI	L	I
9.	LL	L	L

FIGURE 3.3 Wear and friction testing machine.

It took three readings to find the average value. Final weight loss of the samples was measured; weight loss can be correlated to indicate each sample's wear rate. Microhardness was checked for all samples using a Vicker hardness-testing machine. A relative comparison was made between the different hard-faced samples through the derived data. The wear resistance test data taken during experimentation is shown in Table 3.6.

3.3 RESULTS AND DISCUSSION

After conducting tests on the work samples, data was collected, analyzed and compared analytically and graphically. All ten samples were polished first on a disk polishing machine and then the microhardness was tested. On middle position, the

TABLE 3.6
Wear Resistance Test Data

Applied Load (kg)	Disc Diameter (mm)	Time (Minute)	Specimen Size (L*D) (mm)
2	50	5	25*6

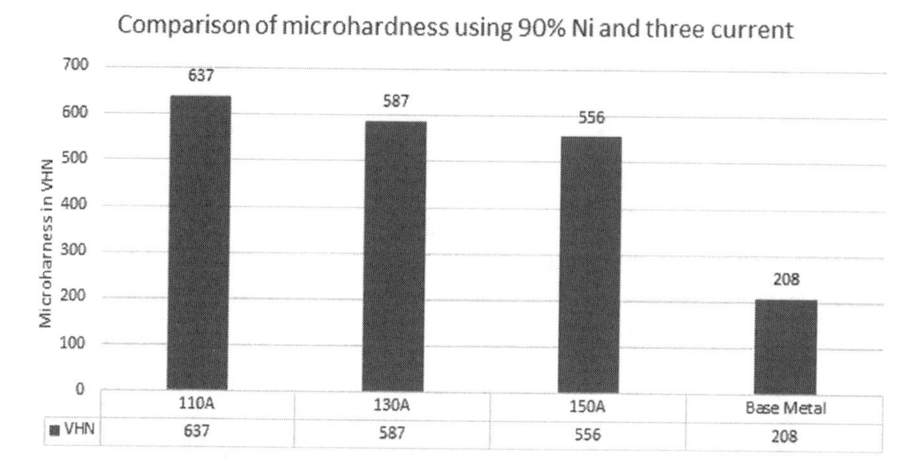

FIGURE 3.4 Microhardness comparison of 90 percent Ni with different currents.

microhardness was tested. One reading had been taken at every study. The microhardness of substrates with 90, 80 and 70 percent nickel under different currents is shown in Figures 3.4–3.6.

It can be easily concluded from the above figures that at a current of 110A the microhardness is at its maximum compared to the microhardness values at 130A and 150A. The key explanation is that all three compositions of the paste decrease with the increase in current hardness because the high current results in slower cooling rates, which results in lower hardness of the softer matrix. Higher cooling rate would yield higher microhardness. It has also been found that the hardness values can be increased by approximately 3.15 times using 90 percent Ni, 3.10 times using 80 percent Ni and 1.90 times using 70 percent Ni powder, as higher concentrations of nickel contribute to increased carbide formation. Moving from the upper to the lower level, the microhardness decreases. The microhardness of all substrates is given in Table 3.7.

Wear rate was calculated by measuring initial and final weights of samples. Loss in weight was calculated for the running wear period of 5 min. Then the wear rate was calculated per hour i.e. for 60 min from loss of weight for 5 min. Loss in weight and wear rate is shown in Table 3.8.

Calculation of wear rate for sample:

In 5 minutes, the sample weight loss $= 0.0051$ g
For 1 minute, the sample weight loss $= 0.0051/5$
For 60 minutes, the sample weight loss $= 0.0051/5 \times 60 = .0612$ g/hr

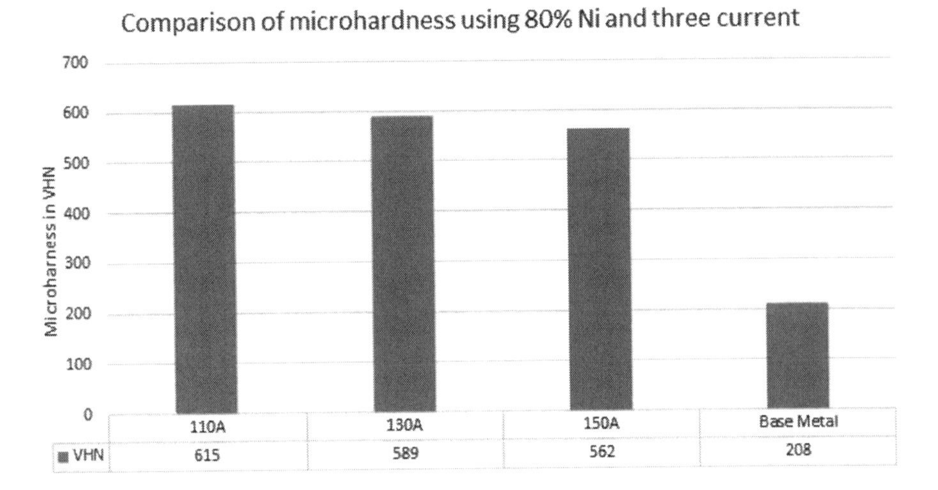

FIGURE 3.5 Microhardness comparison of 80 percent Ni with different currents.

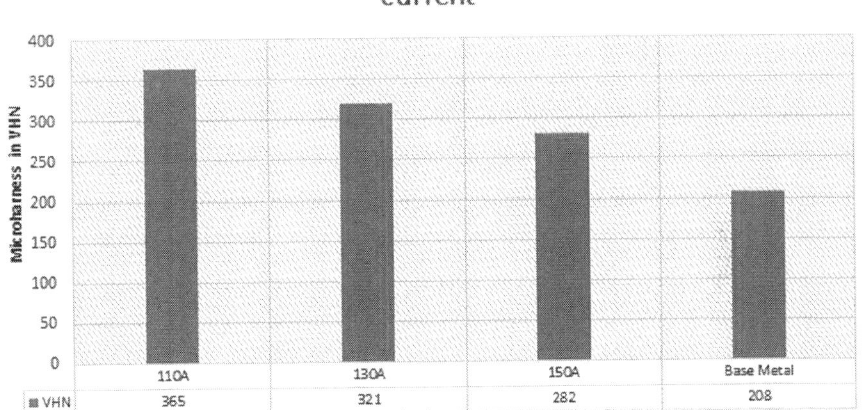

FIGURE 3.6 Microhardness comparison of 70 percent Ni with different currents.

Similarly, wear rate for all samples can be calculated by applying the following formula: Wear rate (g/hr) = loss of weight in (g/5 min) × 60.

The maximum wear rate is for the base metal and minimum for HL sample. Comparison of wear rate of different work samples is shown in Figure 3.7.

Wear is strongly influenced by the amount of nickel in the paste composition. As the amount of wear-resistant elements, such as Ni, increases, carbide formation increases, which results in enhanced wear resistance. Wear resistance can be increased by up to 26 times using 90 percent Ni, 17 times using 80 percent Ni and 12 times using 70 percent Ni as the base metal (mild steel). The line graph comparison of wear rate for different samples is shown in Figure 3.8.

TABLE 3.7
Microhardness of Substrates

Sample No.	Sample Name	Hardness (VHN)
1.	HH	556
2.	HI	587
3.	HL	637
4.	IH	562
5.	II	589
6.	IL	615
7.	LH	282
8.	LI	321
9.	LL	365
10	Base metal	208

TABLE 3.8
Wear Rate Results for All Samples

Sample No.	Sample Name	Initial Weight (g)	Loss Weight (g)	Final Weight (g)	Wear Rate (g/hr)
1	HH	5.6896	0.0051	5.6845	0.0612
2	HI	5.6825	0.0039	5.6786	0.0468
3	HL	5.6876	0.0021	5.6855	0.0252
4	IH	5.5934	0.0051	5.6845	0.0924
5	II	5.6382	0.0049	5.6333	0.0588
6	IL	5.801	0.0046	5.7964	0.0552
7	LH	5.589	0.0108	5.5782	0.1296
8	LI	5.6341	0.0089	5.6252	0.1068
9	LL	5.6559	0.0081	5.6528	0.0972
10	Base Metal	5.5611	0.134	5.4271	1.608

The effect of wear vs. Ni percentage is shown in Figure 3.9. Point 1 along the x axis represents the higher nickel level (90%Ni), hence low wear at that point as compared to Point 0 (80%Ni) and Point −1 (70% Ni). This is because the formation of carbides is higher at Point 1 as compared to the Point 0 and Point −1. It was found that wear also increases for all three compositions of the paste with the rise in current because the high current results in slower cooling rates, which resulting in lower hardness of the softer matrix.

The effect of wear vs. current is shown in Figure 3.10. Point 1 along the x axis denotes the higher level of current, hence lower hardness due to the slower cooling rate, which results in high wear rate as compared to Point 0 and Point −1.

It has been found that, as the current rises, the wear rate often rises because of decreasing hardness and as Ni percent increases, decreasing the wear rate due to

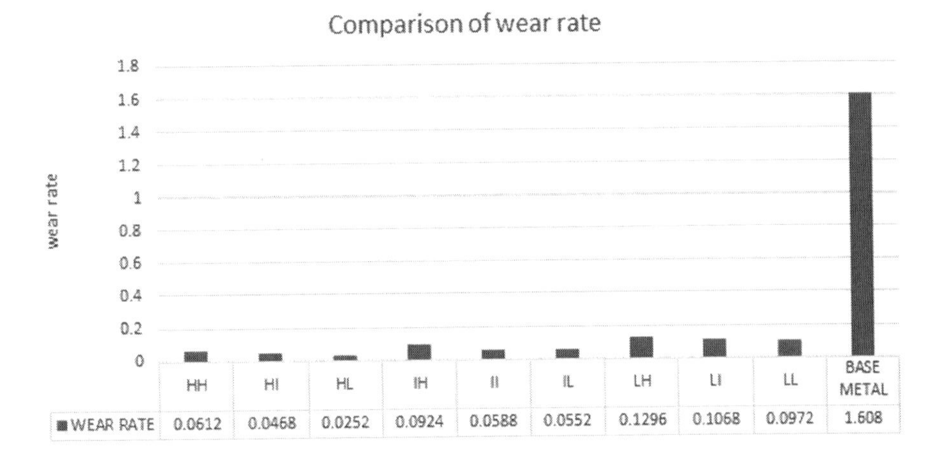

FIGURE 3.7 Comparison of wear rate of different work samples.

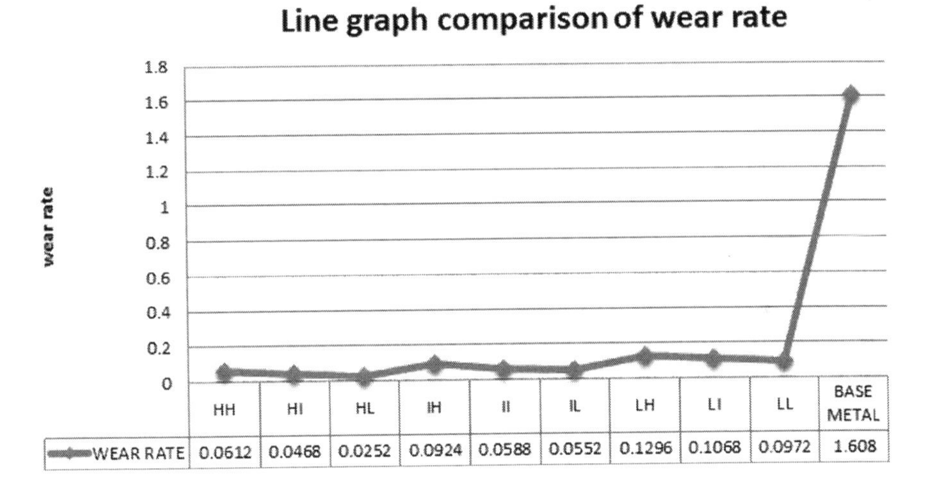

FIGURE 3.8 Comparison line graph of the wear rate for all samples.

higher carbide formation. The same observations were noticed in all three paste compositions. The combined wear-related impact of process parameters is shown is Figures 3.11 and 3.12.

3.4 CONCLUSION

- Using 90 percent Ni, 3.04 times using 80 percent Ni and 1.88 times using 70 percent Ni powder, the hardness values can be increased by approximately 3.10 times.
- Wear resistance can be improved up to 26 times, using 90 percent Ni, 17 times using 80 percent Ni and 12 times using 70 percent Ni (mild steel) than the base metal.

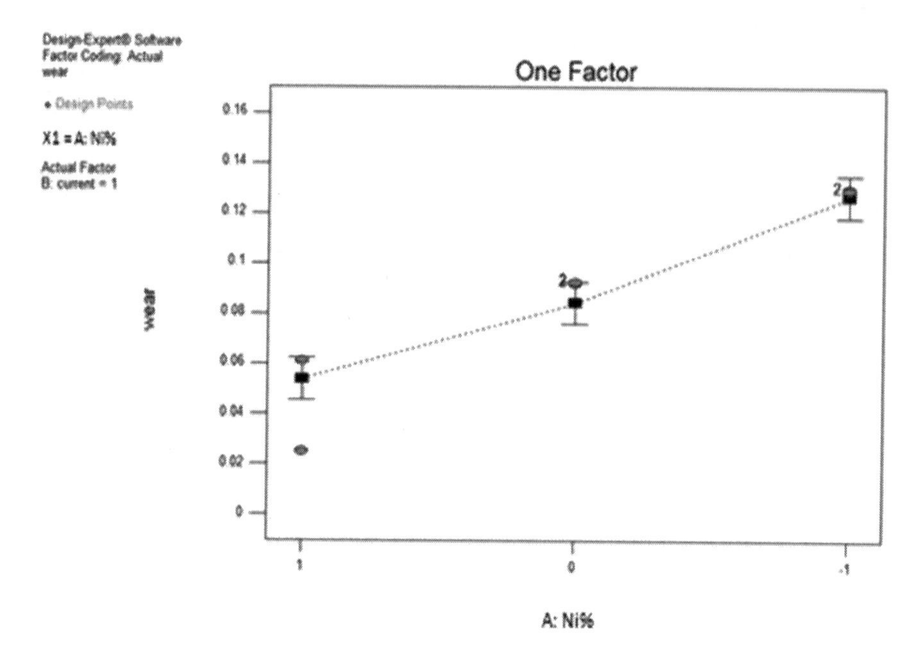

FIGURE 3.9 Interaction between wear and Ni percentage.

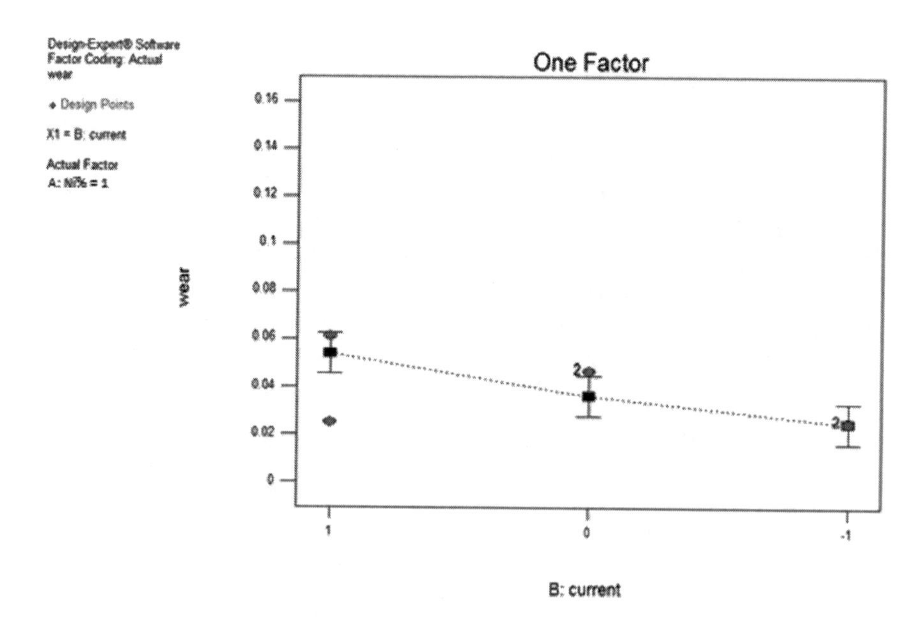

FIGURE 3.10 Interaction between current and wear.

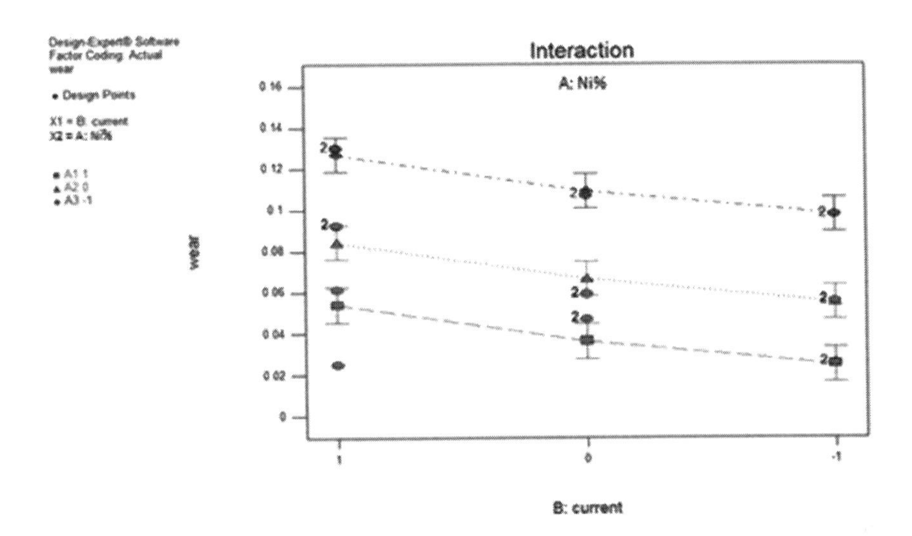

FIGURE 3.11 Combined interaction of process parameters.

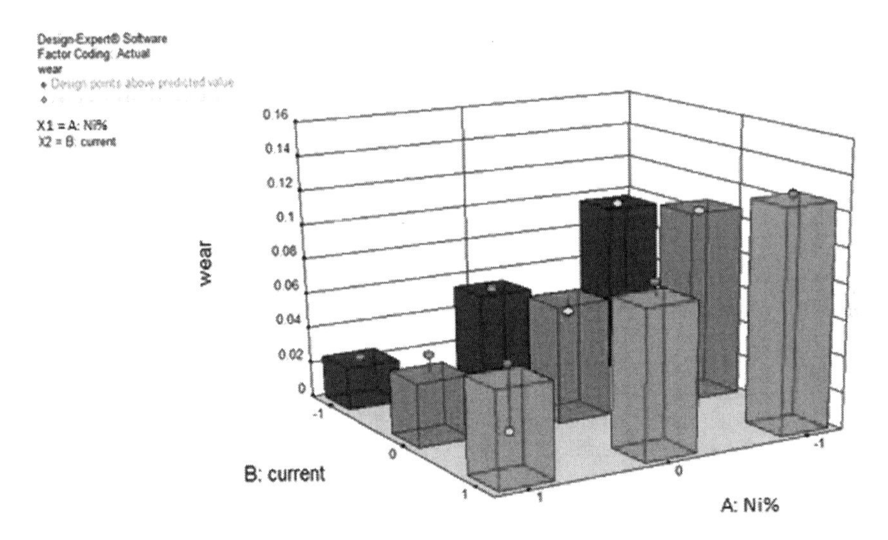

FIGURE 3.12 Combined interaction of process parameters (three-dimensional view).

- Considering all aspects, it can be concluded that 90 percent Ni paste gives better wear and microhardness compared to 80 percent and 70 percent Ni content paste.
- It was found that the current toughness decreases for all three paste compositions because the high current results in slower cooling rates, resulting in a lower hardness matrix.
- Wear studies show that the wear rate also rises because of declining hardness as the current increases. The same observations were noticed in all three paste compositions.

DECLARATION OF CONFLICT OF INTEREST

The authors declare that they have no conflict of interest with respect to the research, authorship and/or publication of this article.

FUNDING

The author(s) did not receive any research funding or grants from any organization.

ETHICAL APPROVAL

This article does not contain any studies with human participants or animals performed by the author.

ACKNOWLEDGEMENTS

The authors wish to acknowledge the Department of RIC, IK Gujral Punjab Technical University, Kapurthala, Punjab, India for providing the opportunity to conduct this research.

REFERENCES

Buchanan, V.E., Shipway, P.H. and McCartney, D.G. 2007. Microstructure and abrasive wear behavior of shielded metal arc welding hard facings used in the sugarcane industry. *Wear*, 263: 99–110.

Chatterjee, S. and Pal, T.K. 2003. Wear behavior of hard facing deposits on cast iron. *Wear*, 255: 417–425.

Coronado, J.J., Caicedo, H.F and Gomez A.L. 2009. Effect of welding processes on abrasive wear resistance for hard facing deposits. *Tribology International*, 42: 745–749.

D'Oliveira, A.S.C.M., Paredes, R.S.C. and Santos, R.L.C. 2006. Pulsed current plasma transferred arc hard facing. *Journal of Materials Processing Technology*, 171:167–174.

Fernandez, J.E, Fernandez, M.D.R, Diaz, R.V. and Navarro, R.T. 2003. Abrasive wear analysis using factorial experiment design. *Wear*, 255: 38–43.

Gualco, A., Svoboda, H.G, Surian, E.S. and Vedia, L.A.D. 2010. Effect of welding procedure on wear behaviour of a modified martensitic tool steel hard facing deposit. *Materials and Design*, 28: 193–213.

Kashani, H., Amadeh, A. and Ghasemi, H.M. 2007. Room and high temperature wear behaviors of nickel and cobalt base weld overlay coatings on hot forging dies. *Wear*, 262(7–8): 800–806.

Kumar, S. and Singh, R. 2019. Investigation of tensile properties of shielded metal arc weldments of AISI 1018 mild steel with preheating process. *Materials Today: Proceedings*, 26(2): 209–222 (Article in press).

Okechukwu, C., Dahunsi, O.A., Oke, P.K., Oladele I.O. and Dauda M. 2017.Review on hard facing as method of improving the service life of critical components subjected to wear in service critical components subjected to wear in service. *Nigerian Journal of Technology*, 36(4):1095–1103.

Rajeev, G.P., Kamaraj, M. and Srinivasa, R.B. 2017. Hard facing of AISI H13 tool steel with Stellite 21 alloy using cold metal transfer welding process. *Surface and Coatings Technology*, 326: 63–71.

Sunil, P. and Kumar, M. 2010. Chromium carbide hard-faced layer made by welding technique using 6013 SMAW electrode.

4 Erosion Response of Martensitic Stainless Steel Subjected to Slurry Flow

Sarfraj Ahmed and Arjun Kundu

CONTENTS

4.1 INTRODUCTION

Slurry transportation systems, such as slurry pumps, nozzles and pipe lines are subjected to slurry erosive wear. Solid minerals are transported in the form of slurry in steel and mining industries. There is a progressive loss of material from the surface of components due to slurry erosion. Because of the impact of solid particles suspended in a liquid medium, slurry erosion takes place. The wear rate due to slurry erosion is affected by slurry concentration, time of run, speed of rotation, particle size, particle hardness and properties of the target material. The effect of wear on industries is significant in terms of environment and finances. The life of the components decreases because of abrasive and erosive wear. Deteriorated or worn out industrial components are repaired by depositing thick, wear-resistant materials. High flow speed of the slurry causes a subclass of slurry erosion known as abrasive slurry erosion. Under such conditions, it is difficult to understand the mechanisms of wear. Abrasive slurry erosion is caused by the high speed of slurry, large particle size and sharp particles. In industries, fine abrasive particles cause low stress slurry erosion, whereas high stress slurry erosion is caused by large abrasive particles (Ojala et al. 2016).

Dube et al. (2009) studied the erosion wear due to alumina and silica slurry in turbulent operating conditions in the solid concentration range of 15–50% using a counter-rotating double-disc erosion tester. Ahmed et al. (2018) studied the slurry abrasive wear behaviour of various steels under experimental conditions. They measured the wear resistance in terms of volume loss and wear performance. The experimental parameters were slurry concentration, sliding distance, particle size, particle properties and normal load. The volume loss was different for different materials depending upon the experimental conditions (Ahmed et al. 2018). Rawat et al. (2017) studied the erosive wear at high concentrations of fly ash. The fly ash slurries were used in the range of 50–70% by mass, and velocity was also varied. They found that erosion wear had a greater dependence on slurry concentration than on velocity. The effect of impact angle was also investigated, and it was noted that at an angle of 45° the erosion rate was at its maximum.

The weld hard-facing technique is employed to improve the wear life of components used in industries such as steel and power plants. Steels have widely been used in slurry transportation systems because of their good wear-resistant properties. Stainless steels have excellent resistance to corrosion, and martensite is the hardest phase of steel. There is a limited scope for change in design of systems because of constraints; therefore, wear resistance can be improved by altering the properties of the material (Tressia et al. 2017; Kishor et al. 2016). Slurry erosion resistance of martensitic stainless steel was investigated under different experimental conditions using a slurry erosion pot tester. A slurry pot tester may be successfully used for simulating in-service applications. Scanning electron microscope (SEM) analysis was carried out to study the mechanisms of material removal.

4.2 MATERIALS AND METHODS

The material used for slurry erosion pot testing was martensitic stainless steel. The material of required composition was deposited on a mild steel plate via a flux-cored arc welding process, known as hardfacing. A mild steel plate, with dimensions of 200 mm × 200mm × 10mm, was used. The welding parameters included the current at 350–380 A, voltage at 28–29 V, stick out at 25–35 mm and weaving at 20–25 mm. The samples were derived from the weld deposited plate, and then it is machined and ground to the appropriate dimensions. The specimens were rectangular in shape, measuring 50 mm in length, 25 mm in width and 6 mm thick. The pack drilling was carried out on specimens to produce a central hole of 8 mm in diameter so that it could be held on spindle (Santa et al. 2007; Sapate and Haque 2010). The chemical composition of the material was determined by a spark emission spectrometer. The specimen used for bulk hardness measurement was ground flat by successive silicon carbide papers upto 1600 grit finish. The bulk hardness of material and silica sand particles was tested using the Vickers hardness tester. The well-polished specimen of the given material was etched with a Kalling's etching solution to reveal the microstructure. The microstructure of martensitic stainless steel as seen through an SEM is shown in Figure 4.1. The chemical composition and hardness values of hard-faced martensitic stainless steel are given in the Tables 4.1 and 4.2.

Experiments were conducted by varying the operating parameters, including slurry concentration, slurry speed, particle size and time duration. The density of

FIGURE 4.1 Microstructure of martensitic stainless steel using SEM.

TABLE 4.1
Chemical Composition of Specimen

Elements	C	Si	Mn	Cr	Ni	Mo	P	S	Fe
Wt.%	0.0483	1.07	0.868	15.3	3.64	0.406	0.0282	0.0115	Balance

TABLE 4.2
Hardness of Specimen and Erodent Particle

Specimen	Avg. Hardness (Vickers)	Erodent	Avg. Hardness (Vickers)
Martensitic Stainless Steel	395 HV	Silica Sand	1180 HV

martensitic stainless steel specimen was 7.88 gm/cm^3. The specimens are rotated inside the slurry pot, and hard abrasive particles strike the surface of the specimen at the maximum tangential velocity, which can be obtained is 4.18 m/sec. The effect of slurry speed, slurry concentration, particle size and time duration on slurry erosion response of martensitic stainless steel was investigated. The volume loss method was adopted, which gave the wear rate. SEM analysis of worn surfaces was conducted to reveal the morphology and mechanisms of material removal (Rao et al. 2016; Lindgren and Perolainen 2014).

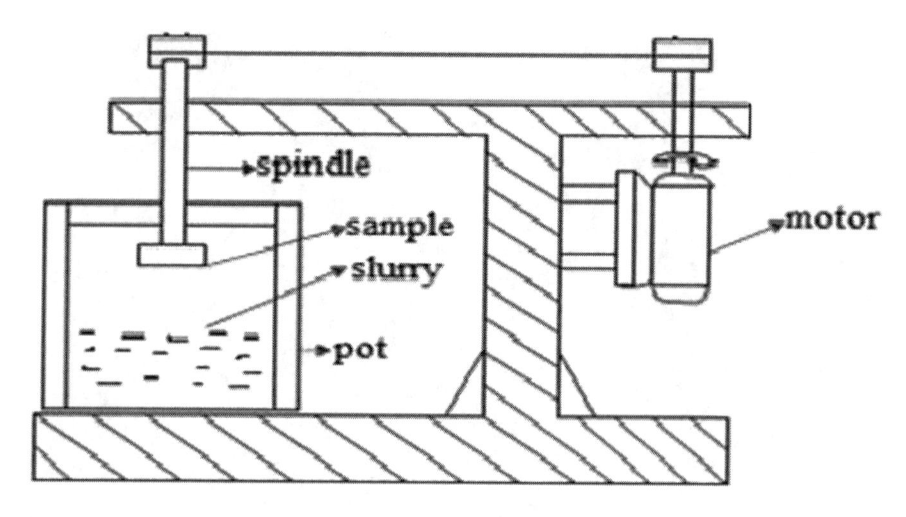

FIGURE 4.2 Schematic of slurry erosion pot testing machine.

4.3 EXPERIMENTAL ANALYSIS

Slurry erosion testing of martensitic stainless steel was conducted using a pot-type slurry erosion tester (Ducom TR-40). The slurry erosion tester consists of six stainless steel slurry pots, and, to create turbulence, the baffles are provided inside the slurry pot. There are six spindles for holding the specimens so that six specimens can be tested at a time. Silica sand particles were used as abrasives and distilled water was added to form slurry. The size of erodent particles was determined by using a SEM. Silica sand particles were selected in the range of 53–73 μm, 125–150 μm, 250–300 μm and 300–425 μm. Silica sand erodent particles were angular in shape (Bhandari et al. 2011).The schematic diagram of the slurry erosion pot tester and SEM images of silica sand erodent particles is shown in Figures 4.2 and 4.3.

The silica sand erodent particles were mixed with distilled water in four different proportions: 10%, 20%, 30% and 40%. The samples were fitted in the specimen holder and dipped into the pot filled with slurry. The specimens were tested at four different speeds: 400 rpm, 800 rpm, 1200 rpm and 1600 rpm. The time duration of testing was varied from 1–4 hrs. The four sets (1–4) of suitably designed experimental plans, in which each set consists of four (a–d) experiments, are shown in Table 4.3.

4.4 RESULTS AND DISCUSSIONS

The slurry erosion response of hard-faced martensitic stainless steel showed different trends depending upon experimental variables. At a speed of 1600 rpm and time duration of 2 hrs, and the slurry concentration was increased from 10% to 40%, the slurry erosive volume loss increased from 2.56 to 8.77 mm^3, which increased the wear rate from 0.0213 to 0.0730 mm^3/min. This occurs because more abrasive particles come in contact with the specimen, increasing the wear rate. The increase in slurry erosive wear rate with respect to slurry concentration is shown in Figure 4.4.

FIGURE 4.3 Silica sand erodent particles of various sizes (a) 53–73 μm (b) 125–150 μm (c) 250–300 μm (d) 300–425 μm.

TABLE 4.3
Sets of Experiments Performed

No. of Sets	Fixed Parameters	Variable Parameters
1.	S – 1600 rpm T – 2 hrs P – 250–300 μm	(a) SC – 10%, (b) SC – 20%, (c) SC – 30% and (d) SC – 40%
2.	SC – 40% P – 250–300 μm	(a) S – 400 rpm and T – 4 hrs,(b) S – 800 rpm and T – 2 hrs (c) S – 1200 rpm and T – 1.2 hrs, (d) S – 1600 rpm and T– 1 hr
3.	SC – 20% S– 1600 rpm P – 250–300 μm	(a) T – 1 hr, (b) T – 2 hrs, (c)T – 3 hrs and (d)T – 4 hrs
4.	S – 1600 rpm SC – 40% T – 1 hr	(a) P– 53–73 μm, (b) P – 125–150 μm, (c) P –250–300 μm and (d) P – 300–425 μm.

Where *S* is speed in rpm, *T* is time in hours, *SC* is slurry concentration in % and *P* is particle size in μm.

When the speed of rotation is increased from 400 rpm to 1600 rpm at 40% slurry concentration, the volume loss increased from 0.4314 to 6.1421 mm^3, and the wear is increased from 0.0018 to 0.102 mm^3/min. It may be expected that at a higher speed, abrasive particles would strike the specimen with a higher impact, leading to a greater indentation on the surface of the specimen (Stack and Pungwiwat 1999;

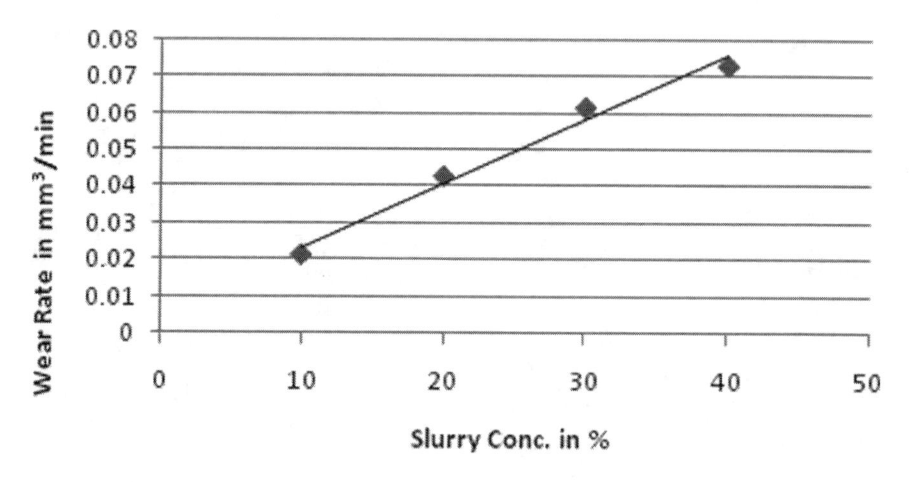

FIGURE 4.4 Slurry concentration vs wear rate at 1600 rpm and 2 hrs.

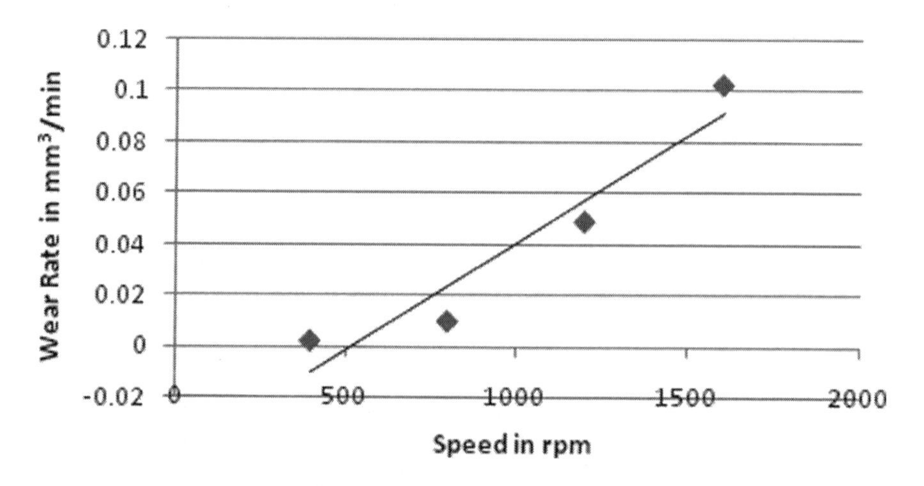

FIGURE 4.5 Speed of rotation vs wear rate at 40%.

Gupta et al. 1995). As shown in Figure 4.5, slurry erosive wear rate with respect to the speed of rotation is not exactly linear. When the particle size was increased from 53–73 μm to 300–425 μm, the wear rate increased from 0.0305 to 0.1492 mm^3/min (see Figure 4.6). The angularity of erodent particles causes material removal at a higher rate, and large-size particles increase the contact area. When the slurry speed is high and particle size is large, abrasive slurry erosion is expected to occur. The slurry erosive wear rate was decreased from 0.054 to 0.036 mm^3/min when the time duration was increased from 1 hr to 4 hrs, which shows an almost linear relationship as seen in Figure 4.7. The SEM analysis of eroded surfaces, as seen in Figure 4.8, shows that the material was mainly removed because of chipping and indentation.

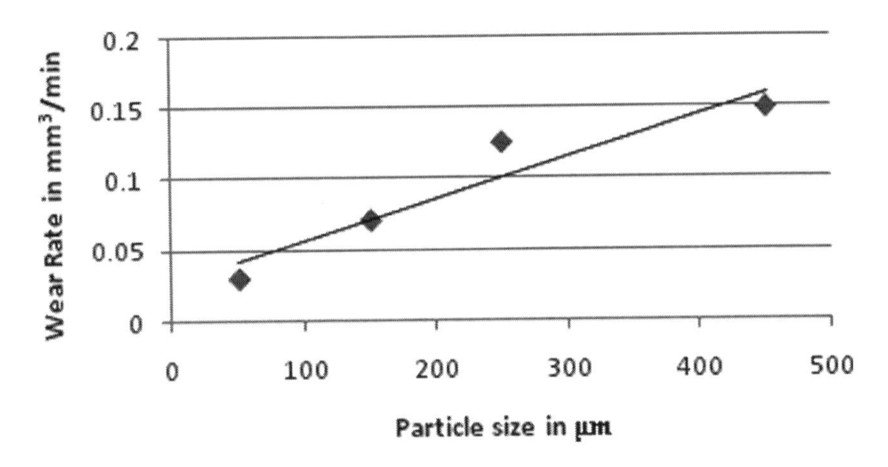

FIGURE 4.6 Particle size vs wear rate at 1600 rpm and 40%.

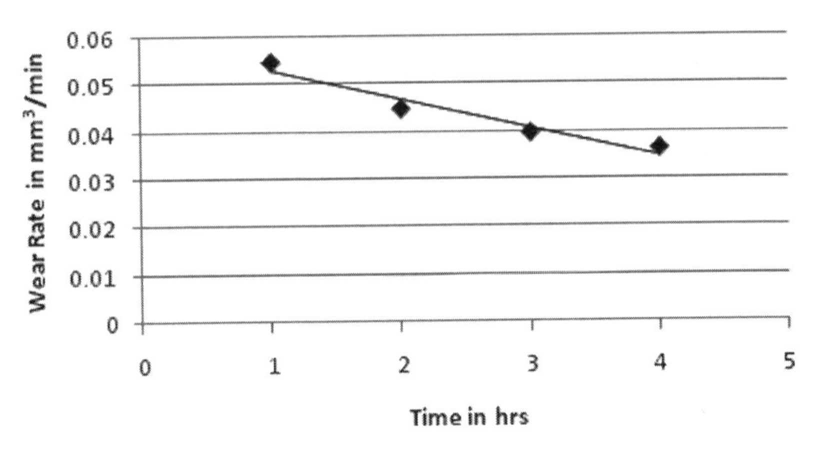

FIGURE 4.7 Time duration vs wear rate.

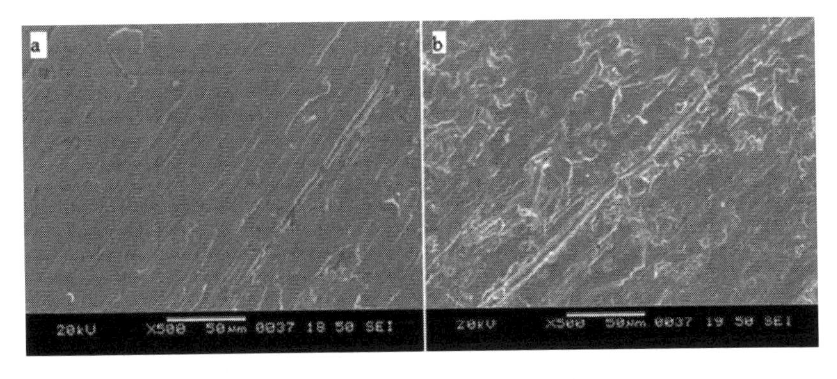

FIGURE 4.8 SEM images of eroded surfaces (a) at 1600 rpm and 40% SC and (b) at particle size of 300–425 µm and 1600 rpm.

4.5 CONCLUSION

The slurry erosion tests were conducted on martensitic stainless steel specimens, and the effect of experimental parameters on slurry erosion response of martensitic stainless steel was investigated. The increase in volume loss was different under each experimental condition. It was found that the wear rate was increased by increasing the slurry concentration, speed of rotation and particle size. Higher slurry speed and large particle size lead to abrasive slurry erosion. Results show that the effect of particle size on slurry erosion was more pronounced as compared to other parameters. It may be because the local flow stress required for deformation of material is reduced because the large-size particles and wear rate is increased. As the experimental investigation shows that increasing the time duration decreases the wear rate, it can be expected that with an increase in time, the material becomes work hardened. Observation of eroded surfaces shows that the material was removed mainly because of chipping and indentation.

REFERENCES

Ahmed, S., Thakare, O.P., Shrivastava, R., Sharma, S. and Sapate, S.G., 2018. A review on slurry abrasion of hardfaced steels. *Materials Today: Proceedings*, 5:3524–3532.

Bhandari, S., Singh, H., Kansal, H.K. and Rastogi, V., 2011. Slurry erosion studies of hydroturbine steels under hydroaccelerated conditions. *Proceedings of the Institution of Mechanical Engineers, Part J: Journal of Engineering Tribology*, 226(3):239–250.

Dube, N.M., Dube, A. and Suman, B.I., 2009. Experimental technique to analyze the slurry erosion Wear due to turbulence. *Wear*, 267:259–263.

Gupta, R., Singh, S. N. and Seshadri, V., 1995. Prediction of uneven wear in a slurry pipeline on the basis of measurements in a pot tester. *Wear*, 184:169–178.

Kishor, B., Chaudhari, G.P. and Nath, S.K., 2016. Slurry erosion of thermo mechanically processed 13Cr4Ni stainless steel. *Tribology International*, 93:50–57.

Lindgren, M. and Perolainen, J., 2014. Slurry pot investigation of the influence of erodent characteristics on the erosion resistance of austenitic and duplex stainless steel grades. *Wear*, 319:38–48.

Ojala, N., Valtonen, K. and Kuokkala, V.T., 2016. Wear performance of quenched wear resistant steels in abrasive slurry erosion. *Wear*, 354–355:21–31.

Rao, K.V.S., Girisha, K.G. and Rakesh, Y.D., 2016. Evaluation of slurry erosive wear of plasma sprayed TiO_2 coated steel. *Materials Science and Engineering*, 149:1–7.

Rawat, A., Singh, S.N. and Seshadri, V., 2017. Erosion wear studies on high concentration fly ash slurries. *Wear*, 378–379:114–125.

Santa, J.F., Baena, J.C. and Toro, A., 2007. Slurry erosion of thermal spray coatings and stainless steels for hydraulic machinery. *Wear*, 263:258–264.

Sapate, S.G. and Haque, N., 2010. Effect of microstructure on slurry erosion behavior of weld hardfacing alloys. *Materials Design and Applications*, 225:49–59.

Stack, M. M. and Pungwiwat, N., 1999. Slurry erosion of metallics, polymers and ceramics: particle size effects. *Materials Science and Technology*, 15:337–344.

Tressia, G., Penagos, J.J. and Sinatora, A. 2017. Effect of abrasive particle size on slurry abrasion resistance of austenitic and martensitic steels. *Wear*, 376–377:63–69.

5 Optimization of Machining Parameters during the Drilling of Natural Fibre-Reinforced Polymer Composites

A Critical Review

Jai Inder Preet Singh, Sehijpal Singh, Vikas Dhawan, Piyush Gulati, Rajeev Kumar, Manpreet Singh, Jujhar Singh, Shubham Sharma and Suresh Mayilswamy

CONTENTS

5.1 INTRODUCTION

It is often observed that humanity strives for betterment and growth. With every development in the field of science and engineering, humans' needs and demands increase. In order to meet society's endless needs, engineers, researchers and scientists work tirelessly to develop new products and improve the performance of existing ones. Modern day applications, especially in aerospace, automotive and chemical industries, require materials with peculiar and unusual properties that are not met by ceramics, metals and polymers alone. In order to overcome the need for new materials for new products and to increase product performance, researchers and designers have resorted to using composites. Composites are a combination of two or more materials to obtain properties that cannot be obtained by individual materials. The components of composites retain their identities, do not dissolve into each other, are physically identifiable and usually present an interface between them. Technically, a composite consists of a matrix material that is reinforced with fibres. Conventionally, polymer matrix composites (PMC) consist of petroleum-based synthetic polymers as matrix material and synthetic fibre, such as carbon, aramid or glass, as reinforcement. These polymers are either thermoplastic (such as polypropylene [PP], nylon, polyvinyl chloride) or thermosets (such as epoxy, phenolic, polyamide, etc.). These conventional polymer composites possess very good mechanical properties, but because of their nonbiodegradable nature and other environmental issues, such as global warming, depletion of fossil resources and rising oil prices, research for sustainable development has increased (Dhawan et al. 2013; Inder et al. 2017). Green composite materials that are derived from natural resources are a possible solution to above mentioned environmental issues and the replacement of conventional polymer composites. Green composites are further classified as partially biodegradable green composites and fully biodegradable green composites. Partially biodegradable green composites are materials in which one of the constituents is derived from a natural resource such as jute fibre/epoxy-based composite, and fully biodegradable green composites are the composites in which both the constituents are derived from a natural resource such as jute fibre/polylactic acid (PLA)-based green composites as reported by Bajpai et al. 2012. Different researchers have used various types of natural fibres (such as banana, sisal, jute, cotton, hemp and coir) as reinforcement in polymers to develop composite materials with improved mechanical properties (Bajpai et al. 2013 Dhawan et al. 2016; Srinivasan et al. 2014; Thakur et al. 2014). Natural fibre composites possess various properties, such as low specific weight, biodegradable nature, low energy consumption and no health hazards (Wambua et al. 2003). Natural fibre composites are being used in a wide range of applications such as automobiles, construction, furniture and so on. A rich application of plant fibres can be found in the Mercedes-Benz E class, as most of the parts of the car, such as the centre console and trim, various damping and insulating parts, seat cushion parts and door trim panels, are made from plant fibre composites (Jalinder 2015).

Generally, polymer composite products are made into their final shape after the primary process, but products with complex shapes also require secondary operations such as machining. Mechanical fastening, i.e. specially bolted joints, are the most common process used for joining of two parts. For the joining of two parts, the location of the hole needed for the bolted joints must first be marked before the polymer composites can be drilled. Drilling polymer composites is completely different from drilling metals and alloys. Mechanisms for the metal removal in the case of drilling metals and alloys are well established, but for polymer composites, it is still under investigation because of the anisotropic nature of polymers. Drilling forces can result in drilling-induced damage on the polymer matrix composites. Delamination and other issues result in poor-quality holes and parts rejection. Today, extensive research is being conducted to define the drill geometry used for drilling holes in polymer composites.

This chapter deals with the drilling aspects of polymer composites in detail. A comparative study has been made for three different drill point geometries with different spindle speeds and feed rates. Different predictive models, such as artificial neural network (ANN), fuzzy logic and adaptive neuro-fuzzy inference system, are used to predict the drilling-induced damage, thrust force and torque induced while drilling the fibre-reinforced polymer (FRP) composites.

5.2 DRILLING OF HYBRID POLYMER COMPOSITES

The drilling of FRP composites has been an area of research of paramount importance for the past two decades. Machining of FRP composites still requires a lot of research input and standardization. The machining conditions required for drilling of anisotropic FRP composites are quite different from those of metal machining. The diverse properties of the fibre and the matrix, combined with different fibre orientations, have a significant effect on the machining process. During drilling, the fibres take on a high proportion of the load, which serves to impair the uniform deformation as would normally be seen in chip formation in metal cutting. The high-strength fibres in composites do not break easily, but rather tend to be pulled by the cutting tool, leading to microcracking and delamination along the cut. The tough fibres are very hard to shear and result in fuzziness along the cut or machined surface. Moreover, overheating at the drilling or cutting zone can heat the resin and locally damage the cured laminate. When fluid is used to cool the work site, the fibres can absorb moisture, also degrading the laminate properties. A high rate of tool wear is usually associated with machining of FRP laminates because of the highly abrasive effect of the reinforced fibre, increasing the total operation time by necessitating frequent tool changes. The drilling operation can be carried out using a radial drilling set-up or computer numerical control (CNC) machining centre. It consists of a vertical column to support the radial arm that is used to give motion to the tool along the x-axis. The machine is able to change the speed and feed rate during the drilling operation. It carries a tool holder to fix different types of drill bits as per the requirements of drilling operation. Generally, cutting speed is set in revolutions per minute (RPM) and feed rate in mm/rev. The composite specimen is set to clamp in the fixture and the machine is programmed to run at the specified cutting speed and feed rate.

Such an experimental set-up requires a radial drilling machine, a workpiece holder, piezoelectric drill dynamometer, charge amplifier, connecting cables, an analog to digital (A/D) converter and a personal computer for data acquisition. The specifications for the radial drilling machine are given in Table 5.1.

Polymer composites laminate were held in the fixture on the top of the dynamometer, which was rigidly mounted with an attachment. Clearance was provided between the workpiece holder and the dynamometer with the help of washers (spacers) to avoid contact between drill and the dynamometer. Highly accurate thrust force and torque signals were acquired using a four-component drill dynamometer. The dynamometer consists of piezoelectric sensors fitted under high preload between a base plate and a top plate. It is compact, robust and capable of measuring three orthogonal thrust forces (Fx, Fy, Fz) and torque (Mz) induced in drilling. In

TABLE 5.1
Specifications of Radial Drilling Machine

Parameter	Specification
Make	Batliboi Pvt. Ltd., Surat, India
Spindle Speed	90–4500 RPM
Feed Rate Range	0.03–0.3 mm/revolution
Drilling Main Motor (Power/Speed)	1.5 KW/1420 RPM
Elevating Motor (Power/Speed)	0.75 KW/1420

drilling, it is used to measure only thrust force (Fz) and torque (Mz) signals. These thrust force and the torque signals are amplified using a charge amplifier. The signal from the amplifier is passed through signal conditioning equipment and is supplied to the personal computer via a data acquisition card.

The challenges encountered while machining of FRP composites are usually not encountered in metal machining. The drilling-induced damage estimation and characterization has been an area of paramount importance. Researchers and scientists worldwide have focused their attention toward the common objective of minimizing damage for the benefit of the composite industry. Several researchers worldwide have tried to observe and model the cutting forces generated during the drilling of FRPs. The optimum properties during the drilling of natural fibre polymer composites are mentioned in Table 5.2.

The study of cutting forces has not only been used to explain the machinability but also to explain the drilling-induced damage. Researchers have conducted many experimental analyzes and have tried to model the cutting forces using different techniques. Cutting forces, especially the thrust force, have been related to the drilling-induced damage in the form of delamination. It has been found that damage occurs when the thrust force exceeds the critical thrust force. Hence, a number of critical thrust-force models based on the linear elastic fracture mechanics and classical plate-bending theory have been proposed to avoid the drilling-induced damage. The study of drilling forces generated by different drill geometries is essential to optimize the operating variables and to reduce drilling-induced damage. This, in turn, will also help to develop dedicated machine tools for machining of FRPs, which can adjust the cutting variables based upon the variation of the drilling forces.

Damage during drilling of an orthotropic composite material depends on:

1. Material of the work piece and orientation of fibres.
2. The stacking sequence and the mechanical characteristics in differently oriented laminate throughout the laminate thickness.
3. Tool geometry and tool material.
4. Operating environment.
5. Process variables.

TABLE 5.2
Optimal Properties during Drilling of Natural Fibre Polymer Composites

Sample No.	Drill Size	Spindle Speed	Feed Rate	Remarks	Source
1	6 mm	3000 rpm	50 mm/min	The optimal drilling parameters suitable for Minimal delamination which in turn improves the hole quality in drilling.	[11]
2	4 mm	900 rpm	0.2 mm/rev	The cutting speed has an insignificant effect on the thrust force when drilling at low feed values. At high feed values, the thrust force decreases with an increased cutting speed.	[10]
3	8 mm	2800 rpm	0.05 mm/rev	With an increase in feed from 0.05 to 0.30 mm/rev, the thrust force and torque increase linearly for a constant value of spindle speed.	[13]
4	6 mm	1500 rpm	0.1 mm/rev	Feed rate was seen to make the largest contribution to the delamination factor. By increasing the feed rate the size of delamination also increased.	[14]

5.3 INPUT AND OUTPUT PARAMETERS

The cutting speed and the feed rate are the two important parameters that influence the drilling process. Chandramohan and Marimuthu (2011) investigated drill size, revolution and feed rate as the input parameters and their effect on thrust force and torque force during the drilling of natural fibres (sisal, banana, roselle), The present work focuses on the prediction of thrust force and torque on the natural fibre-reinforced polymer composite materials and the values compared with the regression model and the scheme of delamination factor/zone using machine vision system. This is also discussed with the help of a scanning electron microscope (SEM). Aravindh and Umanath (2015) studied the delamination in drilling of natural fibre-reinforced polymer composites produced by compression moulding. In the study, drills of three diameters (6, 8, 10 mm) were used with a spindle speed of 1000, 2000 and 3000 rpm and a feed rate of 50, 150 and 200 mm/min. Signal to noise ratio (S/N) response for all the three parameters has been plotted and analyzed. The optimal drilling performance that will minimize the delamination factor was found as 6 mm drill size in Level 1, 3000 rpm spindle speed in Level 3, and 50 mm/1 min feed rate in Level 1. Babu et al. (2012) investigated drilling parameters, such as cutting speed and feed rate, on delamination factors. In their study, three levels of each parameter were selected: 16, 24 and 32 as the cutting speed m/min and 0.10, 0.20 and 0.30 mm/rev as the feed rate. All experiments were conducted on the glass fibre, hemp fibre, jute fibre and banana fibre-reinforced polyester composites.

Debnath et al. (2014) investigated the drilling characteristics of sisal fibre-reinforced epoxy/polypropylene (PP) composites. Three different input parameters used were spindle speed (900, 1800 and 2800 rpm), cutting speed (23, 45 and 70 m/min) and feed rate (0.05, 0.08, 0.12, 0.19 and 0.30 mm/rev). Their effect on thrust force and torque were investigated. Influence of drill geometry on drill forces was also investigated.

5.4 CONCLUSIONS

Natural fibre-reinforced composites offer several advantages such as they are biodegradable, ecofriendly, abundant, renewable, low-density and many more. These are potential materials to replace the synthetic fibre-reinforced composites. This article presents major findings reported in the literature on the study conducted on the optimization of processing parameters for drilling of natural fibre-reinforced polymer composites. Some of the major conclusions drawn from this study are:

1. S/N analysis has been used to find the optimal drilling parameters suitable for minimal delamination, which, in turn, improves the hole quality in drilling.
2. It is observed that the drill size plays a significant role in determining the delamination value.
3. Taguchi design of experiment method provides a systematic and efficient methodology for the design optimization of the process parameters resulting in minimum delamination with far less effect than would be required for most optimization techniques.
4. The feed rate and cutting speed are seen to contribute the most to the delamination effect.

DECLARATION OF CONFLICT OF INTEREST

The authors declare that they have no conflict of interest with respect to the research, authorship and/or publication of this article.

FUNDING

The author(s) did not receive any research funding or grants from any organization.

ETHICAL APPROVAL

This article does not contain any studies with human participants or animals performed by the author.

REFERENCES

Aravindh, S. and Umanath, K. 2015. Delamination in drilling of natural fibre reinforced polymer composites produced by compression moulding. *Trans Tech Publ* 766: 796–800.

Babu, D., Babu, K.S. and Gowd, B.U.M. 2012. Drilling uni-directional fiber-reinforced plastics manufactured by hand lay-up: Influence of fibers. *American Journal of Materials Science and Technology* 1: 1–10.

Bajpai, P.K., Meena, D., Vatsa, S. and Inderdeep, S. 2013. Tensile behavior of nettle fiber composites exposed to various environments. *Journal of Natural Fibers* 10: 244–256.

Bajpai, P.K., Singh, I. and Madaan, J. 2012. Development and characterization of PLA-based green composites: A review. *Journal of Thermoplastic Composite Materials* 27(1): 52–81.

Chandramohan, D. and Marimuthu, K. 2011. Drilling of natural fiber particle reinforced polymer composite material. *International Journal of Advanced Engineering Research and Studies* I(I): 134–145.

Debnath, K., Singh, I. and Dvivedi, A. 2014. Drilling characteristics of sisal fiber-reinforced epoxy and polypropylene composites. *Materials and Manufacturing Processes* 29: 1401–1409.

Dhawan, V., Debnath, K., Inderdeep, S. and Sehijpal, S. 2016. Prediction of forces during drilling of composite laminates using artificial neural network : A new approach. *FME Transactions* 44: 36–42.

Dhawan, V., Sehijpal, S. and Inderdeep, S. 2013. Effect of natural fillers on mechanical properties of GFRP composites. *Journal of Composites* 2013: 1–8.

Inder, J., Singh, P., Singh, S. and Dhawan, V. 2017. Effect of curing temperature on mechanical properties of natural fiber reinforced polymer composites. *Journal of Natural Fibers* 15(5): 1–10.

Jaiinder, P.S., Dhawan, V. and Sehijpal, S. 2015. Development and characterization of green composites. *ELK Asia Pacific Journal* 2: 4–7.

Srinivasan, V.S., Rajendra Boopathy, S., Sangeetha, D. and Vijaya Ramnath, B. 2014. Evaluation of mechanical and thermal properties of banana-flax based natural fibre composite. *Materials and Design* 60: 620–627.

Thakur, V.K., Thakur, M.K. and Gupta, R.K. 2014. Review: Raw natural fiber-based polymer composites. *International Journal of Polymer Analysis and Characterization* 19: 256–271.

Wambua, P., Ivens, J. and Verpoest, I. 2003. Natural fibres: Can they replace glass in fibre reinforced plastics? *Composites Science and Technology* 63: 1259–1264.

6 Mechanical Properties of the Palmyra Fibre Epoxy Composites

S. Magibalan

CONTENTS

6.1 INTRODUCTION

A composite is made out of two elements of which there is a process of reinforcing in the form of fibres that are put into the other step of the matrix process that improves the physical and chemical properties of the composite strengthened by natural fibres (Kahraman et al. 2005; Satyanarayana et al. 1986, 1990; Gowda et al. 1999). Research has discovered that the fibre aspect ratio, volume, proportion, direction, matrix adhesion, etc., influence the mechanical properties of natural fibre-reinforced composites (Pothan et al. 2003; Corbière-Nicollier et al. 2001; Pothan et al. 1997; Gowda et al. 1999) tested the material characteristics of jute fibres reinforced in a polyester matrix. Polyvinyl chloride (PVC), reinforced with natural fibres and plasticization effect, have a strengthening influence and manipulation effects on properties by coupling agents of the composites (Khalil et al. 2013). Abdulmajeed et al. (2011) analyzed the mechanical properties of composites reinforced with unidirectional glass fibre. Garoushi et al. (2007) investigated the effect of reinforcing interpenetrating polymer network (IPN) – a polymer matrix with short E-glass fibres – on the mechanical properties. It was observed that the mechanical properties improved with the use of short E-glass fibres as filler material as compared to conventional restorative composites. Similarly, the thermomechanical and tensile properties of polyamide-6 (PA6) composites reinforced with short carbon fibres were studied by Karsli and Aytac (2013). Chaurasiya et al. (2014) extracted the pulp from the palmyra tree and studied the storage life of this pulp. Velmurugan and Manikandan (2007) conducted an investigation on the mechanical characteristics of a hybrid composite consisting of palmyra and glass fibres reinforced in rooflite resin. In comparing mechanical properties, Venkateshwaran et al. (2011) observed that the tensile strength of palmyra/polyester composites is lower than that of banana fibre/epoxy composites, which is around 43.54% higher than the polyester-reinforced palmyra fibre. Shanmugam and Thiruchitrambalam 2013) investigated the static and dynamic mechanical strength of alkali-treated palm leaf and jute fibres coated in a polyester matrix. The hybrid composites are made with unidirectionally arranged fibres. They found that adding alkali-treated palmyra fibre improved the composite impact power rather than adding jute fibre to the composite. Balakrishna et al. (2013) tested the effects of alkali treatment, fibre volume and fibre length on the tensile strength of Asian palmyra fibre. Further, they developed a mathematical model to predict the tensile strength of the composite under the influence of varying factors. Daud et al. (2015) studied the surface characteristics and mechanical properties of corn stalk and oil palm leaf fibre. Srinivasababu et al. (2014) performed extensive work on the manufacturing processes and characterization of the palmyra palm fibre reinforced in a polyester matrix. Ali et al. (2010) investigated the palmyra palm fruits for its physicochemical properties. Manikandan and Velmurugan (2011) explored the possibility of using saw-dust, coir pitch and palmyra fibre as reinforcement in polyester matrix. Their research revealed that the impact strength, tensile strength and shear strength improved with addition of saw-dust and coir pitch to the sandwiched composite plates. Researchers have also worked to explore the possibilities of using the palmyra palm in areas in which high strength is required. Attom et al. (2009) investigated the effects of palmyra fibre and nylon fibre on the mechanical properties of clay soil. The results revealed that the addition of fibre to the clay soil improved its stiffness, compressive strength and ductility, with palmyra fibre showing a greater improvement in the mechanical properties than nylon fibre. Mahesh et al. (2013) investigated the mechanical as well as thermal properties of chemically treated palmyra fibre reinforced in a polyester matrix. In addition to the palmyra fibre, chalk powder was also used in the matrix as a filler material. Nayak and Mishra (2013) used the palmyra fruit fibre and studied its mechanical and water absorption properties. From the experiments, they found that, with the increase in the fibre content, water absorbing characteristics of the composite also gradually increased. Goulart et al. (2011) investigated the impact strength, flexibility and stiffness of palm fibre reinforced in polypropylene matrix and found that the mechanical properties of composites are enhanced with the addition of coupling agent as compared to pure polymer composites.

6.2 MATERIALS AND METHODS

The composites are produced in the present research work utilizing the basic hand lay-up technique, which is known to be the oldest and cheapest open moulding process of composite manufacture. The compositions of the various composites fabricated include:

A. Epoxy + 4.0 wt.% short palmyra fibre
B. Epoxy + 8.0 wt.% short palmyra fibre
C. Epoxy + 12.0 wt.% short palmyra fibre

The current research work employs palmyra palm as the natural fibre reinforcement in the epoxy matrix to produce a set of fibre-reinforced composites. The structure of palmyra palm leaf stalk itself can be considered as a composite material that consists of long, aligned cellulose fibres soaking in a polymer structure. Palmyra fibre consists mostly of cellulose, hemicellulose, lignin, pectine and some extractives. In this research, short palmyra fibres (SPF) were used in the reinforcing phase for fabricating the composites. The palmyra fibre was extracted after cutting it from the palm tree. The stalks were split into long strips and dipped in water for a duration of 10 days, accompanied by drying in sunlight for 5 days. The fibres were then cut with scissors to an approximate length of 3 mm.

A metal (die steel) mould with 250 mm × 100 mm × 3 mm was made. The tensile test specimen was cut according to ASTM D 3039-76 procedure, and flexural specimens of 100 mm × 10.25 mm × 3 mm were cut as per ASTM D 618 specifications. The impact test sample was made with dimensions of 63.5 mm × 12.7 mm × 12.7 mm, and the notch was prepared in accordance with ASTM D 256 specifications.

6.3 INFLUENCE OF FIBRE LOADING ON COMPOSITE TENSILE STRENGTH

Figure 6.1 describes the impact on composite tensile strength of the weight fraction of the fibre. While the proportion of the fibre weight increased to 88:12 wt., the composite tensile strength improved by a percentage to 135.5 MPa for 3 cm fibre length.

The tensile strength calculated in the research is consistent with earlier investigators; however, the extraction method for palmyra fibre is different in size. Tensile

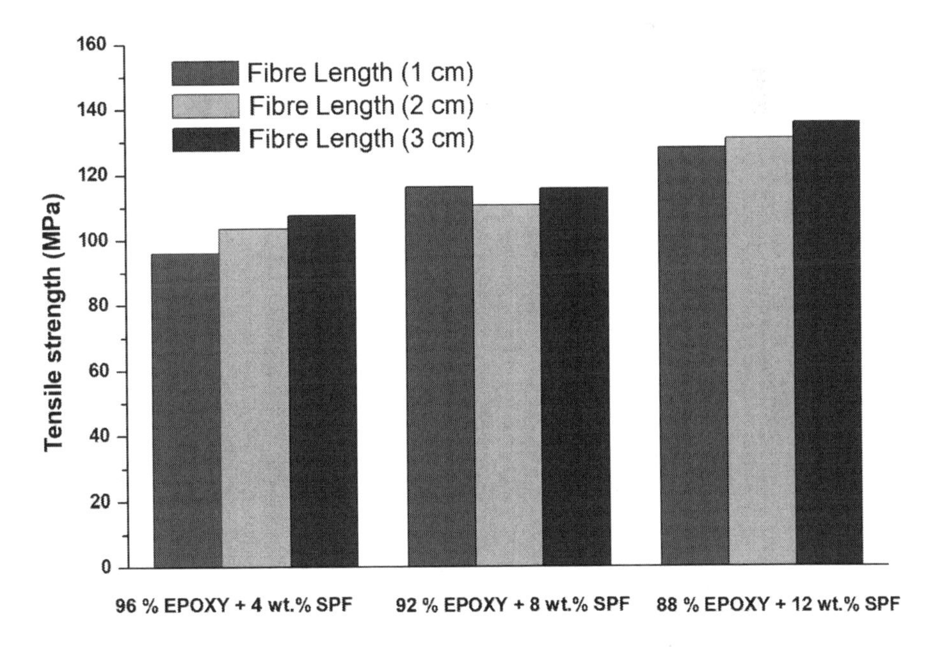

FIGURE 6.1 Impact of fibre loading on composite tensile strength.

strength of the palmyra fibre- epoxy composites is increased with fibre length filled up to 1, 2, 3 cm composites displaying significantly trend increase.

In the first trend, the fibre length was 1 cm, and the fibre loading varied from 96% epoxy with 4% SPF to 88% with 12% SPF. The tensile strength increased from 96 MPa to 128 MPa respectively, as shown in Figure 6.1.

In the second trend, the fibre length was 2 cm, and the fibre loading varied from 96% epoxy with 4% SPF to 88% with 12% SPF. The tensile strength increased from 103.5 MPa to 130.5 MPa respectively, as shown in Figure 6.1.

In the third trend, the fibre length was 3 cm, and the fibre loading varied from 96% epoxy with 4% SPF to 88% with 12% SPF. The tensile strength increased from 107.5 MPa to 135.5 MPa respectively, as shown in Figure 6.1.

6.4 INFLUENCE OF FIBRE LOADING ON COMPOSITE IMPACT STRENGTH

Figure 6.2 explains the impact of the weight fraction of fibre on the composite impact strength. As the fibre weight fraction increased up to 88:12 wt., the composite impact strength increased up to 13.4 J/cm^2 for 3 cm fibre length.

The impact strength calculated in the research is consistent with earlier investigators; however, the extraction method for palmyra fibre is different in size. Impact strength of the palmyra fibre- epoxy composites is increased with fibre length filled up to 1, 2, 3 cm composites displaying significantly trend increase.

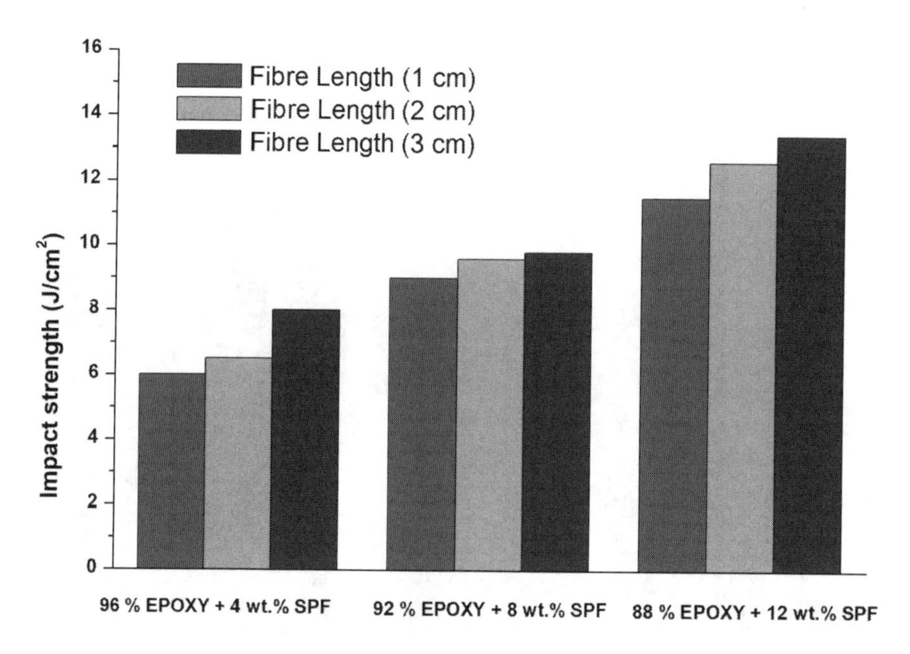

FIGURE 6.2 Impact of fibre loading on composite impact strength.

In the first trend, the fibre length was 1 cm, and the fibre loading varied from 96% epoxy with 4% SPF to 88% with 12% SPF. The impact strength increases from 6 J/cm^2 to 11.5 J/cm^2 respectively, as shown in Figure 6.2.

In the second trend, the fibre length was 2 cm, and the fibre loading varied from 96% epoxy with 4% SPF to 88% with 12% SPF. The impact strength increased from 6.5 J/cm^2 to 12.6 J/cm^2 respectively, as shown in Figure 6.2.

In the third trend, the fibre length was 3 cm, and the fibre loading varied from 96% epoxy with 4% SPF to 88% with 12% SPF. The impact strength increased from 8 J/cm^2 to 13.4 J/cm^2 respectively, as shown in Figure 6.2.

6.5 INFLUENCE OF FIBRE LOADING ON COMPOSITE FLEXURAL STRENGTH

Figure 6.3 explains the impact of the weight fraction of fibre on the composite flexural strength. As the fibre weight fraction increased up to 88:12 wt., the composite flexural strength increased up to 151 MPa for 3 cm fibre length.

The flexural strength calculated in the research is consistent with earlier investigators; however, the extraction method for palmyra fibre is different in size. Flexural strength of the palmyra fibre- epoxy composites is increased with fibre length filled up to 1, 2, 3 cm composites displaying significantly trend increase.

In the first phase, the fibre length was 1 cm, and the fibre loading varied from 96% epoxy with 4% SPF to 88% with 12% SPF, as seen in Figure 6.3. The flexural

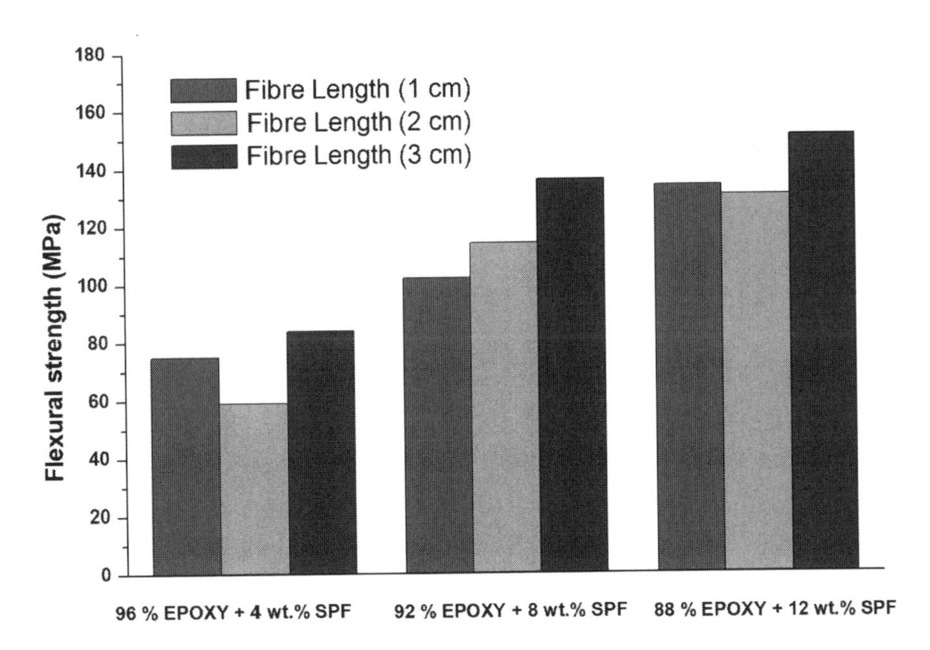

FIGURE 6.3 Impact of fibre loading on composite flexural strength.

strength increased from 75 MPa to 134 Mpa. In the second phase, the fibre length was 2 cm, and the fibre loading varied from 96% epoxy with 4% SPF to 88% with 12% SPF, as seen in Figure 6.3. The flexural strength increased from 78 MPa to 130.5 Mpa. In the third phase, the fibre length was 3 cm, and the fibre loading varied from 96% epoxy with 4% SPF to 88% with 12% SPF, as seen in Figure 6.3. The flexural strength increased from 81 MPa to 151 MPa.

6.6 CONCLUSIONS

1. Successful manufacturing of epoxy composites reinforced with short palmyra fibres (SPF) using easy hand lay-up technique is essential.
2. Tensile strength of fibre length at 3 cm and fibre loading of 88% with 12% SPF increased the tensile strength to 135.5 MPa.
3. Impact strength of fibre length at 3 cm and fibre loading of 88% with 12% SPF increased the impact strength to 13.4 J/cm^2.
4. Flexural strength of fibre length at 3 cm and fibre loading of 88% with 12% SPF increased the flexural strength 151 MPa.

REFERENCES

Abdulmajeed, A. A., Närhi, T. O., Vallittu, P. K. and Lassila, L. V. 2011. The effect of high fibre fraction on some mechanical properties of unidirectional glass fibre-reinforced composite. *Dental materials* 27(4): 313–321.

Ali, A., Alhadji, D., Tchiegang, C. and Saïdou, C. 2010. Physico-chemical properties of palmyra palm (*Borassus aethiopum* Mart.) fruits from Northern Cameroon. *African Journal of Food Science* 4(3): 115–119.

Attom, M. F., Al-Akhras, N. M. and Malkawi, A. I. 2009. Effect of fibres on the mechanical properties of clayey soil. *Proceedings of the Institution of Civil Engineers: Geotechnical Engineering* 162(5): 277–282.

Balakrishna, A., Rao, D. N. and Rakesh, A. S. 2013. Characterization and modeling of process parameters on tensile strength of short and randomly oriented *Borassus flabellifer* (Asian Palmyra) fibre reinforced composite. *Composites Part B: Engineering* 55: 479–485.

Chaurasiya, A. K., Chakraborty, I. and Saha, J. 2014. Value addition of palmyra palm and studies on the storage life. *Journal of Food Science and technology* 51(4): 768–773.

Corbière-Nicollier, T., Laban, B. G., Lundquist, L., Leterrier, Y., Månson, J. A. and Jolliet, O. 2001. Life cycle assessment of biofibres replacing glass fibres as reinforcement in plastics. *Resources, Conservation and recycling* 33(4): 267–287.

Daud, Z., Hatta, M. Z. M. and Awang, H. 2015. Oil palm leaf and corn stalk–Mechanical properties and surface characterization. *Procedia: Social and Behavioral Sciences* 195: 2047–2050.

Garoushi, S., Vallittu, P. K. and Lassila, L. V. 2007. Short glass fibre reinforced restorative composite resin with semi-inter penetrating polymer network matrix. *Dental materials* 23(11): 1356–1362.

Goulart, S. A. S., Oliveira, T. A., Teixeira, A., Miléo, P. C. and Mulinari, D. R. 2011. Mechanical behaviour of polypropylene reinforced palm fibres composites. *Procedia Engineering* 10: 2034–2039.

Gowda, T. M., Naidu, A. C. B. and Chhaya, R. 1999. Some mechanical properties of untreated jute fabric-reinforced polyester composites. *Composites – Part A: Applied Science and manufacturing* 30(3): 277–284.

Kahraman, R., Abbasi, S. and Abu-Sharkh, B. 2005. Influence of epolene G-3003 as a coupling agent on the mechanical behavior of palm fibre-polypropylene composites. *International Journal of Polymeric Materials* 54(6): 483–503.

Karsli, N. G. and Aytac, A. 2013. Tensile and thermomechanical properties of short carbon fibre reinforced polyamide 6 composites. *Composites Part B: Engineering* 51: 270–275.

Khalil, H. A., Tehrani, M. A., Davoudpour, Y., Bhat, A. H., Jawaid, M. and Hassan, A. 2013. Natural fibre reinforced poly (vinyl chloride) composites: A review. *Journal of Reinforced Plastics and Composites* 32(5): 330–356.

Mahesh, C., Kondapanaidu, B., Govindarajulu, K. and Murthy, V. B. 2013. Experimental investigation of thermal and mechanical properties of palmyra fibre reinforced polyster composites with and without chemical treatment and also addition of chalk powder. *International Journal of Engineering Trends and Technology* 5(5): 259–271.

Manikandan, V. and Velmurugan, R. 2011. Utilization of bioresources such as coir-pith, saw dust and palmyra fibre as reinforcement material in polyester matrix. *Materials Science: An Indian Journal* 7(2): 94–99

Nayak, N. C. and Mishra, A. 2013. Development and mechanical characterization of palmyra fruit fibre reinforced epoxy composites. *Journal of Production Engineering* 16(2): 69–72.

Pothan, L. A., Oommen, Z. and Thomas, S. 2003. Dynamic mechanical analysis of banana fibre reinforced polyester composites. *Composites Science and technology* 63(2): 283–293.

Pothan, L. A., Thomas, S. and Neelakantan, N. R. 1997. Short banana fibre reinforced polyester composites: mechanical, failure and aging characteristics. *Journal of Reinforced Plastics and Composites* 16(8): 744–765.

Satyanarayana, K. G., Sukumaran, K., Kulkarni, A. G., Pillai, S. G. K. and Rohatgi, P. K. 1986. Fabrication and properties of natural fibre-reinforced polyester composites. *Composites* 17(4): 329–333.

Satyanarayana, K. G., Sukumaran, K., Mukherjee, P. S., Pavithran, C. and Pillai, S. G. K. 1990. Natural fibre-polymer composites. *Cement and Concrete composites* 12(2): 117–136.

Shanmugam, D. and Thiruchitrambalam, M. 2013. Static and dynamic mechanical properties of alkali treated unidirectional continuous palmyra palm leaf stalk fibre/jute fibre reinforced hybrid polyester composites. *Materials and Design* 50: 533–542.

Srinivasababu, N., Kumar, J. S. and Reddy, K. V. K. 2014. Manufacturing and characterization of long palmyra palm/*Borassus flabellifer* petiole fibre reinforced polyester composites. *Procedia Technology* 14: 252–259.

Velmurugan, R. and Manikandan, V. 2007. Mechanical properties of palmyra/glass fibre hybrid composites. *Composites – Part A: Applied Science and manufacturing* 38(10), 2216–2226.

Venkateshwaran, N., ElayaPerumal, A., Alavudeen, A. and Thiruchitrambalam, M. 2011. Mechanical and water absorption behaviour of banana/sisal reinforced hybrid composites. *Materials and Design* 32(7): 4017–4021.

7 ANFIS-Based Prediction of MRR and Surface Roughness in Electrical Discharge Machining of HAMMC

D. Mala, N. Senthilkumar, B. Deepanraj and T. Tamizharasan

CONTENTS

7.1 INTRODUCTION

The conventional method of removing desired material from a metal matrix composite (MMC) is a challenging job because their cells and cell edges are damaged and/or collapsed during the machining processes, thereby deteriorating their fundamental properties. The much-improved mechanical, thermal and chemical properties of newly developed engineering materials makes it possible to machine them with the help of traditional/conventional machining processes, such as abrasion and cutting (Ilio and Paoletti 2012), as conventional machining methods basically rely on subtraction of unwanted materials using harder cutting tools than the newly developed engineering materials. The higher machining cost associated with composite materials and ceramics, along with the developed damage during material removal, adds complexity and constraints to these engineering materials. Apart from advanced materials, low-rigidity structures, more complicated profiles and components that require micromachining with close tolerances and a superfinish surface texture are usually desirable. Conventional methods of machining that involve metal cutting of parts from advanced materials are ineffective, which illustrates the need to develop novel methods in die making and the automotive, aerospace and moulding industries (El-Hofy 2005). This problem can be overcome, to a certain extent, by machining this material using the electrodischarge machining (EDM) process (Uthayakumar and Parameswaran 2019).

Mandal et al. (2018) mapped the relationship existing between the inputs and outputs using adaptive neuro-fuzzy inference system (ANFIS) and forecast the responses with a mean square error (MSE) of 0.0416 in wire-cut EDM (WEDM) process for different membership function (MF). Singh et al. (2019) compared Buckingham pie-theorem, artificial neural networks (ANN) and ANFIS for envisaging roughness on the machine surface that was attained during argon assisted EDM machining of D3 steel. They found that fitting of surface roughness (SR) by ANFIS is better than the other two methods with a lower MSE of 2.44×10^{-6}. Shehabeldeen et al. (2019) performed friction stir welding for joining dissimilar aluminium alloys (AA2024 and AA5083) and mapped the relationship among the inputs and ultimate tensile strength by adopting ANFIS and response surface methodology (RSM). This obtained an MSE of 2.33589 for ANFIS against 53.50798 for RSM, thereby developing an adequate ANFIS model. Fazlollahtabar and Gholizadeh (2020) investigated the influence of EDM parameters on corrosion of tool electrodes, surface roughness and metal removal rate (MRR) by engaging ANFIS, and they were able to predict the optimal levels during simultaneous optimization with a lower mean error.

Muthuramalingam et al. (2019) adopted an ANFIS-based model for forecasting white layer thickness while machining silicon steel by mapping duty factor, peak current and circuit voltage over white layer thickness considering gaussian MF with an accuracy of 96.8%. Manikandan et al. (2019) performed WEDM on LM6/SiC/Dunite hybrid composite and developed an intelligent decision-making model ANFIS toward predicting the desired output characteristic. They found that the ANFIS network more correctly envisages the responses than the results obtained from experimental outcomes and evolved models.

From literature survey it was identified that, least work was performed on EDM machining of hybrid aluminium metal matrix composites (HAMMC), prepared through P/M route and prediction of output responses through the adoption of ANFIS methodology. Hence, in this present study, a HAMMC Al+4%Cu+7.5%SiC, fabricated through P/M technique was considered for EDM process, and the input parameter and output response mapping were done with ANFIS-based fuzzy theory to develop an adequate model for forecasting the output responses with fewest errors.

7.2 MATERIALS SELECTION AND METHODOLOGY

7.2.1 Selection of Base Materials

Aluminium is the foremost choice for researchers and scientists because of its ability to resist corrosion well in extreme conditions with passivation phenomenon and with lower density, which also makes it an ideal material for use in conventional and novel applications. Aluminium powders can be formed by different techniques and approaches that were consolidated into larger structural components by several powder metallurgy processes (Davis 1999). Copper in pure form and as alloys, such as bronzes, brasses and cupronickels, finds extensive applications in heat exchangers, domestic heating, automobile radiators, solar panels and numerous applications that require extreme conduction of heat across a medium. Because of its corrosion resistance and strength, copper-based materials are used for fabricating pipes, valves and fittings used to transport water, potable water, process water and other aqueous fluids. Silicon carbide (SiC) ceramic particles possess high hardness, high oxidation resistance, high electrical conductivity and excellent thermal shock resistance. SiC is attractive for several applications because of its unique combination of low thermal expansion and density, higher thermal conductivity, elastic modulus and flexural strength. It decomposes without melting at >2500°C under an atmospheric pressure.

7.2.2 Powder Metallurgy

Powder metallurgy techniques utilize the powders of alloying elements to convert it into a specific component or product through the sequence of operations that comprises of mixing the ingredients, compacting, sintering and finishing (Eisenet al. 1998; Upadhyaya and Upadhyaya 2011). The initial procedure is to measure the primary alloying powders in appropriate amounts by volume or weight fractions and then thoroughly mix it for homogeneity. The next step is to compact them into the specific shape and size as per the die, where the properly mixed powders are pressed using a mechanical or hydraulic press with a specific loading (Tsukerman 1965). For this study, the powders were compacted with a 20-ton load. Sintering process comprises of heating the compacted green mould for 3 hours at 500°C so that proper bonding of powders takes place, during which the furnace was kept under the atmosphere of nitrogen supplied at the rate of 0.5 litres per minute.

7.2.3 Electrical Discharge Machining

EDM is a nonconventional type of machining technique in which the required shape of the workpiece is made in the tool electrode and by eroding the workpiece material around the electrode using intense sparks (electrical discharges) that are produced between the tool and workpiece (Jameson 2001). With a series of sparks that occurs between the tool and workpiece, the material is eroded between the negative and positive plattens that are parted by a fluid medium, called dielectric, which is nonconductive in nature (Lauwers et al. 2012). Generally, kerosene is the dielectric fluid used, and copper is the tool electrode material, which is also used in this experimentation. Specifications for the EDM machine used in this experiment were: maximum current of 32 Amps, maximum holding size of workpiece as 74×45×27 cm, axis travel x/y/z of 35×25×20 cm, maximum workpiece weight of 200 kg and maximum tool electrode weight of 50 kg. Generally in EDM, the effect of process parameters, such as applied current, pulse on-time, pulse off-time, gap voltage and flushing pressure, can be studied to determine the machinability behaviour of the processed material over surface roughness, tool wear rate, metal removal rate and thickness of recast layer (Yang et al. 2017).

7.2.4 Taguchi's Technique for Experimental Design

Design of experiments (DoE) is a procedure of experimentation that is applied to processes that are complex in nature, with the aim of optimizing the outcomes (Tamizharasan et al. 2019). DoE refers to the procedure of planning the experiment, design and analysis so that a valid and concrete conclusion may be reached efficiently and effectively, which forms the basis for Taguchi's technique (Senthilkumar and Tamizharasan 2014). Taguchi laid an experimental array called orthogonal arrays (OAs) for different combinations of inputs and their interactions to conduct experiments as per Latin square design (Ross 2005a; Roy 2001). Taguchi's design is unique because of its simplicity and robustness while requiring fewer experimental trials. To perform the machining studies, Taguchi's DoE was adopted to formulate the orthogonal array for three parameters, considering three level values for each, and a suitable L_{27} orthogonal array was considered. The EDM electrical parameters considered for machining the HAMMC are current, pulse on-time and pulse off-time. Table 7.1 presents the selected input parameters and their control values.

7.2.5 Adaptive Neuro-Fuzzy Inference System (ANFIS)

Fuzzy logic is a seamless procedure to denote ambiguity and inadequate expert information (Ross 2005). The vital difficulty in fuzzy systems is formulating different sets of if-then rules to map the input and output space. Two methods might be applied: initially, converting experience and knowledge of human expertise and then to generate rules automatically by using an expert system (Czogala and Leski 2000). By combining fuzzy inference systems (FIS) and ANNs, real-life problems can

TABLE 7.1
Process Conditions and Values

Parameters/Levels	Units	Level 1	Level 2	Level 3
Current	Amp	3	5	7
Pulse ON-Time (T_{on})	μs	50	100	150
Pulse OFF-Time (T_{off})	μs	20	30	40

be solved using adaptive intelligence in several scientific and engineering sectors, which has fascinated the collective attention of academicians and scientists. In ANN, the developed network acquires knowledge from scratch by fine-tuning the weight of interconnection and their strengths between different layers of information sharing. An examination exposes that the weaknesses of these tactics appear complementary and, therefore, it is sensible to ponder constructing a combined system that connects different concepts into one, thereby utilizing the advantages of both methods. The ANN's capability to learn from the provided data during training can be used for generating if-then rules automatically in the fuzzy system (Kumar et al. 2019; Bhiradi et al. 2020). The linking of ANN with FIS is referred to as neuro-fuzzy, which utilizes an adaptive learning technique. The advantage of FIS is its capability to learn, while the ANN's advantage is the automatic establishment of a linguistic rule base (Nedjah and Mourelle 2005).

7.3 RESULTS AND DISCUSSIONS

Before performing the machining studies, the micrographs and mechanical characterization of the fabricated Al-4%Cu-7.5%SiC P/M HAMMC were carried out. The microstructure of the fabricated HAMMC, presented in Figure 7.1, displays the optical and SEM micrographs. Some free copper can be seen in the unfused condition in the aluminium matrix by a volume fraction of 0.6%. Remaining portions of the micrograph show some Cu-Al_2 intermetallic in a fused form in the solid solution of aluminium. Homogeneous and uniform dispersal of ceramic particles of SiC were observed as dark grey particulates, and a SiC particle is higher in the matrix (Selvakumar et al. 2016).

The properties of the fabricated hybrid composite determined through different characterization techniques as per American Society for Testing and Materials (ASTM) standard were: ultimate compression strength of 0.062 kN/mm^2 (ASTM E9), breaking load of 19.955 kN, Rockwell hardness of 42.93 (ASTM E18), wear rate of

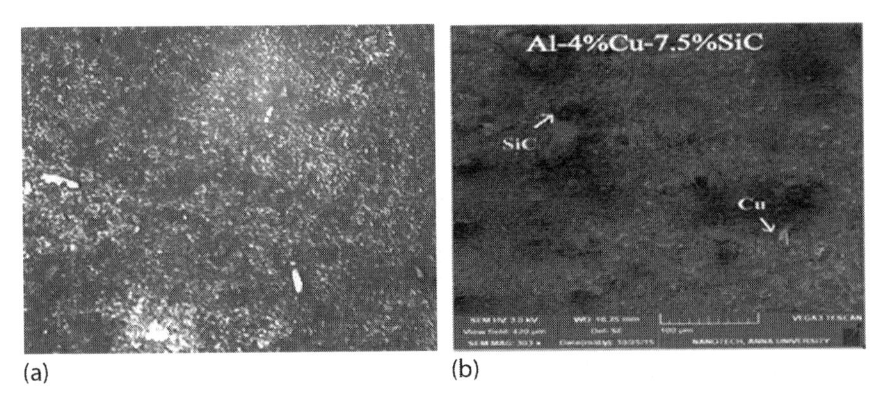

FIGURE 7.1 Optical and scanning electron microscope micrographs of 7.5%SiC in Al-4%Cu matrix.

6.82×10^{-8} g/cm (ASTM G99) and thermal conductivity of 174 W/mK (Selvakumar et al. 2017).

Experiments were conducted in a computer numerical controlled (CNC) EDM machine, as per the designed L_{27} OA. A separate workpiece was considered for each experiment, and a copper electrode was used. The measured output responses after performing the experiments are given in Table 7.2. Two replications were considered, i.e. experiments were conducted twice to avoid any experimental deviation (error), and the mean values are considered for experimental analysis.

Observations made from experimental investigation illustrates that, when current is varied from 3A to 5A, a steep increase in MRR is noticed (Kumar et al. 2018).A significant increase in MRR by 272.63% is seen. When the current is further increased to 7A, MRR is further increased by 87.49%, which makes it obvious that current

TABLE 7.2
Inner and Outer Array of Designed Experiment

Exp. No.	Current (A)	Pulse ON-Time (μs)	Pulse OFF-Time (μs)	MRR (g/min)	SR (microns)
1	3	50	20	0.00432	4.798
2	3	50	30	0.00364	4.23
3	3	50	40	0.00311	3.576
4	3	100	20	0.00351	7.459
5	3	100	30	0.00175	6.011
6	3	100	40	0.00164	5.437
7	3	150	20	0.03914	6.177
8	3	150	30	0.01802	5.93
9	3	150	40	0.01038	5.384
10	5	50	20	0.00652	9.408
11	5	50	30	0.01194	8.573
12	5	50	40	0.02875	7.271
13	5	100	20	0.04485	5.763
14	5	100	30	0.05027	6.191
15	5	100	40	0.04886	7.571
16	5	150	20	0.03563	6.367
17	5	150	30	0.0482	7.407
18	5	150	40	0.04339	8.251
19	7	50	20	0.04803	14.666
20	7	50	30	0.05044	11.281
21	7	50	40	0.05499	10.479
22	7	100	20	0.06532	10.819
23	7	100	30	0.06693	9.019
24	7	100	40	0.0673	9.222
25	7	150	20	0.06311	11.458
26	7	150	30	0.08381	11.498
27	7	150	40	0.09743	11.101

applied is the influential input factor that contributes toward MRR (Hourmand et al. 2019). When T_{on} is changed from one level value of 50 μs to another level value of 100 μs, a considerable increase in material removal rate by 36.31% is observed. When it is further increased to150 μs from 100 μs, a significant improvement in MRR is observed by 49.02%. While considering T_{off}, a decrease in MRR is observed. When T_{off} is changed from 20 to 30 μs, MRR is reduced by 8.81%. By increasing the T_{off} further to 40 μs, MRR is reduced by 2.63% (Aghdeab and Ahmed 2016).

While analyzing surface roughness, it is found that when current increases, SR increases, whereas it decreases with an increase in T_{off}. With an increase in T_{on}, SR initially decreases and then increases with further increase in T_{on}. Observations made from the results show that MRR and surface roughness have a significant correlation. When current is increased from 3A to 5A, SR increases by 36.31%, and a further increase of 49.02% in SR is observed when the current is increased from 5A to 7A. This result can be correlated with the MRR results, as MRR increases, SR also increases, i.e. when more material is unbonded from workpiece surface, a rough surface is generated. When $T_{on}$is varied from 50 μs to 100 μs, SR is decreased by 9.15%. When it is further increased from 100 μs to150 μs, an increase of 9.01% in SRis observed. Hence, it is obvious that a moderate value of T_{on} is necessary to maintain the SR of the workpiece surface. When T_{off} is varied from 20 μs to 30 μs, a 8.81% in SR is observed. Further increasing T_{off} reduces SR by 2.63%. The effect of this parameter is inevitable because even though MRR increases with an increase in level values, SR is decreased, matching with the aim and objective of this work. Figure 7.2 presents the influence of input factors over the output parameters MRR and SR.

Use of analysis of variance (ANOVA) to identify the variance among the selected input factors allowed for the screening of the total variance identified in the study into different components. The total variance identified will be the variance of the individual scores obtained when it is determined for the dependent variable. Dependent variables are output responses, whereas independent variables are input factors (Gamst et al. 2008). Table 7.3 presents the ANOVA table formulated for MRR and SR, considering 95% confidence interval (CI) (Thirumalvalavan and Senthilkumar 2019). It was observed that a substantial model was developed for both of the outputs. Current is the highest influential factor for MRR, which contributes by 73.14%, followed by T_{on} by 14.62% and the influence of T_{off} is negligible (Pradhan and Biswas 2011). For SR, current contributes by 77.15% and the influence of T_{on} and T_{off} is negligible.

The lower order polynomial model developed for MRR and SR using regression modelling (Senthilkumar and Tamizharasan 2015) is given in Equation 7.1 and Equation 7.2, which is used to predict the outputs. With R^2 predicted value of 78.74% for MRR and 65.74% for SR, the level of prediction will be lower for both of these cases.

$$\text{MRR} = -0.06684 + 0.014218 \times I + 0.000253 \times T_{\text{on}} + 0.000252 \times T_{\text{off}} \tag{7.1}$$

$$\text{SR} = 2.472176 + 1.403917 \times I - 0.00079 \times T_{\text{on}} - 0.04791 \times T_{\text{off}} \tag{7.2}$$

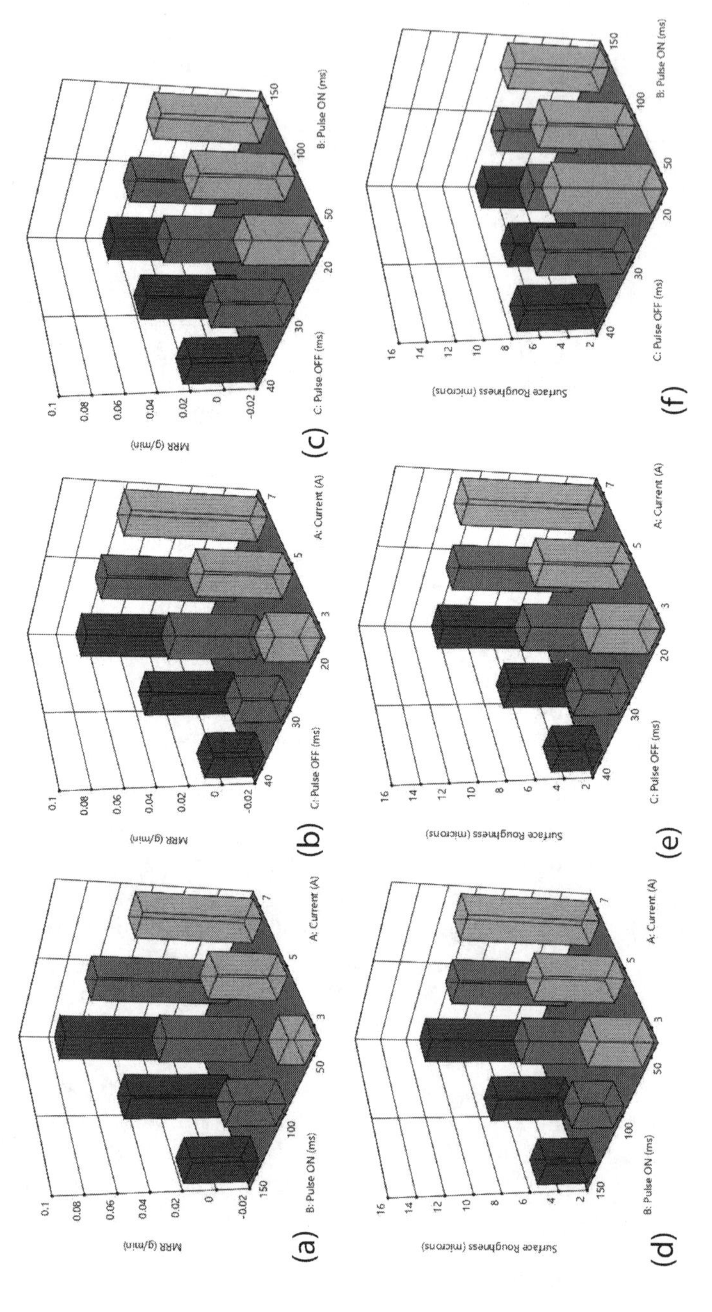

FIGURE 7.2 Effect of input parameters on output responses (material removal rate and surface roughness).

TABLE 7.3
ANOVA Table for Outputs

Output	Source	Sum of Squares	df	Mean Square	F-value	p-value	% Contribution
Material Removal Rate	Model	0.0175	6	0.0029	25.24	< 0.0001	88.34%
	Current	0.0146	2	0.0073	62.70	< 0.0001	73.14%
	T_{on}	0.0029	2	0.0015	12.54	< 0.0001	14.62%
	T_{off}	0.0001	2	0.000057	0.49	0.618	0.58%
	Residual	0.0024	23	0.000116			11.67%
	Cor Total	0.02	26				100.00%
	R^2	0.8834		R^2 (Adj.)	0.8484	R^2 (Pred.)	0.7874
Surface Roughness	Model	153.721	6	25.62	14.398	2.5E-06	81.20%
	Current	146.045	2	73.022	41.04	8.34E-08	77.15%
	T_{on}	3.096	2	1.548	0.87	0.434	1.64%
	T_{off}	4.58	2	2.290	1.29	0.298	2.42%
	Residual	35.589	20	1.779			18.80%
	Cor Total	189.310	26				100%
	R^2	0.812		R^2 (Adj.)	0.7556	R^2 (Pred.)	0.6574

One of the downsides and limitations of this EDM procedure is the creation of a remelt/recast layer that is formed on the machined surface and another layer called heat-affected zone (HAZ) that forms underneath the workpiece surface, which affects the surface finish and integrity and lowers fatigue strength. The white layer is formed by heating the workpiece to a molten state and is not in a condition to be flushed away through the gap and get ejected out. The EDM process has essentially reformed the metallurgical characteristics and general structure of the material. The white layer structure is mainly comprised of discharge craters and microcracks that cause poor surface quality and are very difficult to eliminate owing to higher hardness and bonding characteristics when compared with the base material (Wang et al. 2009).

The EDM drilled surface of the HAMMC shown in Figure 7.3 for experiment conditions 3, 5, 10 and 24 shows two distinct zones. The first zone shows the clear metal matrix edge in the absence of composite SiC particles, whereas the second layer with the composite particles showed resistance to drilling, leading to the deviation of the drilled edge. The composite particles are unaffected by the process. A very fine recast layer was observed, which got etched. Fine particles of composite are observed at the layer. The main metal matrix showed the uniform distribution of SiC particles. The grain size corresponds to fewer than 100 microns at the edge.

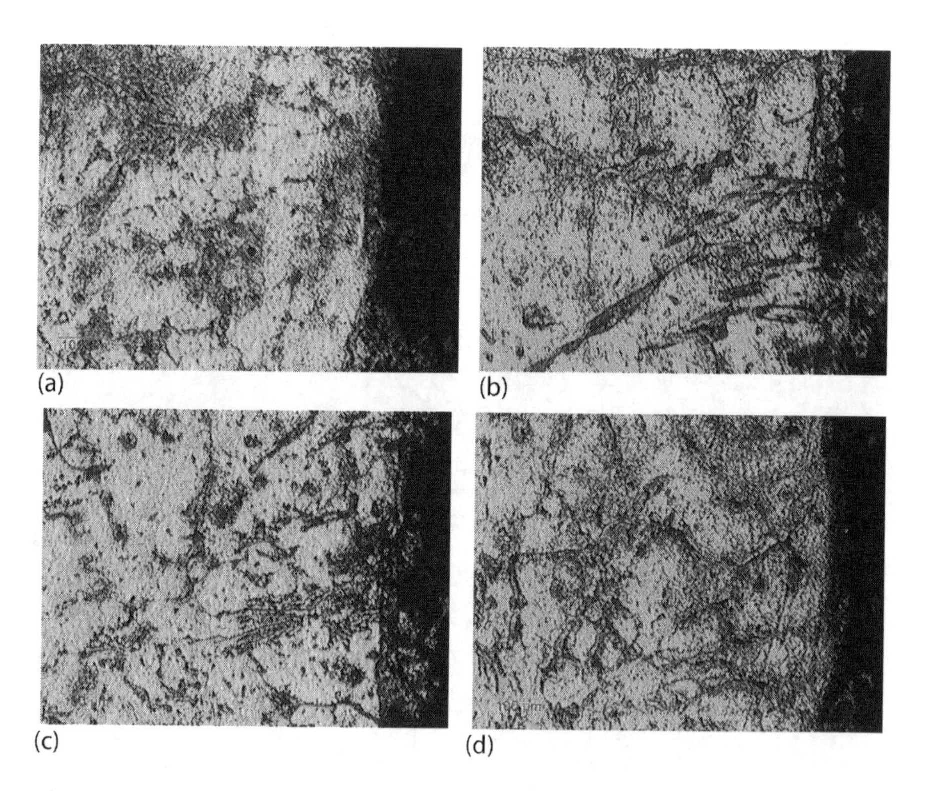

FIGURE 7.3 White layer formation and heat-affected zone in work piece. (a) Experiment No. 3; (b) Experiment No. 5; (c) Experiment No. 10; (d) Experiment No. 24.

The purpose of ANFIS is to predict the responses by developing a good mathematical model by reducing the fuzziness available in the experimental outputs by means of fuzzy inference engine and if-else-then rules framed between the input and output. During training in ANFIS, for both MRR and SR, trapezoidal membership function (MF) is considered with four linguistic variables (small, low, medium and high) for all the three inputs. The AND method considered is prod (product), which scales the output fuzzy set and the probor (probabilistic OR) approach, commonly referred to as an algebraic sum, which can be determined as the equation: probor $(a,b)=a+b-ab$ (Bansal et al.2009). Table 7.4 shows the parameters considered in this study for Sugeno-based ANFIS system. Figure 7.4 shows the relationship between the training data provided to the Sugeno FIS and the output obtained. Observation shows that both the input and output data match accurately, i.e. a perfect prediction is made through the rule base generated using the ANFIS model.

TABLE 7.4
Parameters in Sugeno-Based FIS

Parameter	Value
AND	Prod
OR	Probor
Aggregation	Max
Fuzzification	Sugeno
Defuzzification	Wtaver
Optimization method	Hybrid
Number of Epochs	300
Input membership function	Trapezoidal
Output membership function	Linear

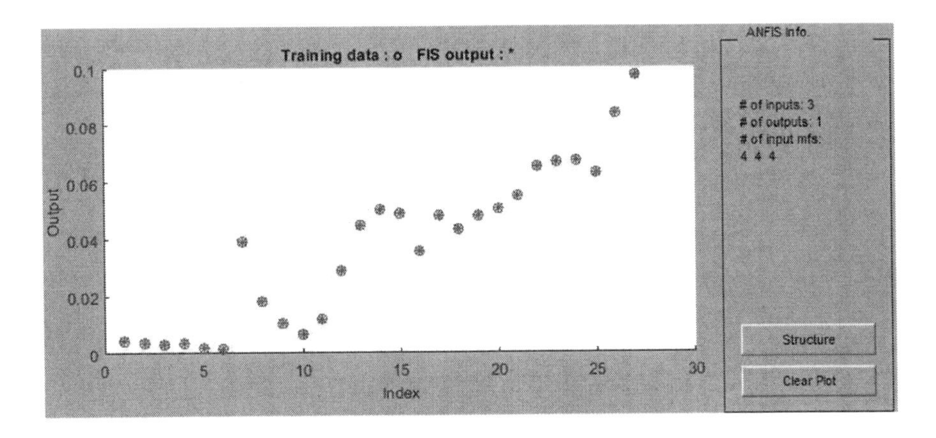

FIGURE 7.4 Plot during training data and fuzzy inference system predicted data.

Figure 7.5 presents the neural network architecture model developed during the ANFIS. Three inputs were provided, and only one output (either MRR or SR) is considered during the model development. The input MF (trapezoidal) considered was with four linguistic variables, represented by four neurons for each input, which is connected with the rule base (if-else-then) and the linear MF of output, providing the predicted output value. During model development, the MRR error was 3.81×10^{-8}, whereas the SR error was 5.3984×10^{-6}, which is closer to the ideal value of zero.

The surface plots (3D) obtained during the ANFIS modelling are presented in Figure 7.6 for all the inputs and for the outputs MRR and SR. These surface plots represent the relationship that exists between the input and output based on the if-else-then rules framed (Senthilkumar et al. 2015) based on the training function of the ANN (Naresh et al. 2020). Based on the rule, the rule editor is formed (Figure 7.7), from which the outputs can be predicted for a given set of input parameters.

From the rule editor, prediction of outputs MRR and SR was performed for the given input condition and is compared with the experimental output and regression model outputs. Figure 7.8 presents the comparison of values for MRR, and Figure 7.9 presents the comparison of SR values obtained from ANFIS and regression model with the experimental output. It is found that a perfect prediction is more possible with the ANFIS model than the regression model, as a perfect fit is observed between the experimental and ANFIS models, whereas the residuals were higher from the regression model, which is due to the lower R^2 predicted value of MRR and SR.

Table 7.5 presents the comparison of mean average error, mean square error and root mean square error obtained from ANFIS and RSM models. Observation shows that the errors obtained with the ANFIS prediction model is closer to the ideal value

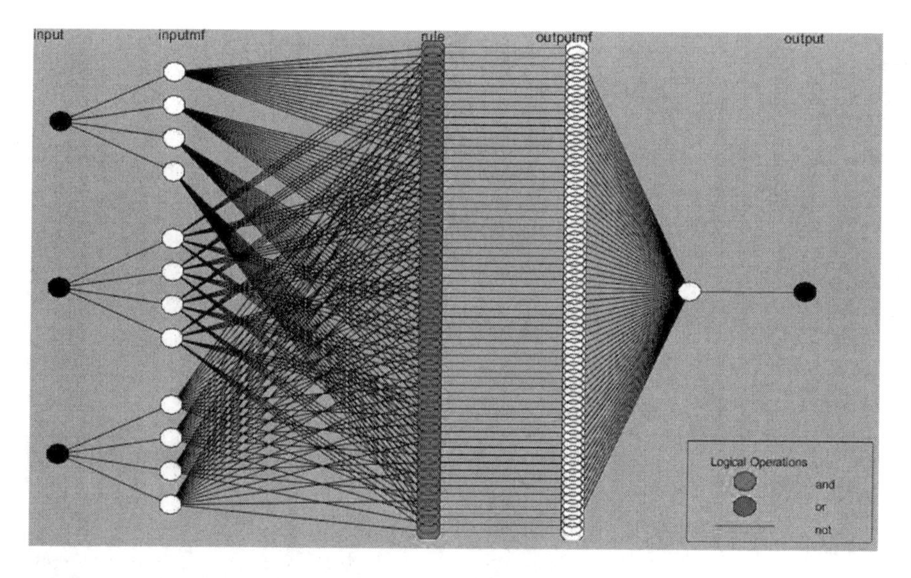

FIGURE 7.5 Neural network architecture obtained during adaptive network-based fuzzy inference system.

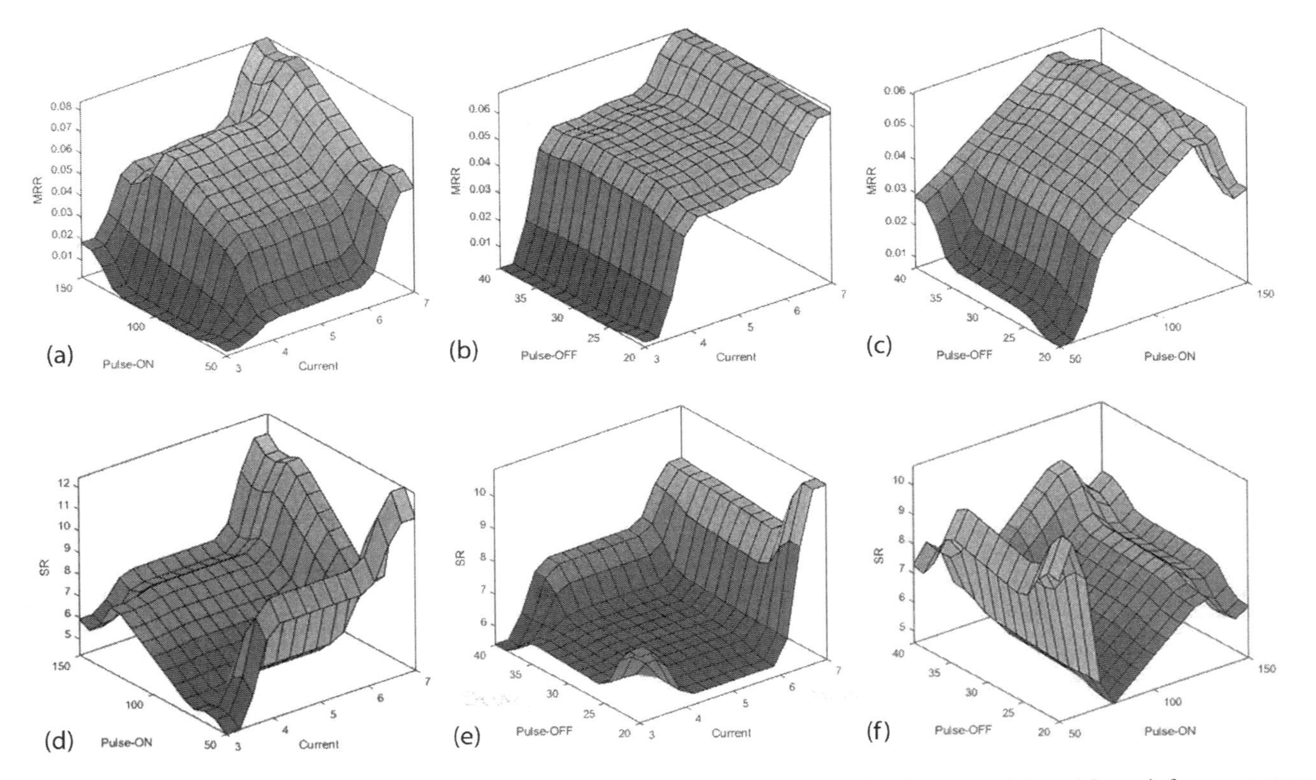

FIGURE 7.6 Surface plots obtained for material removal rate and surface roughness during adaptive network-based fuzzy inference system.

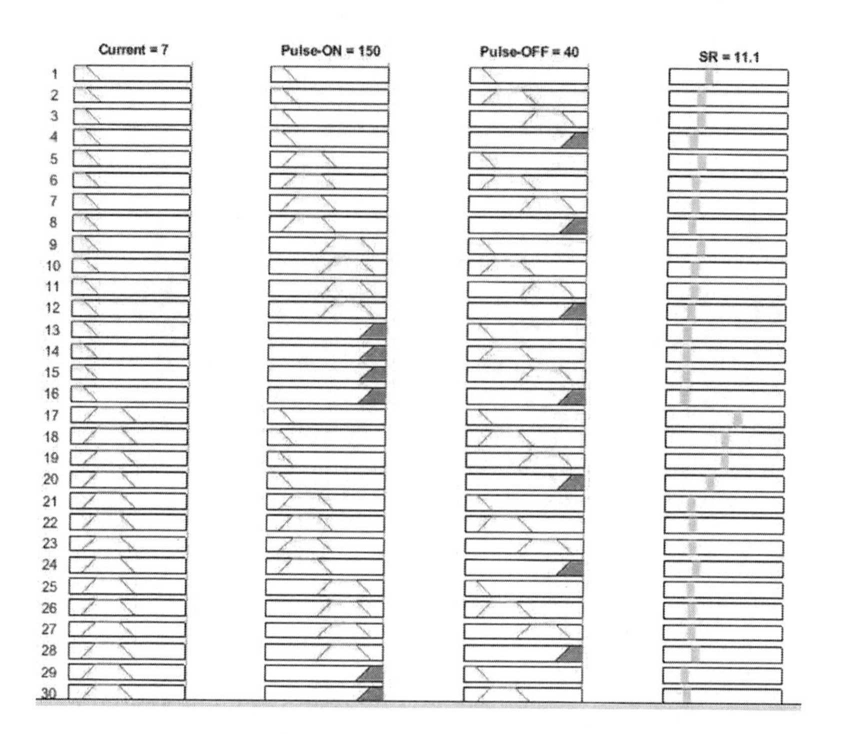

FIGURE 7.7 Rule editor for prediction of outputs in adaptive network-based fuzzy inference system

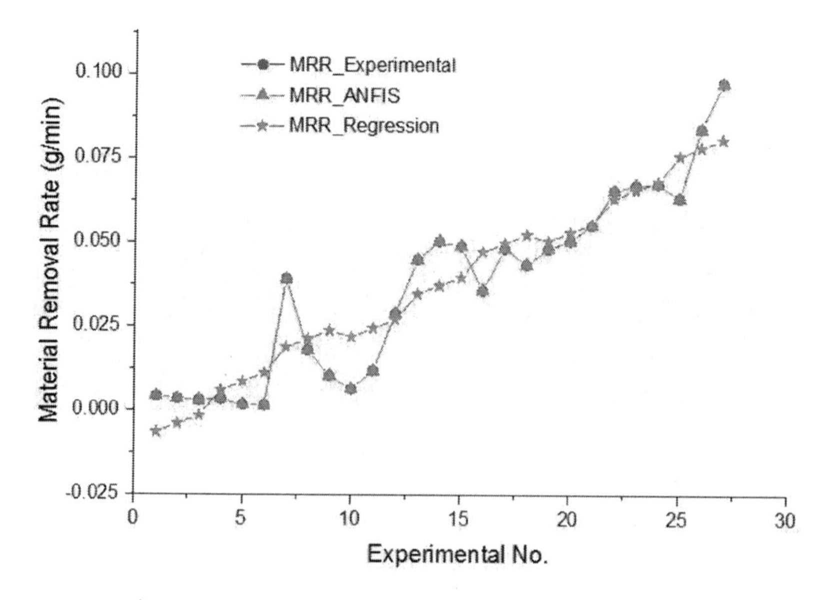

FIGURE 7.8 Comparison of material removal rate between experimental, regression and adaptive network-based fuzzy inference system.

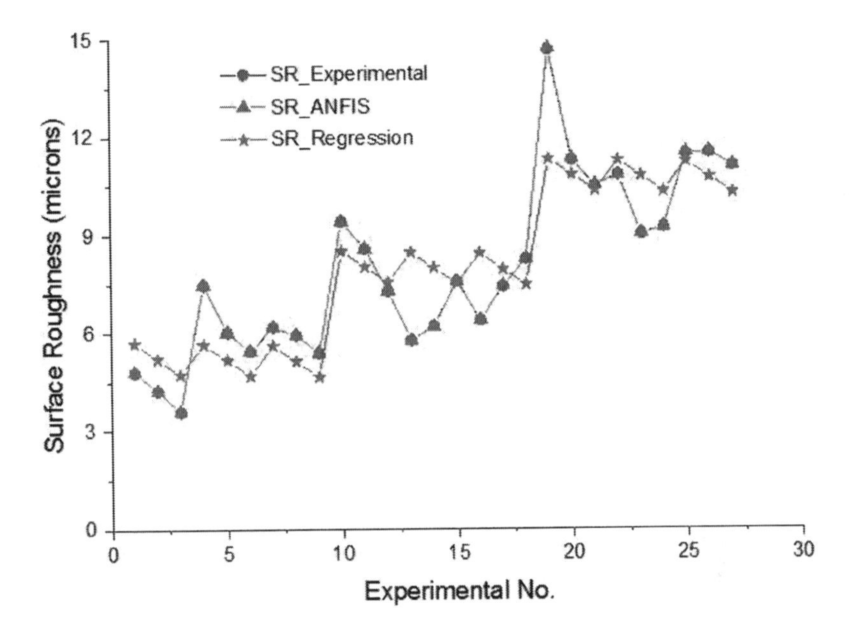

FIGURE 7.9 Comparison of surface roughness between experimental, regression and adaptive network-based fuzzy inference system.

TABLE 7.5
Comparison of Errors between ANFIS and RSM Model

Type of Error	Formulae	MRR		SR	
		ANFIS	RSM	ANFIS	RSM
Mean Average Error	$MAE = \frac{1}{n}\sum_{i=1}^{n}\lvert x_d - x_p \rvert$	0.0000189	0.0077	0.00693	1.00954
Root Mean Square Error	$RMSE = \sqrt{\frac{1}{n}\sum_{i=1}^{n}(x_d - x_p)^2}$	0.000219	0.077	0.0745	10.78989
Mean Square Error	$MSE = \frac{1}{n}\sum(y - \hat{y})^2$	0.000000048	0.00593	0.00554	116.42165

of zero, and it is comparatively much lower than the RSM-based error values, which proves the competence of the established model using ANFIS.

Apart from the prediction of outputs for optimizing the output parameters of MRR and SR simultaneously, desirability analysis is adopted. During the methodology of implementing desirability, values that fall in between the probable values of 1 and 0 will be assigned by the desirability function, where 1 represents the desirable or ideal value, and 0 represent the undesirable outcome (Ponnuvel and Senthilkumar 2019).

The formulae to calculate desirability for minimization and maximization of responses are provided in Equation 7.3, where A and B are extreme limits of selected inputs and the exponent s governs the weightage toward attaining the value of target; the input vector is X, and f_r is the prediction model used, and R is the number of desirability functions (responses) considered in the study.

$$d_r^{\max} = \begin{cases} 0 & \text{if } f_r(X) < A \\ \left(\dfrac{f_r(X) - A}{B - A} \right)^S & \text{if } A \le f_r(X) \le B \\ 1 & \text{if } f_r(X) > B \end{cases}$$

$$d_r^{\min} = \begin{cases} 1 & \text{if } f_r(X) < A \\ \left(\dfrac{f_r(X) - B}{B - A} \right)^S & \text{if } A \le f_r(X) \le B \\ 0 & \text{if } f_r(X) > B \end{cases} \tag{7.3}$$

$$D = \left(\prod_{r=1}^{R} d_r \right)^{1/R}$$

The optimum input value obtained is current of 5.79 A, T_{on} of 150 μs and T_{off} of 40 μs. The predicted outputs are MRR of 0.0635 g/min and SR of 8.565 microns, as seen from the ramp plot displayed in Figure 7.10. With the obtained optimal inputs, a validation trial was conducted for validating the outputs, for which MRR was 0.06184 g/min and SR was 8.275 microns, which is 2.61% and 3.39% lower than the predicted values.

7.4 CONCLUSIONS

In this work, fabrication and characterization of P/M based HAMMC (Al+4%Cu+7.5%SiC) was carried out and machining studies were carried out in EDM with an objective of optimizing current, T_{on} and T_{off} over MRR and SR. For mapping EDM input factors and the measured output responses (MRR and SR), ANFIS methodology was adopted to develop a model for predicting the output responses. The predicted results obtained from the lower order polynomial model developed for output responses were compared with the ANFIS-based predictions. ANFIS was found to provide a better prediction model with a lower error percentage. For MRR and SR, the influence of current is higher than that of T_{on} and T_{off}. Desirability based multiobjective optimization produces an optimal condition of 5.79 A of current, T_{on} of 150 μs and T_{off} of 40 μs. Validation of optimum conditions were conducted through a confirmation experiment that provided a least deviation of 2.61% and 3.39% for MRR and SR than the predicted outputs.

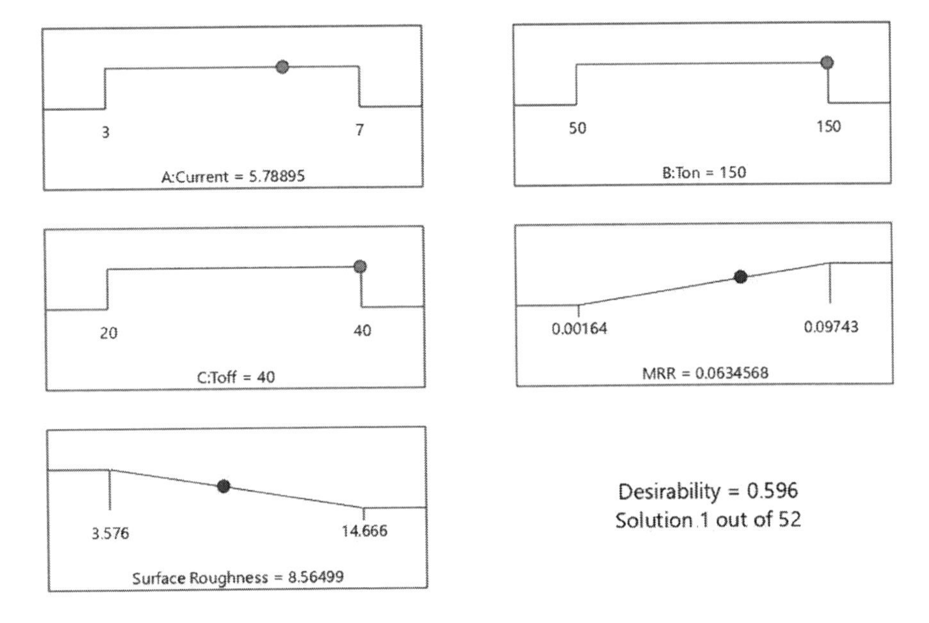

FIGURE 7.10 Ramp plot indicating optimal condition.

REFERENCES

Aghdeab, S.H. and Ahmed, A.I. 2016. Effect of pulse on time and pulse off time on material removal rate and electrode wear ratio of stainless steel AISI 316L in EDM. *Engineering and Technology Journal*, 34(15 Part (A) Engineering), 2940–2949.

Bansal, R.K., Goel, A.K. and Sharma, M.K. 2009. *MATLAB and Its Applications in engineering*. India: Pearson Education.

Bhiradi, I., Raju, L. and Hiremath, S.S. 2020. Adaptive neuro-fuzzy inference system (ANFIS): modelling, analysis, and optimisation of process parameters in the micro-EDM process. *Advances in Materials and Processing Technologies*, 6(1), 133–145.

Czogala, E. and Leski, J. 2000. *Fuzzy and Neuro-Fuzzy Intelligent Systems*. New York, NY: Physica-Verlag, A Springer-Verlag company.

Davis, J.R. ed. 1999. *Corrosion of aluminum and Aluminum Alloys*. Materials Park, OH: ASM International.

DiIlio, A. and Paoletti, A. 2012. Machinability aspects of metal matrix composites. In *Machining of Metal Matrix Composites* (pp. 63–77). London: Springer.

Eisen, W.B., Ferguson, B.L., German, R.M., Iacocca, R., Lee, P.W., Madan, D., Moyer, K., Sanderow, H. and Trudel, Y. 1998. *Powder Metal Technologies and applications*. Materials Park, OH: ASM International.

El-Hofy, H.A.G. 2005. *Advanced Machining Processes: Nontraditional and Hybrid Machining Processes*. USA: McGraw Hill Professional.

Fazlollahtabar, H. and Gholizadeh, H. 2020. Fuzzy possibility regression integrated with fuzzy adaptive neural network for predicting and optimizing electrical discharge machining parameters. *Computers and Industrial Engineering*, 140, 106225.

Gamst, G., Meyers, L.S. and Guarino, A.J. 2008. *Analysis of Variance Designs: A conceptual and Computational Approach with SPSS and SAS*. Cambridge, London, UK: Cambridge University Press.

Hourmand, M., Sarhan, A.A., Farahany, S. and Sayuti, M. 2019. Microstructure characterization and maximization of the material removal rate in nano-powder mixed EDM of Al-Mg_2 Si metal matrix composite – ANFIS and RSM approaches. *International Journal of Advanced Manufacturing Technology*, 101(9–12), 2723–2737.

Jameson, E.C. 2001. *Electrical Discharge Machining.* Michigan, USA: Society of Manufacturing Engineers.

Kumar, R., Roy, S., Gunjan, P., Sahoo, A., Sarkar, D.D. and Das, R.K. 2018. Analysis of MRR and surface roughness in machining Ti-6Al-4V ELI titanium alloy using EDM process. *Procedia Manufacturing*, 20, 358–364.

Kumar, S., Dhanabalan, S. and Narayanan, C.S. 2019. Application of ANFIS and GRA for multi-objective optimization of optimal wire-EDM parameters while machining Ti–6Al–4V alloy. *SN Applied Sciences*, 1(4), 298.

Lauwers, B., Vleugels, J., Malek, O., Brans, K. and Liu, K. 2012. Electrical discharge machining of composites. In *Machining Technology for Composite Materials* (pp. 202–241). UK:Woodhead Publishing Limited.

Mandal, S., Pramanick, A., Chakraborty, S. and Dey, P.P. 2018. ANFIS based model to forecast the wire-EDM parameters for machining an ultra high temperature ceramic composite. *IOP Conference Series: Materials Science and Engineering377*: 012088. doi: 10.1088/1757-899X/377/1/012088.

Manikandan, N., Balasubramanian, K., Palanisamy, D., Gopal, P.M., Arulkirubakaran, D. and Binoj, J.S. 2019. Machinability analysis and ANFIS modelling on advanced machining of hybrid metal matrix composites for aerospace applications. *Materials and Manufacturing Processes*, 34(16), 1866–1881.

Muthuramalingam, T., Saravanakumar, D., Babu, L.G., Phan, N.H. and Pi, V.N. 2020. Experimental investigation of white layer thickness on EDM processed silicon steel using ANFIS approach. *Silicon*, 12, 1905–1911–7.

Naresh, C., Bose, P.S.C. and Rao, C.S.P. 2020. ANFIS based predictive model for wire edm responses involving material removal rate and surface roughness of Nitinol alloy. *Materials Today: Proceedings*, Doi:10.1016/j.matpr.2020.03.216.

Nedjah, N. and de Macedo Mourelle, L. 2005. *Fuzzy Systems Engineering: theory and practice.* Berlin/Heidelberg/The Netherlands: Springer-Verlag.

Ponnuvel, S. and Senthilkumar, N. 2019. A study on machinability evaluation of Al-Gr-B4C MMC using response surface methodology-based desirability analysis and artificial neural network technique. *International journal of Rapid Manufacturing*, 8(1–2), 95–122.

Pradhan, M.K. and Biswas, C.K. 2011. Effect of process parameters on surface roughness in EDM of tool steel by response surface methodology. *International Journal of Precision Technology*, 2(1), 64–80.

Ross, P.J. 2005a. *Taguchi Techniques for Quality Engineering.* New Delhi, India: Tata McGraw Hill Publishing Company Ltd.

Ross, T.J. 2005b. *Fuzzy logic with engineering Applications.* West Sussex, UK: John Wiley &Sons.

Roy, R.K. 2001. *Design of experiments using the Taguchi approach: 16 steps to product and process improvement.* New York, USA: John Wiley &Sons

Selvakumar, V., Muruganandam, S. and Senthilkumar, N. 2017. Evaluation of mechanical and tribological behavior of Al–4% Cu–x% SiC composites prepared through powder metallurgy technique. *Transactions of the Indian Institute of Metals*, 70(5), 1305–1315.

Selvakumar, V., Muruganandam, S., Tamizharasan, T. and Senthilkumar, N. 2016. Machinability evaluation of Al–4% Cu–7.5% SiC metal matrix composite by Taguchi–Grey relational analysis and NSGA-II. *Sādhanā*, 41(10), 1219–1234.

Senthilkumar, N. and Tamizharasan, T. 2014. Experimental investigation of cutting zone temperature and flank wear correlation in turning AISI 1045 steel with different tool geometries, *Indian Journal of Engineering & Materials Sciences*, 21(2), 139–148

Senthilkumar, N. and Tamizharasan, T. 2015. Flank wear and surface roughness prediction in hard turning via artificial neural network and multiple regressions. *Australian Journal of Mechanical Engineering*, 13(1), 31–45.

Senthilkumar, N., Sudha, J. and Muthukumar, V. 2015. A grey-fuzzy approach for optimizing machining parameters and the approach angle in turning AISI 1045 steel. *Advances in Production Engineering and Management*, 10(4), 195–208.

Shehabeldeen, T.A., Zhou, J., Shen, X., Yin, Y. and Ji, X. 2019. Comparison of RSM with ANFIS in predicting tensile strength of dissimilar friction stir welded AA2024-AA5083 aluminium alloys. *Procedia Manufacturing*, 37, 555–562.

Singh, N.K., Singh, Y., Kumar, S. and Sharma, A. 2020. Predictive analysis of surface roughness in EDM using semi-empirical, ANN and ANFIS techniques: a comparative study. *Materials Today: Proceedings*, 25(4), 735–741.

Tamizharasan, T., Senthilkumar, N., Selvakumar, V. and Dinesh, S. 2019. Taguchi's methodology of optimizing turning parameters over chip thickness ratio in machining P/M AMMC. *SN Applied Sciences*, 1(2), 160.

Thirumalvalavan, S. and Senthilkumar, N. 2019. Experimental investigation and optimization of HVOF spray parameters on wear resistance behaviour of Ti-6Al-4V Alloy. *Comptesrendus de l'Académiebulgare des Sciences*, 72(5), 665–674.

Tsukerman, S.A.1965. *Powder metallurgy*. Oxford: Pergamon Press.

Upadhyaya, A. and Upadhyaya, G.S. 2011. *Powder Metallurgy: Science, Technology and Materials*. Hyderabad, India: Universities Press.

Uthayakumar, M. and Parameswaran, P. 2019. Evaluation of electrical discharge machining performance on Al (6351)–SiC–B4C composite. In *Non-Conventional Machining in Modern Manufacturing Systems* (pp. 109–124). Hershey, PA: IGI Global.

Wang, C.C., Chow, H.M., Yang, L.D. and Lu, C.T. 2009. Recast layer removal after electrical discharge machining via Taguchi analysis: a feasibility study. *Journal of Materials Processing Technology*, 209(8), 4134–4140.

Yang, W.S., Chen, G.Q., Wu, P., Hussain, M., Song, J.B., Dong, R.H. and Wu, G.H. 2017. Electrical discharge machining of Al2024-65 vol% SiC composites. *Acta MetallurgicaSinica (English Letters)*, 30(5), 447–455.

8 Experimental Investigation and Optimization of EDM Process Parameters on Ti-6Al-4V Alloy Using Taguchi-Based GRA and Modelling by RSM

A. Palanisamy, V. Sivabharathi,
S. Sivasankaran and S. Navaneethakrishnan

CONTENTS

8.1 INTRODUCTION

Today, advanced machining processes, such as electrical discharge machining (EDM) (spark erosion technique), are being used mandatorily in most of the manufacturing industries for machining hard metals with specific applications (e.g., aerospace, automotive, nuclear, tool and die and medical) to produce various cutting, and forming tools (Shabgard et al. 2013). Several features, namely, no mechanical stresses, chatters, and vibrations, can be obtained in the EDM process, as there is no direct contact between the tool and the workpiece. Any hard metals can also be machined by EDM (provided it should conduct electricity) in which titanium-based hard alloy can be easily machined, as it possesses lower thermal conductivity with a high strength-to-weight ratio (Magabe et al. 2019). Particularly medical components, steam turbine blades and parts for aircraft engine compressors, submarines and race cars are manufactured from titanium-based hard alloy using EDM. The EDM process falls into a machining category that favours producing any kind of complicated parts that cannot be produced any other machining technique. Therefore, it is mandatory to reduce the machining time to achieve high throughput (Shabgard et al, 2013; Magabe et al. 2019). To achieve this, several EDM process parameters are to be optimized to maximize the material removal rate (MRR) and minimize the electrode wear rate (EWR) and surface roughness (R_a). The selection of EDM process parameters could define the product quality directly, which needs to be selected based on machining handbooks and or operator's experiences; however, this does not guarantee EDM parameters. The research community has developed numerous mathematical models to correlate the EDM process parameters to its responses using the design of experiments (Montgomery, 2017; Lin et al. 2002; Verma and Sahu 2017).To select the best process parameters by considering several responses, optimization techniques can be used with the help of objective functions with certain constraints. For this, substantial scientific knowledge and various practices are necessary to apply the latest approach to real-time industries. Using conventional methods, numerous experiments are to be carried out, such as one-to-one, which consumes a lot of time and cost for developing the mathematical models to predict the system behaviour. Recently, the Taguchi technique has been used as an effective tool for designing the experiments by which the system performance can be controlled to produce high-quality products (Magabe et al. 2019; Lin et al. 2002). In the past, the Taguchi system was used to optimize single response characteristics. The steps involved in the Taguchi design include selecting a suitable orthogonal array (OA), carrying out the experiment trails founded on OA, analyzing the facts, identifying the most favourable order, and executing confirmation trials (Lin et al. 2002). Several researchers have used the Taguchi method to optimize different machining processes, such as drilling, end milling and turning, using various alloys (Panda et al. 2015; Jeykrishnan et al. 2016). Using the Taguchi method requires a more scientific knowledge for handling the multiresponse characteristics. For instance, the enhancement of one response characteristic may not consider the importance of other performance characteristics. This meant that the optimization of multiresponse characteristics is highly challenging compared to optimizing single-response characteristics. Grey relational analysis (GRA), proposed by Deng (Julon 1989), often is the best method for multiresponse optimization problems, as it considers several input parameters. Currently, Taguchi-based GRA is the best approach for finding optimal machining conditions for multiresponse problems in various machining processes (Angappan et al. 2017; Sahu and Mahapatra, 2019; Yuvaraj and Suresh 2019). Hence, the Taguchi-based GRA approach was attempted in this work to optimize the EDM parameters by considering various responses (EWR, MRR, and R_a) (Hasçalık and Çaydaş 2007; Alshemary et al 2018).The most influencing parameters that affect the EDM performance are the current (*I*), pulse ON time and pulse OFF time (Santos et al. 2016; Urso et al 2015). The average value of discharged current intensity from EDM indicates the pulse current; the period of time that the current is permitted per cycle represents the pulse ON time, which directly influences the material removal (the peak current and the duration define the energy input); the time taken between two consecutive on time, or two EDM sparks, indicate the pulse OFF time, which helps to allow the electrode cool and flush away the chips and to control the speed and stabilize the cut (Pachaury and Tandon 2017). The main goals of this research study are to examine the EDM of hard titanium alloy (Ti-6Al-4V) by adopting the Taguchi-based GRA optimization and modelling by RSM with Design-Expert 12 software. The most influencing parameters were examined using ANOVA analysis with the help of Minitab 16.0 version software (Bahgat et al. 2019; Palanisamy et al 2014; Tsai et al. 2018; Palanisamy and Selvaraj 2019). The topography of the machined (EDM) surfaces was examined using white-light interferometer (WLI).

8.2 MATERIALS AND METHODS

Ti-6Al-4V, with a rectangular cross-section (200 × 15 × 5 mm) was used for machining the EDM. The chemical composition of as-received Ti-6Al-4V was tested by metal spark analyzer (Test Point Company, Coimbatore, India), which was aluminium (6.08 wt.%), vanadium (4.02 wt.%), iron (0.22 wt.%), oxygen (0.18 wt.%), carbon (0.02 wt.%), nitrogen (0.01 wt.%), hydrogen (0.0053 wt.%), and titanium (89.46 wt.%). The photographs of the specimen and the rectangular electrodes (15 × 15 × 3 mm) are shown in Figure 8.1a. The schematic diagram of EDM process is shown in Figure 8.1b. The EDM experiments, conducted on the Sparkonix EDM machine with a maximum capacity of operating current of 15 Amps, are shown in Figure 8.1c. Finely ground graphite electrodes (54 electrodes, 3 replicas × 18 Nos, corresponding to each experiment) were used to machine a rectangular slot of 15 × 3 × 0.5 mm. To avoid dimensional variations, the graphite electrode was used only one time for each experiment. Three replicas were used in each experiment to examine exact EDM behaviour. The metal was removed from the workpiece by electric spark erosion action, which is the main principle of EDM (El-Hofy 2005). The workpiece was connected to the positive terminal, and the electrode was connected to the negative terminal of the DC power supply. The spark gap of 0.025 mm was constantly maintained with the help of the servo-controlled mechanism for all experiments. Kerosene oil with a flushing pressure of 0.2 kg/cm^2 was used as a dielectric medium (Bai et al 2018). The side-flushing technique was used to remove the debris formed

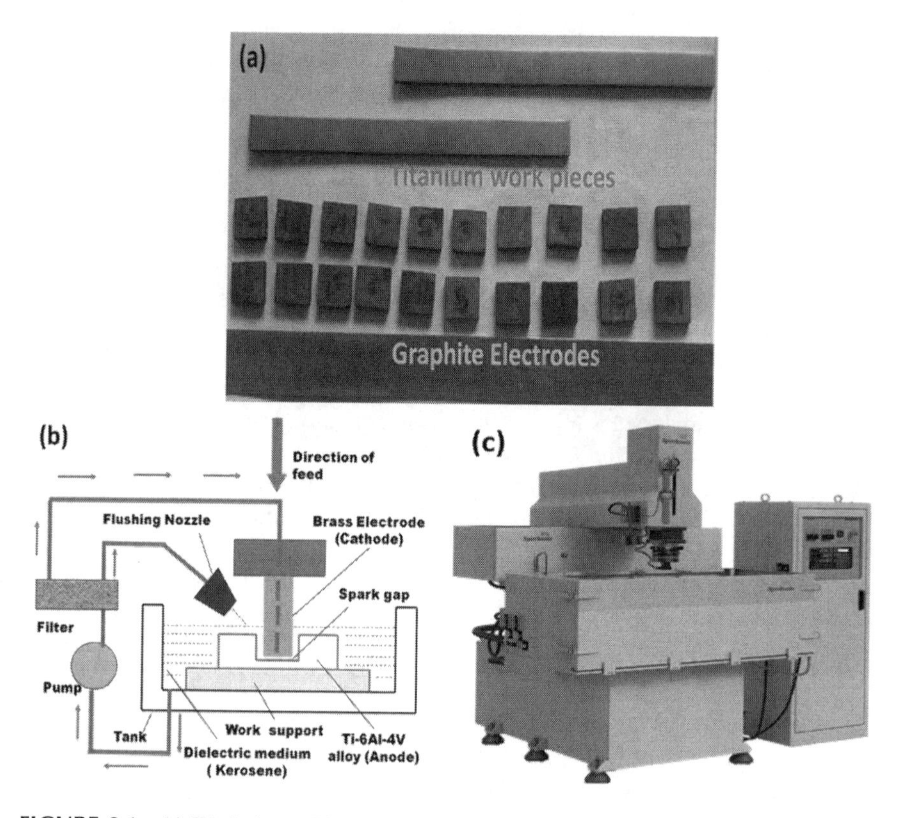

FIGURE 8.1 (a) Workpiece (Ti – 6Al – 4V alloy) and graphite electrodes; (b) schematic diagram of the EDM process; (c) photograph of the electric discharge machine.

during the EDM process. The L_{18} OA was chosen to design and to conduct the experiments (Senthil et al 2014; Selvarajan et al 2015). An electronic weighing balance (Mitutoyo Shimadzu, Japan, BL 220H), with a precision of 0.001 g and a capacity of 220 g, was used to weigh the workpiece and electrodes before and after machining. Surface roughness of machined (EDM) workpieces was measured by a surftest instrument (Mitutoyo, SJ-310) with a cut-off length of 0.8 mm and a traverse length of 5 mm. Also, the machined surface topography was examined by the WLI (Rtech Instruments, USA). The EWR and MRR were calculated using the standard formula, by means of weight difference, before and after EDM by machining time (Eq. 8.1, and Eq. 8.2):

$$\text{EWR}\left(\frac{\text{g}}{\text{min}}\right) = \frac{\text{Electrode removal weight}}{\text{Machining time}} \tag{8.1}$$

$$\text{MRR}\left(\frac{\text{g}}{\text{min}}\right) = \frac{\text{Workpiece removal weight}}{\text{Machining time}} \tag{8.2}$$

TABLE 8.1
Experimental Layout Based on L_{18} OA with the Input EDM Parameters and the Measured Responses

	Input EDM parameters				Output Responses		
Ex. No.	Pulse Current, Amps (A)	Pulse ON Time, μs (B)	Pulse OFF Time, μs (C)	Machining Time (min)	EWR, g/min	MRR, g/min	R_a, μm
1	4	200	20	40.1	0.00175	0.00197	5.77
2	4	400	30	14.1	0.00284	0.00560	5.90
3	4	600	40	4.12	0.01214	0.01915	6.30
4	8	200	20	19.2	0.00156	0.00411	7.46
5	8	400	30	18.24	0.00274	0.00433	7.84
6	8	600	40	8.2	0.00732	0.00962	8.61
7	12	200	30	19.1	0.00209	0.00413	8.89
8	12	400	40	14.3	0.00420	0.00552	9.12
9	12	600	20	6.11	0.00818	0.01291	10.20
10	4	200	40	28.44	0.00246	0.00277	5.20
11	4	400	20	8.3	0.00445	0.00950	6.14
12	4	600	30	3.31	0.01208	0.01750	6.99
13	8	200	30	34.12	0.00205	0.00231	7.66
14	8	400	40	18.51	0.00162	0.00426	7.99
15	8	600	20	8.3	0.00576	0.00950	8.87
16	12	200	40	13.9	0.00432	0.00568	8.89
17	12	400	20	22.23	0.00225	0.00355	9.54
18	12	600	30	8.2	0.00854	0.00962	10.25

The experimental layout based on L_{18} OA with the EDM input parameters and their measured experimental results is illustrated in Table 8.1.

8.3 RESULTS AND DISCUSSIONS

8.3.1 Optimization of Process Parameters by Taguchi Method and GRA

In the Taguchi method (Palanisamy et al 2018; Payal et al. 2019), optimization is the main step to determine the best input parameters for achieving high-quality products at lower cost. This method is based on OAs by which the system behaviour over the space can be examined effectively. In this method, the signal-to-noise (S/N) ratio term is used to investigate the system behaviour. This S/N ratio has three categories, namely, smaller is better (SB), larger is better (LB), and nominal is better (NB). All the experimental results obtained as per Table 8.1 were investigated using S/N ratio values based on the objective functions of each response. Among the responses, the EWR and the R_a are to be minimized, hence, SB characteristic was used in these objective functions. In contrast, MRR is to be maximized to enhance the productivity, hence the LB characteristic was used in this objective function. Eq. 8.3 and Eq. 8.4 represent the objective functions for minimization and maximization, respectively.

$$\text{Smaller is Better; } \ {}^{S}\!/_{N}\ \text{ratio} = -10 * \log\left(\frac{1}{n}\right)\sum_{i=1}^{n} y_{ij}^{2} \tag{8.3}$$

$$\text{Larger is Better; } \ {}^{S}\!/_{N}\ \text{ratio} = -10 * \log\left(\frac{1}{n}\right)\sum_{i=1}^{n} 1 / y_{ij}^{2} \tag{8.4}$$

Here, n indicates the replicas and y_{ij} indicates the output values. From these objective functions (Eq. 8.3 and Eq. 8.4), the highest S/N ratio value will give us the optimal solution for each response.

GRA technique can be used to optimize the multiresponse characteristics by converting all the objective functions into a single one. This method is based on allocating ranks to each experimental part, and, because of this, the highest grey-relational grade (GRG) is the optimal solution. In GRA, all the responses (EWR, MRR and R_a) were normalized from 0 to 1 using SB characteristics for EWR and R_a, whereas LB characteristic was used for MRR. After that, the grey relational coefficient (GRC) was determined, and the overall GRG was calculated. Based on GRG, the highest GRG will be near to the optimal region, and the corresponding input parameters can be identified from OA. The steps to be followed for optimizing the EDM parameters in GRA are:

Step: 1

The obtained S/N ratios from the Eqs. 8.3 and 8.4 were normalized using Eqs. 8.5 and 8.6 for SB and LB characteristics, respectively.

$$N_i^*(k) = \frac{\max y_i(k) - y_i(k)}{\max y_i(k) - \min y_i(k)} \tag{8.5}$$

$$N_i^*(k) = \frac{y_i(k) - \min y_i(k)}{\max y_i(k) - \min y_i(k)} \tag{8.6}$$

Where $i = 1 \ldots m$, $k = 1, 2, 3 \ldots n$, m = no. of replicas, n = number of parameters, $y_i(k)$ = original sequence, $N_i^*(k)$ value after GRG and $min_{yi}(k)$ and $max_{yi}(k)$ are the minimum and maximum value of $y_i(k)$, respectively.

Step: 2

GRC is calculated to map the relationship between actual input and normalized responses. The GRC was calculated using Eq. 8.7.

$$\epsilon_i(k) = \frac{\Delta \min + \tau\Delta \max}{\Delta_{oi}(k) + \tau\Delta \max} \tag{8.7}$$

Where, $\epsilon_i(k)$ is the GRC, Δ_{oi} is deviation among $N_o^*(k)$ and $N_i^*(k)$, $N_o^*(k)$ = ideal (reference) sequence; Δmax = highest value of $\Delta_{oi}(k)$, Δmin = least value of $\Delta_{oi}(k)$ and τ is assumed to 0.5 in this case (distinguishing coefficient).

Step: 3

After averaging the GRC, the GRG was calculated using Eq. 8.8.

$$GRG = \left(\frac{1}{m}\right)\Sigma\epsilon_i(k) \tag{8.8}$$

Where m indicates the number of response variables. The highest GRG indicates the closer to the ideal solution (i.e. optimum), which meant the deviation between the actual experimental result and the ideal value is minimum.

Table 8.2 shows the S/N ratio of the responses and their corresponding normalized values. Table 8.3 shows the analysis of mean (ANOM) for the S/N ratio of the responses by L_{18} OA. Also, the last column of Table 8.3 shows the corresponding optimal parameter combinations for an individual response. Based on ANOM with respect to L_{18} OA for EWR, the pulse current at 4 Amps, 600 µs pulse ON time and

TABLE 8.2
S/N Ratio of the Responses and Their Normalized Values

Ex.	S/N ratio			Normalized values		
No.	EWR	MRR	R_a	EWR	MRR	R_a
1	55.16	−54.12	−15.22	0.0541	0.0000	0.1533
2	50.94	−45.04	−15.42	0.2910	0.4180	0.1861
3	38.32	−34.36	−15.99	1	1	0.2828
4	56.12	−47.73	−17.45	0.0000	0.3236	0.5318
5	51.24	−47.28	−17.89	0.2742	0.3462	0.6050
6	42.71	−40.34	−18.70	0.7532	0.6975	0.7431
7	53.58	−47.68	−18.98	0.1429	0.3259	0.7902
8	47.54	−45.17	−19.20	0.4819	0.4531	0.8279
9	41.74	−37.78	−20.17	0.8078	0.8268	0.9928
10	52.18	−51.14	−14.32	0.2217	0.1510	0.0000
11	47.04	−40.44	−15.76	0.5102	0.6922	0.2449
12	38.36	−35.14	−16.89	0.9979	0.9604	0.4359
13	53.76	−52.72	−17.68	0.1328	0.0710	0.5708
14	55.81	−47.41	−18.05	0.0179	0.3397	0.6330
15	44.79	−40.44	−18.96	0.6365	0.6922	0.7869
16	47.30	−44.92	−18.98	0.4957	0.4656	0.7902
17	52.96	−49.00	−19.59	0.1777	0.2593	0.8942
18	41.37	−40.34	−20.21	0.8284	0.6975	1

TABLE 8.3
Analysis of Mean (ANOM) for the S/N Ratio of the Responses by L_{18} OA

Responses	Levels	Current (Amps) A	Pulse ON time (μs) B	Pulse OFF time (μs) C	Optimal Parameters
EWR	Level 1	46.9989	53.0160	49.6357	$A_1B_3C_3$
	Level 2	50.7387	50.9220	48.2086	
	Level 3	47.4159	41.2155	47.3092	
	Max-Min	3.7398	11.8005	2.3265	
	Rank	2	1	3	
MRR	Level 1	−43.3737	−49.7177	−44.9182	$A_1B_3C_3$
	Level 2	−45.9850	−45.7229	−44.6998	
	Level 3	−44.1468	−38.0650	−43.8876	
	Max-Min	2.6113	11.6527	1.0306	
	Rank	2	1	3	
R_a	Level 1	−15.6001	−17.1065	−17.8605	$A_3B_3C_1$
	Level 2	−18.1225	−17.6514	−17.8450	
	Level 3	−19.5222	−18.4869	−17.5393	
	Max-Min	3.9222	1.3804	0.3212	
	Rank	1	2	3	

pulse OFF time 40 μs were the optimal input parameters ($A_1B_3C_3$). Based on the rank obtained by L_{18} OA, pulse ON time was the most significant parameter followed by pulse current and pulse OFF time for the response of EWR. For the MRR, the lowest level of pulse current and the highest levels of pulse ON time and pulse OFF time ($A_1B_3C_3$) were the optimal parameters, i.e. pulse current of 4 Amps, pulse ON time of 600 μs and pulse OFF time of 40 μs. Pulse ON time was the most significant parameter followed by pulse current and pulse OFF time for the response of EWR. For R_a, the obtained optimal parameter combinational set was at the lowest levels of pulse current, pulse ON time and pulse OFF time ($A_1B_1C_3$), i.e. pulse current of 4 Amps, pulse ON time of 200 μs and pulse OFF time of 40 μs. The pulse current has more influence than the pulse ON time and the last pulse OFF time (based on rank, see Table 8.3). Table 8.4 shows the GRC, GRG and their corresponding rank for each combinational set. The ninth experiment was the largest GRG (0.817) value; it may be nearer to the optimum parameters based on GRG. The best optimum combinational parameter set, calculated based on the mean response analysis as per L_{18} OA, is $A_3B_3C_3$. The corresponding optimum values are 0.001967 g/min of EWR, 0.012930 g/min of MRR and 10.22 μm of R_a. The pulse ON time is the most significant parameter followed by pulse current and pulse OFF time (based on rank, see Table 8.5). The mean of the GRG for each level of the EDM parameters can be calculated by averaging the GRG as per the L_{18} OA. The mean response values of the GRG are shown in Table 8.5. The main effects plot for the responses of EWR, MRR, R_a and GRG are shown in Figure 8.2 (a) – (c) and (d), respectively.

TABLE 8.4
Grey Relational Coefficient, GRG and Their Ranks

Ex. No.	Grey Relational Coefficient (GRC)			GRG	Rank
1	0.3458	0.3333	0.3713	0.350	18
2	0.4135	0.4621	0.3805	0.419	15
3	1.0000	1.0000	0.4108	0.804	2
4	0.3333	0.4250	0.5164	0.425	14
5	0.4079	0.4333	0.5587	0.467	12
6	0.6695	0.6231	0.6606	0.651	5
7	0.3684	0.4259	0.7045	0.500	11
8	0.4911	0.4776	0.7439	0.571	7
9	0.7223	0.7427	0.9858	**0.817**	**1**
10	0.3911	0.3706	0.3333	0.365	17
11	0.5052	0.6190	0.3984	0.508	10
12	0.9959	0.9267	0.4699	0.797	3
13	0.3657	0.3499	0.5381	0.418	16
14	0.3373	0.4309	0.5767	0.448	13
15	0.5791	0.6190	0.7012	0.633	6
16	0.4979	0.4834	0.7045	0.562	8
17	0.3781	0.4030	0.8254	0.535	9
18	0.7445	0.6231	1.0000	0.789	4

TABLE 8.5
Mean Response Table for the Grey Relational Grade (GRG)

	Parameters		
Levels	**Pulse Current, A**	**Pulse ON Time, B**	**Pulse OFF Time, C**
Level 1	0.5404	0.4366	0.5447
Level 2	0.5070	0.4913	0.5649
Level 3	0.6290*	0.7486*	0.5668*
Max-Min	0.1220	0.3120	0.0221
Rank	2	1	3

* Indicates the optimum level of EDM parameters i.e. $\mathbf{A_3B_3C_3}$

8.3.2 Examination of EDM Process Parameters on the EWR

The EWR with different input parameters were explained in this section. Based on the results (Table 8.1, Table 8.2, Table 8.3, Table 8.4, Table 8.5, and Figure 8.2(a)), the lowest EWR of 0.0016 g/min was obtained at the lowest levels of all the three input parameters i.e., pulse current (4 Amps), pulse ON time (200 μs) and pulse OFF time

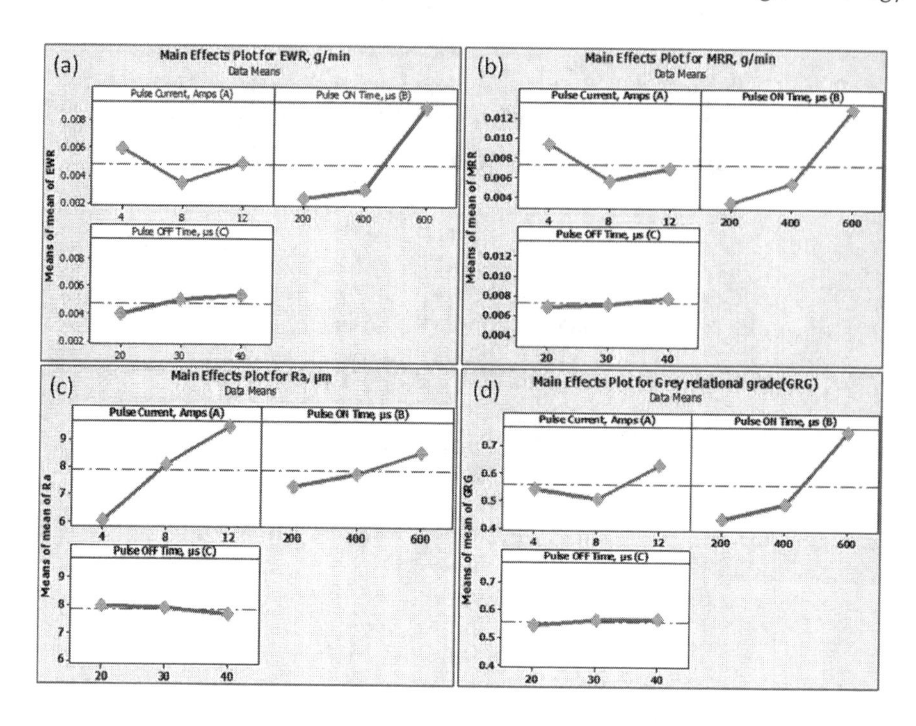

FIGURE 8.2 Main effects plot for the responses (a) EWR, (b) MRR, (c) R_a and (d) GRG.

(20 µs). This was due to the lowest value of applied pulse current, pulse ON time and pulse OFF time, which introduces lower thermal effects on the machined surface though more machining time occurred at this condition. However, the EWR was increased when the pulse current increased from a low level (4 Amps) to a high level (12 Amps) due to more thermal erosion on both the workpiece and the graphite electrodes. This result was attributed to an increase of molten metal and expelled metal from the workpiece. The high value of EWR (0.01214 g/min) is obtained at the lowest level of pulse current (4 Amps), highest level of pulse ON time (600 µs) and highest level of pulse OFF time (40 µs). The ranges of EWR values are 0.00156–0.01214 g/min. The optimal value of the EWR was 0.00818 and was obtained at the level of $A_3B_3C_1$; and the corresponding main effects plot of EWR is shown in Figure 8.2(a). The curve showing larger amounts of inclination is the most significant curve (pulse ON time and pulse current), and the curve exhibiting horizontal to the mean line mean, it has less significant over the EWR (Pulse OFF time). With increasing pulse ON time, the higher amount of positively charged metal particles from the workpiece would impinge/hit the graphite electrodes, leading to more EWR through the use of a dielectric medium (Singaravel et al. 2019).

8.3.3 Examination of EDM Process Parameters on the MRR

MRR is to be maximized to increase the throughput of the manufacturing industries. The results of MRR are given in Table 8.1, Table 8.2, Table 8.3, Table 8.4,

Table 8.5 and Figure 8.2(b). The results show that low-value MRR of 0.00197 g/min was attained at the lowest level of all the three input parameters, i.e. pulse current (4 Amps), pulse ON time (200 μs) and pulse OFF time (20 μs), ie., ($A_1B_1C_1$); this is due to lower thermal energy acting on the machining area at lower levels of input parameters, and, as a result the machining time for completion of the workpiece increased compared to other combinations of input parameters. The high value of MRR of 0.001915 g/min was obtained at the highest level of pulse current (12 Amps), highest level of pulse ON time (600 μs) and lowest level of pulse OFF time (20 μs). In contrast, more thermal spark erosion has occurred when the applied pulse current and pulse ON time increased from a lower level to a higher level which leads to the removal of more material from the workpiece consequently the electrode get in more over the workpiece (Ahmed et al. 2019; Klocke et al. 2013). The ranges of MRR values are 0.00197–0.001915 g/min. The observed optimal MRR value was 0.01291 g/min at $A_3B_3C_1$; (12 Amps of current, 600 μs of pulse ON time and 20 μs of pulse OFF time). Figure 8.2(b) shows the corresponding main effect plot of MRR; the curve showing larger amount of inclination is the most significant curve (pulse ON time and pulse current), whereas, the curve behaving horizontal to the mean line has less significant over the MRR (pulse OFF time).

8.3.4 Examination of EDM Process Parameters on R_a

The examination of R_a with machining parameters can be performed based on the results illustrated in Table 8.1, Table 8.2, Table 8.3, Table 8.4, Table 8.5 and Figure 8.2(c). From Table 8.1, it is observed that the lowest measured R_a value was 5.77 μm at the lowest levels of the input parameters (4 Amps of pulse current, 200 μs of pulse ON time and 20 μs of pulse OFFF time). This was due to the lower thermal effects on the surface of the workpiece (machining time is very high compared with other combinational sets) by the lowest current, pulse ON time and pulse OFF time. While the pulse current and pulse ON time increased from a low level to a high level, the R_a values were increased on the machined surfaces because of more thermal erosion and higher material removal from the work surface during the EDM process (Klocke et al 2013; Sultan et al. 2014)]. The highest measured R_a value was 10.25 μm at the highest levels of pulse current (12 Amps), pulse ON time (600 μs) and the middle level of pulse OFF time (30 μs). The optimal EDM parameters for getting better surface finish for the multiresponses was at $A_3B_3C_1$, i.e. at 12 Amps of current, 400 μs of pulse ON time, and 20 μs of pulse OFF time. The optimum R_a value was 10.2 μm. The corresponding main effects plots for the R_a are shown in Figure 8.2(c). The curve showing larger amount of inclination is the most significant curve (Pulse current followed by Pulse ON time), whereas, the curve behaving horizontal to the mean line has less significant over the MRR (Pulse OFF time). Also, the corresponding three-dimensional (3D) surface topography of WLI images was shown in Figure 8.3(a)–(c) for the initial condition (current 4 Amps, Ton 200 μs, Toff 20 μs), optimal condition (current 12 Amps, Ton 600 μs, Toff 40 μs) and for the final condition (current 12 Amps, Ton 600 μs, Toff 20 μs).

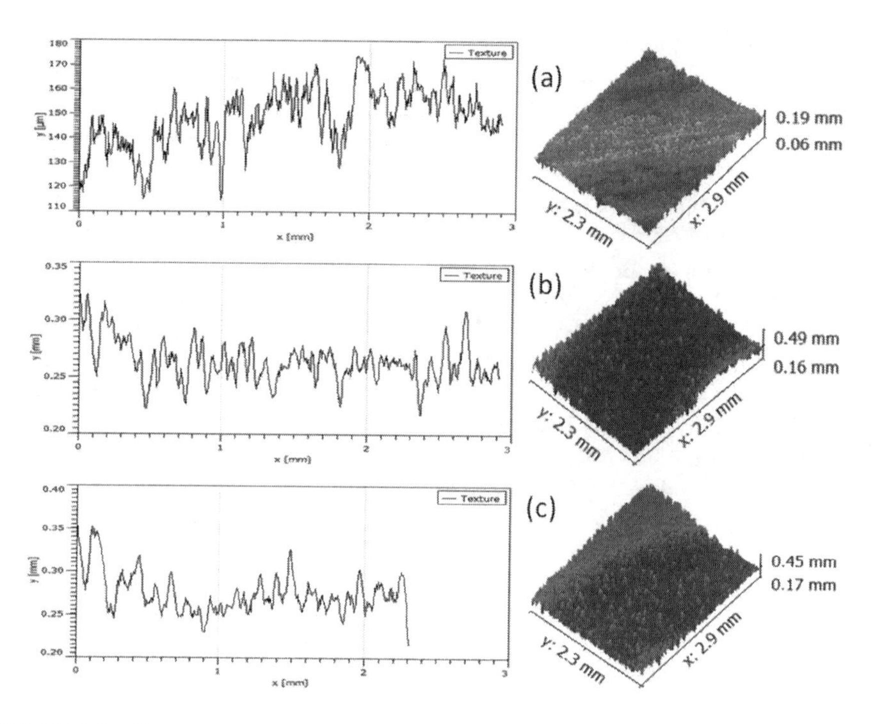

FIGURE 8.3 3D surface texture image of the (a) initial condition, (b) optimal condition and (c) final condition.

8.3.5 Response Surface Methodology (RSM)

RSM is also one of the statistical modelling techniques by which the mapping of input parameters over the responses in terms of mathematical regression equations, two-dimensional (2D) contour, and surface plots can be examined (Bahgat et al. 2019). In this work, the development of mathematical models is done by RSM for the four responses (SR, EWR, RCLT and MRR). The relationship between input variables (Current (*A*), pulse ON time (*B*) and pulse OFF time (*C*)) and the output responses (electrode wear ratio, MRR, R_a or (*Y*)) can be given as:

$$Y = F(A, B, C) \tag{8.9}$$

Where, *F* is the response function.

$$\begin{aligned} EWR = {} & +0.009069 - 0.001532 * A - 0.000026 * B + 0.000041 * C \\ & - 1.32284E-06(A * B) + 0.000121(A * A) + 6.67708E-08(B * B) \end{aligned} \tag{8.10}$$

$$\begin{aligned} MRR = {} & +0.008546 - 0.001555 * A - 7.77629E-06 * B - 0.000016 * C \\ & - 3.07328E-06(A * B) + 0.000154(A * A) + 7.03125E-08(B * B) \end{aligned} \tag{8.11}$$

$$\mathrm{Ra} = +3.22306 + 0.734792 * A - 0.000321 * B - 0.015583 * C$$
$$-0.019115(A * A) + 4.22917E - 06(B * B) \tag{8.12}$$

$$\mathrm{GRG} = +0.585239 - 0.045623 * A - 0.000824 * B + 0.000032 * C$$
$$-0.000053(A * B) + 0.004870(A * A) + 2.53542E - 06(B * B) \tag{8.13}$$

8.3.6 Analysis of Variance (ANOVA)

Minitab 16.0 Software was used to determine the statistical contribution of each EDM parameter for the experimental outcome and for the GRG in terms of ANOVA, which was done at a 5% significance level (i.e. 95% confidence), and the same is illustrated in Table 8.6. Table 8.6 demonstrates the individual parameter effect using F-ratio and p-value. If the F-ratio is greater than 4 (referred from the statistical table) and p-value is less than 0.05 mean, the corresponding input parameter would have a significant effect on response. The most influencing parameter is pulse ON time followed by pulse current and pulse OFF time for the responses and for the GRG. From Table 8.6 shows the effect of individual parameters the EWR: the pulse ON time (78.25%) had more influence ($p < 0.000$, F-ratio of 43.03 >4) followed by pulse current with 8.82%; ($p < 0.031$, F-ratio of 4.84 >4) and pulse OFF time (2.97%). For MRR, the pulse ON time (70.86%) had more influence ($p < 0.000$, F-ratio of 21.08 >4) followed by pulse current with 10.04% and pulse OFF time was not significant. For the R_a, the pulse current (87.18%) had more influence ($p < 0.000$, F-ratio of 622.4 >4) followed by pulse ON time with 11.27% ($p < 0.000$, F-ratio of 80.48 >4) and pulse OFF time was not significant. The corresponding R-Sq and R-Sq (adj.) values for the EWR, MRR and R_a were 90% and 84.55%, 81.51% and 71.42%, and 99.23% and 98.81%, respectively. After converting multiresponse characteristics into a single objective function based on GRG, the ANOVA was carried out, which is shown in Table 8.6. The pulse ON time (79.92%) had more influence ($p < 0.000$, F-ratio of 53.94 >4) followed by pulse current with 11.49%; ($p < 0.008$, F-ratio of 7.72 >4) and pulse OFF time. The corresponding R-Sq and R-Sq (adj.) values for the GRG are 91.82% and 87.36% respectively. Based on these results, the control of pulse ON time and pulse current would give improved results on EDM of Ti-6Al-4V. ANOVA was also conducted for the developed RSM models by Design-Expert 12 software to determine the most influencing parameter effect, which is shown in Table 8.7. While constructing the ANOVA Table 8.7, 95% confidence level (i.e. 5% significance) was set for the examination. F-values more than 4 indicate that the model terms are significant. There was only a 0.01% chance that a model F-value this large could occur because of noise. Also, p-values less than 0.0500 indicate that the model terms are significant for all the responses. Values greater than 0.1000 indicate that the model terms are not significant. Based on F value and p-value, all the developed models by RSM of EWR, MRR, R_a and GRG are significant. In the case of EWR, A, B, AB, A^2 and B^2 are significant model terms. For MRR, A, B, AB, A^2 and B^2 are significant. For R_a, A, B, C, A^2 and B^2 are significant model terms. For

TABLE 8.6
ANOVA for (a) EWR, (b) MRR, (c) R_a and (d) GRG Using Adjusted SS for the Tests

Parameters	DF	Sum of Squares	Mean Square	*F*-Value	*p*- Value	PC %	Remarks
(a) EWR							
Current	2	0.0000181	0.0000090	4.84	0.031	8.82	S=0.00136560 R-Sq=90.00%
Pulse ON time	2	0.0001605	0.0000802	43.03	0.000	78.25	R-Sq(adj)=84.55%
Pulse OFF time	2	0.0000061	0.0000030	1.63	0.240	2.97	
Error	11	0.0000205	0.0000019			10.0	
Total	17	0.0002051				100	
(b) MRR							
Current	2	0.0000433	0.0000217	2.99	0.092	10.04	S=0.00269264 R-Sq=81.51%
Pulse ON time	2	0.0003056	0.0001528	21.08	0.000	70.86	R-Sq(adj)=71.42%
Pulse OFF time	2	0.0000025	0.0000013	0.18	0.841	0.58	
Error	11	0.0000798	0.0000073			18.50	
Total	17	0.0004313				100	
(c) R_a							
Current	2	35.7031	17.8516	622.4	0.000	87.18	S=0.169351 R-Sq=99.23%
Pulse ON time	2	4.6163	2.3082	80.48	0.000	11.27	R-Sq(adj)=98.81%
Pulse OFF time	2	0.3175	0.1588	5.54	0.022	0.78	
Error	11	0.3155	0.0287			0.77	
Total	17	40.9525				100	

(Continued)

TABLE 8.6 (CONTINUED)
ANOVA for (a) EWR, (b) MRR, (c) R_a and (d) GRG Using Adjusted SS for the Tests

Parameters	DF	Sum of Squares	Mean Square	*F*-Value	*p*- Value	PC %	Remarks
(d) GRG							
Current	2	0.047961	0.023980	7.72	0.008	11.49	S=0.0557224 R-Sq=91.82%
Pulse ON time	2	0.333725	0.166863	53.74	0.000	79.92	R-Sq(adj)=87.36%
Pulse OFF time	2	0.001753	0.000877	0.28	0.759	0.42	
Error	11	0.034155	0.003105			8.18	
Total	17	0.417594				100	

TABLE 8.7
Analysis of Variance (ANOVA) for the Developed RSM Models

Source	DF	Sum of Squares	Mean Square	*F*-value	*p*-value	remarks
(a) Model of EWR	6	0.0002	0.0000	27.12	< 0.0001	Significant
A-Pulse Current	1	3.142E-06	3.142E-06	2.66	0.1311	R^2 = 0.9367
B-Pulse ON Time	1	0.0001	0.0001	111.73	< 0.0001	Adj R^2 = 0.9021
C-Pulse OFF Time	1	1.839E-06	1.839E-06	1.56	0.2379	Pred R^2 = 0.8403
AB	1	8.120E-06	8.120E-06	6.88	0.0237	Adeq Precision = 16.17
A^2	1	0.0000	0.0000	12.66	0.0045	
B^2	1	0.0000	0.0000	24.16	0.0005	
Residual	11	0.0000	1.181E-06			
Cor Total	17	0.0002				
(b) Model of MRR	6	0.0004	0.0001	20.11	< 0.0001	Significant
A-Pulse Current	1	0.0000	0.0000	5.79	0.0349	R^2 = 0.9165
B-Pulse ON Time	1	0.0003	0.0003	83.64	< 0.0001	Adj R^2 = 0.8709
C-Pulse OFF Time	1	2.772E-07	2.772E-07	0.0846	0.7765	Pred R^2 = 0.7757
AB	1	0.0000	0.0000	13.38	0.0038	Adeq Precision = 13.95
A^2	1	0.0000	0.0000	7.45	0.0196	
B^2	1	0.0000	0.0000	9.66	0.0100	
Residual	11	0.0000	3.275E-06			
Cor Total	17	0.0004				

(Continued)

TABLE 8.7 (CONTINUED)
Analysis of Variance (ANOVA) for the Developed RSM Models

Source	DF	Sum of Squares	Mean Square	*F*-value	*p*-value	remarks
(c) Model of R_a	5	40.61	8.12	285.31	< 0.0001	Significant
A-Pulse Current	1	35.33	35.33	1241.02	< 0.0001	$R^2 = 0.9917$
B-Pulse ON Time	1	4.50	4.50	158.14	< 0.0001	Adj $R^2 = 0.9882$
C-Pulse OFF Time	1	0.2914	0.2914	10.24	0.0076	Pred $R^2 = 0.9811$
A^2	1	0.3741	0.3741	13.14	0.0035	Adeq Precision = 51.003
B^2	1	0.1145	0.1145	4.02	0.0680	
Residual	12	0.3416	0.0285			
Cor Total	17	40.95				
(d) Model of GRG	6	0.3959	0.0660	33.66	< 0.0001	Significant
A-Pulse Current	1	0.0237	0.0237	12.08	0.0052	$R^2 = 0.9483$
B-Pulse ON Time	1	0.2923	0.2923	149.13	< 0.0001	Adj $R^2 = 0.9202$
C-Pulse OFF Time	1	1.126E-06	1.126E-06	0.0006	0.9813	Pred $R^2 = 0.8672$
AB	1	0.0130	0.0130	6.64	0.0257	Adeq Precision = 14.535
A^2	1	0.0243	0.0243	12.39	0.0048	
B^2	1	0.0411	0.0411	20.99	0.0008	
Residual	11	0.0216	0.0020			
Cor Total	17	0.4175				

GRG, A, B, AB, A^2 and B^2 are significant model terms. Moreover, "Adeq Precision" indicates the S/N ratio in which large value represent the less effect produced by the uncontrollable parameters (e.g. ventilation, lighting, atmosphere, etc.). A ratio greater than 4 is desirable. A reasonable "Adeq Precision" value was obtained in all the responses (EWR, MRR and R_a) as well as for the GRG. In Table 8.8, the coefficients of "Pred R-Squared" and "Adj R-Squared" values were well agreed to experimental results and these result indicate higher accuracy of the developed models. Figure 8.4(a)–(d) indicates that all the predicted values of the EWR, MRR, R_a and GRG are close to the experimental values, confirming that the developed models could predict the response precisely (Surekha et al. 2017, Shabgard et al. 2017, Zainal et al. 2017, Dhupal et al. 2018). Figure 8.5(a)–(c) and Figure 8.6(a)–(b) shows the corresponding contour and 3D surface plot of the EWR, MRR, R_a and GRG, respectively.

8.3.7 Confirmation Experiment

The improved performance of each response based on the optimal conditions was checked by the confirmation experiments. The estimated GRG ($\gamma_{\text{estimated}}$) at the optimal condition can be computed by the following:

$$\gamma_{\text{estimated}} = \gamma n + \sum_{i=1}^{n} \left(\gamma i - \gamma n \right) \tag{8.14}$$

Where γn indicates the total mean of GRG, γi represents the mean of GRG at the optimal condition (i.e. significant parameters of A, B, and C); n is the number of input variables that have considerable effect on the responses. Confirmation experiments were performed with two replicas at the optimal conditions. The obtained results are shown in Table 8.8. This result demonstrated the improvement of responses when compared to the initial condition.

TABLE 8.8
Results of the Confirmation Experiment

		Optimal cutting conditions	
	Initial conditions	Prediction	Experiment
Levels	*A1B1C1*	*A3B3C3*	*A3B3C3*
EWR, g/min	0.017		0.001967
MRR, g/min	0.0082		0.012930
$R_{a,}$ µm	5.77		10.22
GRG	0.350	0.827	0.819

The percentage improvement in GRG = 57.26%.

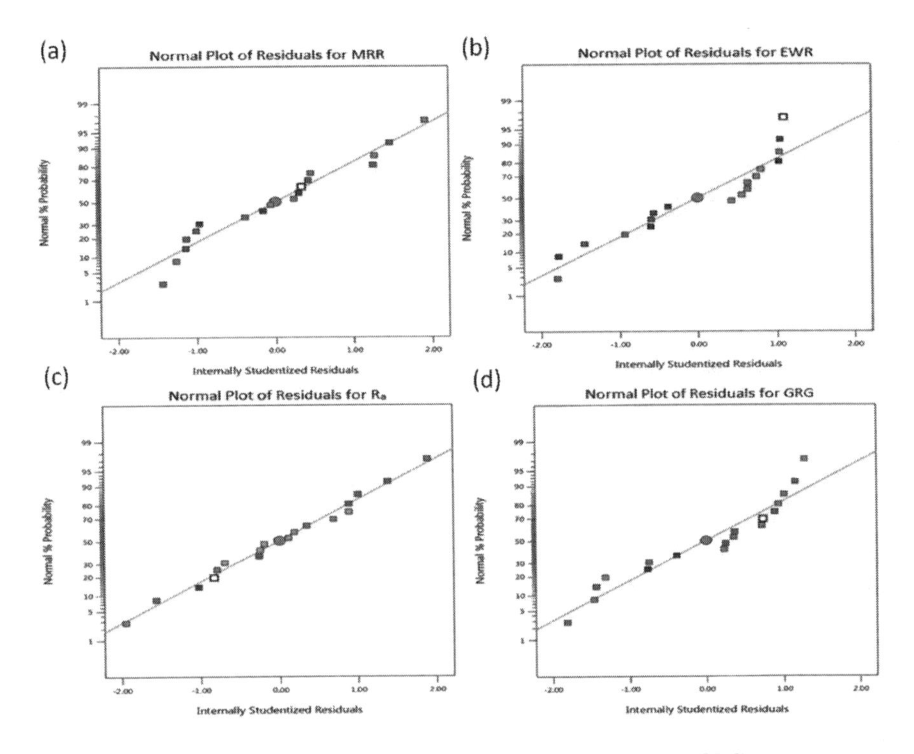

FIGURE 8.4 Normal plot of residuals (a) MRR; (b) EWR; (c) R_a; (d) GRG.

8.4 CONCLUSIONS

The EDM of hard Ti-6Al-4V was conducted with different input parameters (pulse current, pulse ON time, and pulse OFF time) for measuring several responses (EWR, MRR and R_a) in the EDM process. The experiments conducted using Taguchi L_{18} OA optimized the single and multiresponse characteristics using Taguchi-based GRA modelled by RSM. The optimized input parameters were the pulse current of 12 Amps, pulse ON time of 600 μs, and pulse OFF time of 40 μs ($A_3B_3C_3$) for the multiresponse characteristics. At the optimal input parameters, the improved the response value was 0.001967 g/min, 0.012930 g/min and 10.22 μm for EWR, MRR and $R_{a,}$ respectively. Further, the improvement of performance of around 57.26% by considering all the responses was obtained at optimal conditions when compared to the initial condition, which was identified by the confirmation test. In addition, the pulse ON time is the parameter with the most influence, which affected more responses than the pulse current and pulse OFF time as identified by ANOVA based on GRG. The controlling of these parameters suggested ranges of pulse current, pulse OFF time and pulse ON time could give us improved EDM behaviour of Ti-6Al-4V. Mathematical prediction models obtained from RSM have produced a well-agreed result with the experimental one. Through this research work, the obtained results could be used as a reference database in manufacturing industries.

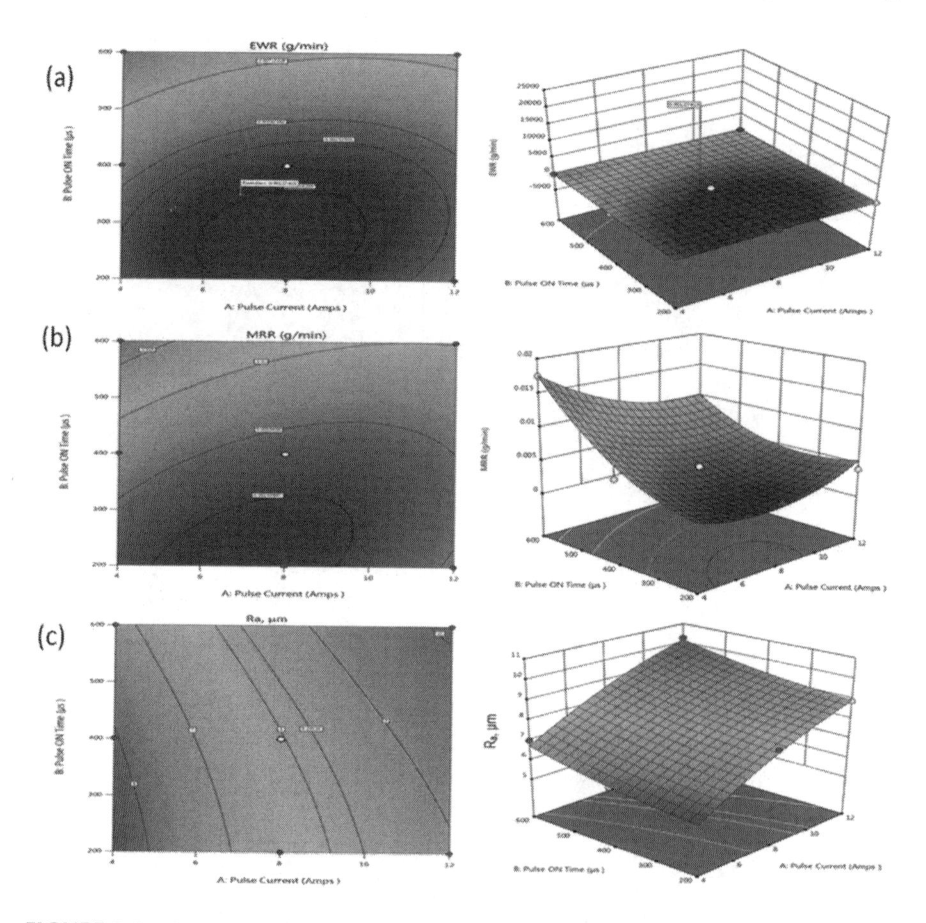

FIGURE 8.5 Contour and 3D surface plot (a) EWR, (b) MRR, (c) R_a.

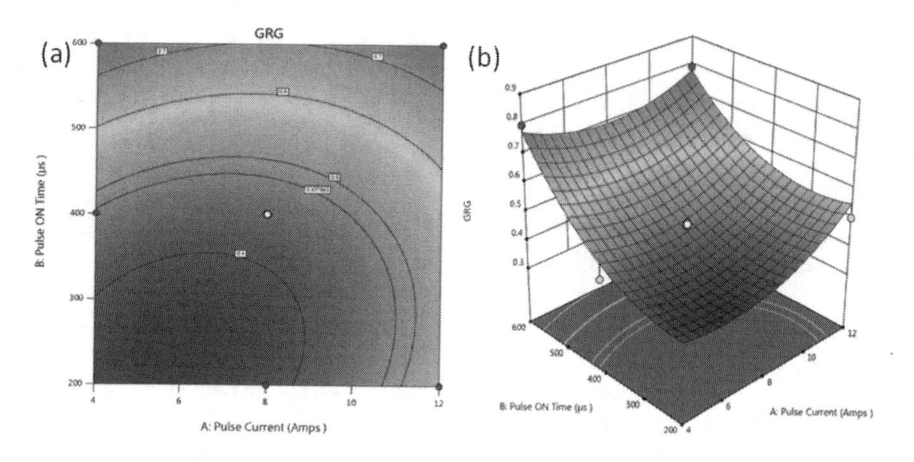

FIGURE 8.6 (a) Contour plot of GRG (b) 3D surface plot of GRG.

Dr. A. Palanisamy and **Dr. V. Sivabharathi:** conceptualization, methodology, formal analysis, investigation, and writing of the original draft. **Dr. S. Navaneethakrishnan** and **Dr. S. Sivasankaran:** data curation, supervision, software, review and editing

DECLARATION OF CONFLICT OF INTEREST

The authors declare that they have no conflict of interest with respect to the research, authorship, and/or publication of this article.

FUNDING

The author(s) did not receive any research funding or grants from any organization.

ETHICAL APPROVAL

This article does not contain any studies with human participants or animals performed by the author.

REFERENCES

Ahmed, N., Ishfaq, K., Rafaqat, M., Pervaiz, S., Anwar, S. and Salah, B. 2019. EDM of Ti-6Al-4V: Electrode and polarity selection for minimum tool wear rate and overcut. *Materials and Manufacturing Processes*, 34(7): 769–778.

Alshemary, A., Pramanik, A., Basak, A.K. and Littlefair, G. 2018. Accuracy of duplex stainless steel feature generated by electrical discharge machining (EDM). *Measurement*, 130 137–144. doi:10.1016/j.measurement.2018.08.013.

Angappan, P., Thangiah, S. and Subbarayan, S. 2017. Taguchi-based grey relational analysis for modeling and optimizing machining pa- rameters through dry turning of Incoloy. *Journal of Mechanical Science and Technology*, 31(9): 4159–4165. doi:10.1007/s12206-017-0812-y.

Bahgat, M.M., Shash, A.Y., Abd-Rabou, M. and El-Mahallawi, I.S. 2019. Influence of process parameters in electrical discharge machining on H13 die steel. *Heliyon*, 5(6): e01813.

Bai, X., Yang, T. and Zhang, Q. 2018. Experimental study on the electrical discharge machining with three-phase flow dielectric medium. *The International Journal of Advanced Manufacturing Technology*, 96(5–8): 2003–2011.

Dhupal, D., Naik, S. and Das, S.R. 2018. Modelling and optimization of Al–SiC MMC through EDM process using copper and graphite electrodes. *Materials Today: Proceedings*, 5(5): 11295–11303.

El-Hofy, H.A.G. 2005. *Advanced Machining Processes: Nontraditional and Hybrid Machining Processes*. USA: McGraw Hill Professional.

Hasçalık, A. and Çaydaş, U. 2007. Electrical discharge machining of titanium alloy (Ti–6Al–4V). *Applied Surface Science*, 253: 9007–9016. doi:10.1016/j.apsusc.2007.05.031.

Jeykrishnan, J., Ramnath, B.V., Akilesh, S. and Kumar, R.P. 2016. Optimization of process parameters on EN24 tool steel using Taguchi technique in electro-discharge machining (EDM). In *IOP Conference Series: IOP Publishing, Materials Science and Engineering*, 149 (1): 012022. doi:10.1088/1757-899X/149/1/012022.

Julong, D. 1989. *Introduction to Grey System Theory*, vol. 1, England and China Petroleum Industry Press, P.R. China, 1–24.

Klocke, F., Schwade, M., Klink, A., and Veselovac, D. 2013. Analysis of material removal rate and electrode wear in sinking EDM roughing strategies using different graphite grades. *Procedia CIRP*, 6: 163–167.

Lin, J.L. and Lin, C.L. 2002. The use of the orthogonal array with grey relational analysis to optimize the electrical discharge machining process with multiple performance characteristics. *International Journal of Machine Tools and Manufacture*, 42: 237–244.

Magabe, R., Sharma, N., Gupta, K., and Davim, J.P. 2019. Modeling and optimization of wire-EDM parameters for machining of Ni 55.8 Ti shape memory alloy using hybrid approach of Taguchi and NSGA-II. *The International Journal of Advanced Manufacturing Technology*, 102: 1703–1717.

Montgomery, D.C. 2017. *Design and analysis of experiments*. Singapore Pte.Ltd., John Wiley & Sons.

Pachaury, Y. and Tandon, P. 2017. An overview of electric discharge machining of ceramics and ceramic based composites. *Journal of Manufacturing Processes*, 25: 369–390.

Palanisamy, A., Rekha, R., Sivasankaran, S. and Sathiya Narayanan C. 2014. Multi-objective optimization of EDM parameters using grey relational analysis for titanium alloy (Ti–6Al–4V). *Applied Mechanics and Materials*, 592: 540–544. doi:10.4028/www.scientific.net/amm.592-594.540.

Palanisamy, A. and Selvaraj, T. 2019. Optimization of turning parameters for surface integrity properties on Incoloy 800H superalloy using cryogenically treated multi-layer CVD coated tool. *Surface Review and Letters*, 26 (02): 1850139.

Palanisamy, A., Selvaraj T. and Sivasankaran, S. 2018. Optimization of turning parameters of machining incoloy 800H superalloy using cryogenically treated multilayer CVD-coated tool. *Arabian Journal for Science and Engineering*, 43: 4977–4990.

Panda, S., Mishra, D., Biswal, B.B. and Nanda, P. 2015. Optimization of multiple response characteristics of EDM process using taguchi-based grey relational analysis and modified PSO. *Journal of Advanced Manufacturing Systems*, 14(03): 123–148.

Payal, H., Maheshwari, S., Bharti, P.S. and Sharma, S.K. 2019. Multi-objective optimisation of electrical discharge machining for Inconel 825 using Taguchi-fuzzy approach. *International Journal of Information Technology*, 11(1): 97–105.

Sahu, A.K. and Mahapatra, S.S. 2019. Optimization of electrical discharge machining of titanium alloy (Ti6Al4V) by grey relational analysis based firefly algorithm. In AlMangour B. (eds.), *Additive Manufacturing of Emerging Materials*. Cham: Springer, 29–53.

Santos, R.F., Silva, E.R., Sales, W.F. and Raslan, A.A. 2016. Analysis of the surface integrity when nitriding AISI 4140 steel by the sink electrical discharge machining (EDM) process. *Procedia CIRP*, 45: 303–306. doi:10.1016/j.procir.2016.01.197.

Selvarajan, L., Sathiya Narayanan, C. and Jeyapaul, R. Optimization of EDM hole drilling parameters in machining of $MoSi_2$-SiC intermetallic/composites for improving geometrical tolerances. 2015. *Journal of Advanced Manufacturing Systems*, 14(04): 259–272.

Senthil, P., Vinodh, S. and Singh, A.K. 2014. Parametric optimisation of EDM on Al-Cu/TiB_2 in-situ metal matrix composites using TOPSIS method. *International Journal of Machining and Machinability of Materials*, 16(1): 80–94.

Shabgard, M., Ahmadi, R., Seyedzavvar, M., and Oliaei, S.N.B. 2013. Mathematical and numerical modeling of the effect of input-parameters on the flushing efficiency of plasma channel in EDM process. *International Journal of Machine Tools and Manufacture*, 65: 79–87.

Shabgard, M. and Khosrozadeh, B. 2017. Investigation of carbon nanotube added dielectric on the surface characteristics and machining performance of Ti–6Al–4V alloy in EDM process. *Journal of Manufacturing Processes*, 25: 212–219.

Singaravel, B., Shekar, K.C., Reddy, G.G. and Prasad, S.D. 2019. Experimental investigation of vegetable oil as dielectric fluid in electric discharge machining of Ti-6Al-4V. *Ain Shams Engineering Journal*, 11(1): 143–147.

undefinedSultan, T., Kumar, A. and Gupta, R.D. 2014. Material removal rate, electrode wear rate, and surface roughness evaluation in die sinking EDM with hollow tool through response surface methodology. *International Journal of Manufacturing Engineering*. Volume 2014, Article ID 259129, 16 pages, http://dx.doi.org/10.1155/2014/259129.

Surekha, B., Swain, S., Suleman, A.J. and Choudhury, S.D. 2017. Performance capabilities of EDM of high carbon high chromium steel with copper and graphite electrodes. In *AIP Conference Proceedings*, 1859(1): 020070.

Tsai, M.Y., Fang, C.S. and Yen, M.H. 2018. Vibration-assisted electrical discharge machining of grooves in a titanium alloy (Ti-6A-4V). *The International Journal of Advanced Manufacturing Technology*, 97(1–4): 297–304.

Urso, G.D., Maccarini, G. and Ravasio, C. 2015. Influence of electrode material in micro-EDM drilling of stainless steel and tungsten carbide. *The International Journal of Advanced Manufacturing Technology*, 85(9–12), 2013–2025. doi:10.1007/s00170-015-7010-9.

Verma, V. and Sahu, R. 2017. Process parameter optimization of die-sinking EDM on Titanium grade–V alloy (Ti6Al4V) using full factorial design approach. *Materials Today: Proceedings*, 4(2): 1893–1899.

Yuvaraj, T. and Suresh, P. 2019. Analysis of EDM process parameters on Inconel 718 using the Grey-Taguchi and Topsis methods. *Strojniski Vestnik – Journal of Mechanical Engineering*, 65(10): 557–564.

Zainal, N., Zain, A.M., Sharif, S., Hamed, H.N.A. and Yusuf, S.M. 2017. An integrated study of surface roughness in EDM process using regression analysis and GSO algorithm. *Journal of Physics: Conference Series, IOP Publishing*, 892(1): 012002.

9 High-Temperature Tribological Behaviour of Surface-Coated Tool Steel

B. Guruprasad and K. Murugan

CONTENTS

9.1 INTRODUCTION

Tribology is defined as the science and technology of interacting surfaces in relative motion and the related practices. Tribology is primarily concerned with the study of friction, lubrication and wear, and was developed to reduce friction force [1]. In the following years, surface engineering was developing. Instead of searching for materials to satisfy some requirement, efforts were applied toward developing materials, coatings and lubricants [2]. The science of tribology began to anticipate the needs and many new products came from recently acquired knowledge in material science, organic chemistry, molecular dynamics and so on. In addition, enduring questions about adhesion and friction were explored in terms of the behaviour of atoms and subatomic particles [3]. Tribology in the 20th century has been developed by a combination from a wide variety of backgrounds: surface science, chemistry of lubricants, machine design and behaviour, material science, rheology and fluid mechanics. The researchers successfully produced coating layers by Chemical Vapor Deposition (CVD), Physical Vapor Deposition (PVD); alloying surfaces based on alloy elements such as Ti, W, N, Mo, Ni, Cr, C, S, etc.; and creating new materials with good tribology properties [4, 5]. Tribology, the science of friction, wear and lubrication, is all about surfaces and is a new field of study compared to other engineering subjects such as thermodynamics, mechanics and plasticity [6]. Wear is a phenomenon of removal and lost materials on sliding surfaces. Friction is the resistance to movement of one body over another body and a principal cause of wear and energy dissipation [7, 8]. The wear phenomenon appears in many ways depending on the working conditions of contacting surfaces in relative motion. There are four types of wear: adhesive wear, abrasive wear, corrosive wear and fatigue wear [9].

Lubrication is an effective means of controlling wear and reducing friction. The lubricants are thin, low shear-strength layers of gas, liquid or solid that are interposed between two surfaces in order to reduce friction and prevent damage [10-11]. In general, the thickness of these films ranges from 1 μm to 100 μm or thinner that many kinds of materials in different ways. Liquid and gaseous films are usually termed "hydrodynamic" and "hydrostatic lubrication", while lubrication by solid is termed "solid lubrication". In hydrodynamic or hydrostatic lubrication, both surface layers are separated by a thin, liquid film, and the friction present in a hydrodynamic lubrication system is the friction only due to the shearing of the lubricant itself [12.-13]. Hydrodynamic lubrication is an excellent method of lubrication, as it is able to achieve coefficients of friction as low as 0.001 ($\mu = 0.001$), and there is no wear between the moving parts. In solid lubrication, a thin film made of a single solid or a combination of solids is introduced between two rubbing surfaces for the purpose of reducing friction and wear. Graphite and molybdenum disulphide (MoS_2) are the most frequently used inorganic solid lubricants [14-17].

Surface coating and surface engineering, current methods of improving surface properties, aim to deal with tribology problems. The methods are based on theories of physics, chemistry and material science and are applied to create special surface layers to improve tribological properties [18]. These include thermal spray coatings, electroplated coatings, PVD and CVD coatings, ion implantation, laser surface processing, carburizing, nitriding and nitrocarburizing. The main advantages of these techniques are to form layers with more hardness and resistance to erosion, corrosion and abrasion at low and high temperatures as well as preventing adhesion, reducing friction and increasing life of working components. The objective of the present work is to study the low- and high-temperature (up to 800°C) tribological characteristics of tool steels coated with different coatings of TiAlN, TiSiN and CrN sliding against uncoated high-strength boron steel [19-21].

The friction and wear characteristics of the coatings were evaluated using an Optimol SRV machine. The goal of this study is to gain a better understanding of friction and wear of TiAlN, TiSiN and CrN coatings at low and high temperatures under dry sliding conditions.

9.2 EXPERIMENTAL MATERIALS

The study was aimed at investigating the tribological behaviour of the coated tool steels sliding against high-strength boron steel. The tool steel has been used in many applications of plastic moulding dies, swaging tools, wear mouldings, sheet metal press dies and machine components. In order to strengthen the working life, surface engineering has been used, such nitriding, nitrocarbirizing, boronizing, coatings and oxidation. In this study, the tool steels have been coated with TiSiN, TiAIN and CrN coatings by PVD technique, and the research aimed to investigate tribological properties of the coatings. Chemical compositions are shown in Table 9.1.

Crystal structure of the ceramic compounds depends on the contents of the elements. Some papers have noted that the crystal structures are cubic and hexagonal, and the hardness, therefore, changes with structure accordingly. In this investigation, the structure of the coatings has not been studied; only hardness of coatings is mentioned.

9.3 CONFIGURATION AND SLIDING CONTACT

The sliding contact configuration features pins with spherical ends that oscillate against the flat surface of the disc. The radius of the pin is 5 mm round with a 10 mm total length, while the disc is 24 mm in diameter with an 8 mm thickness. The pin is fixed in the upper holder, which can move forward and backward, and the lower disc is fixed and mounted onto a heating block that can control temperature up to 900°C during testing. The contact configuration is shown as in Figure 9.1.

The contact between the pin and the disc is a point contact under 50 N loading. At room temperature, it seems that stress-strain behaviour causes elastic reaction. Within elastic deformation, the contact is referred to as Hertz's contact. Based on topographic analysis, some parameters, such as contact radius and indentation, could be evaluated and are shown in Table 9.2.

9.4 TEST SETTING AND PROCEDURES

Within the work, we studied the tribological properties of three different coatings of TiAlN, TiSiN and CrN sliding on high-strength boron steel. The performance was evaluated at room temperature and at high temperature (800°C) using standard equipment. The topographic parameters of the coating layers of the tool steel were characterized using a Wyko NT 1100 apparatus. Thus, the roughness was determined and profile pictures were taken with 5X or 10X magnification.

Before testing, all the samples have been cleaned ultrasonically and rinsed by ethanol to remove all the dust particles that affect tribological properties and fixed on the right position in Optimol SRV tribometer. Within the apparatus, the pin and disc were slid relatively to control speed, load, time and temperature. The experiments

TABLE 9.1
Chemical Compositions of High-Strength Boron Steel and Tool Steels

Materials	C	Mn	Cr	Si	B	P	S	Ni	Mo	V	HV300g
Boron steel	0.2–0.5	1.0–1.3	0.14–0.26	0.2–0.35	0.05	>0.03	>0.01	–	–	–	155
Tool steel uncoated	0.31	0.09	1.35	0.6	–	100 ppm	40 ppm	0.7	0.8	0.145	–
Tool steel TiAlN coating	0.31	0.09	1.35	0.6	–	100 ppm	40 ppm	0.7	0.8	0.145	1250
Tool steel TiSiN coating	0.31	0.09	1.35	0.6	–	100 ppm	40 ppm	0.7	0.8	0.145	1200
Tool steel CrN coating	0.31	0.09	1.35	0.6	–	100 ppm	40 ppm	0.7	0.8	0.145	1600

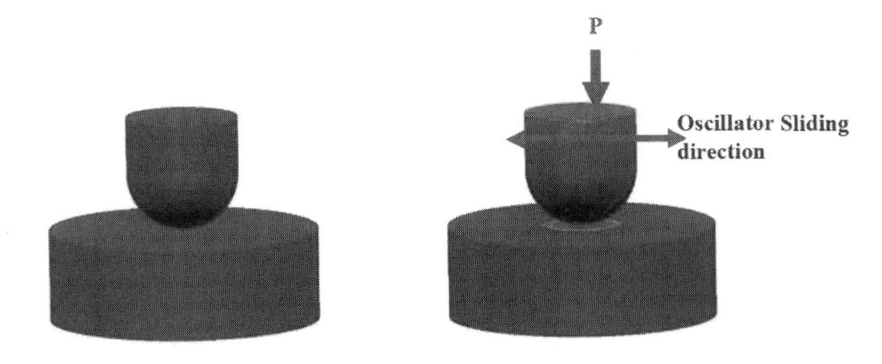

FIGURE 9.1 Pin-disc contact configuration.

TABLE 9.2
Typical Contact Parameters of Pin-Disc at 800°C

Coatings penetrate	Contact radius (mm)	Indentation δ (mm)	Mean contact pressure (GPa)
TiAlN	0.19	3.61 E-3	0.44
TiSiN	0.16	2.56 E-3	0.62
CrN	0.19	3.61 E-3	0.44

TABLE 9.3
Parameters at Room Temperature

Parameters	Values
Load	50N
Temperature	40°C
Stroke length	1 mm
Frequency	50 Hz
Duration	10 min.

were carried out at room and high temperature [22-23]. The testing parameters are shown in Table 9.3 and Table 9.4.

Testing procedure at high temperature was a little different compared to the one at room temperature. At high temperature, the upper pin was separated with a gap small enough to avoid contact while the lower disc was heating up [24]. We kept the temperature stable for 3 minutes before making the contact and running the test to make sure that the sample was at the same temperature. After testing, tested surfaces were characterized by using scanning electron microscopy (SEM) integrating energy dispersive spectroscopy (EDS). Thus, surface morphology of the test specimens was

TABLE 9.4
Parameters at 800°C

Parameters	Values
Load	50N
Temperature	800°C
Stroke length	1 mm
Frequency	50 Hz
Duration	10 min.

examined with high resolution, and the coating composition was analyzed at the same time.

9.5 RESULTS AND DISCUSSION

9.5.1 Characterization of Surface Topography

All the tested specimens were preprocessed before test running. In order to avoid factors that affect the results, such as geometries, all surfaces of samples were polished and consequently coated by TiAlN, TiSiN and CrN with respect to tool steels. The roughness of the disc surface was controlled and ranged in the interval R_a=0.3–0.5 µm, but the roughness of coatings depends on coating processes and they were not uniform. The roughness of CrN coatings was not as good compared to others, ranging from R_a=0.3–1.5 µm [25].

9.5.2 Friction Results

9.5.2.1 Friction of Coating Initially

In this study, the friction coefficient of the pairs was measured over a long period of time or long distance of sliding; however, the friction was reported during 1 min. (app. 6 meters sliding) at 40°C; the friction coefficient is shown in the Figure 9.2.

It was observed that the friction coefficients increased rapidly in this period because of roughness. Initially, surface peaks were sharp, and they interlocked during the contact, therefore, when two surfaces slide relatively, more force is needed to overcome the obstacles [26-27]. It was also observed that the coating layers became smoother compared to the original, the maximum distance between the highest and lowest peaks Rt was reduced, which may have been caused by a reduction of friction coefficients subsequently. At 800°C, the friction coefficients behaved as at 40°C initially, but the increase was lower.

The phenomenon was due to the larger contact area between the pin and disc, and the pin penetrated deeper into the disc surface. The friction coefficient was governed by time/sliding distance as shown in Figure 9.3. There were at least three

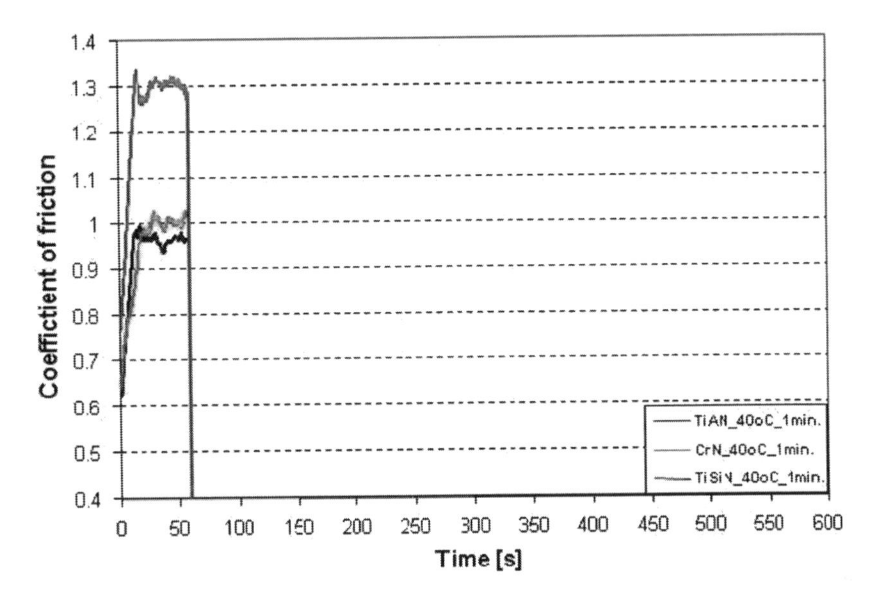

FIGURE 9.2 Friction coefficient of TiAlN, CrN and TiSiN coatings sliding against high-strength boron steel at 40°C for 1 minute.

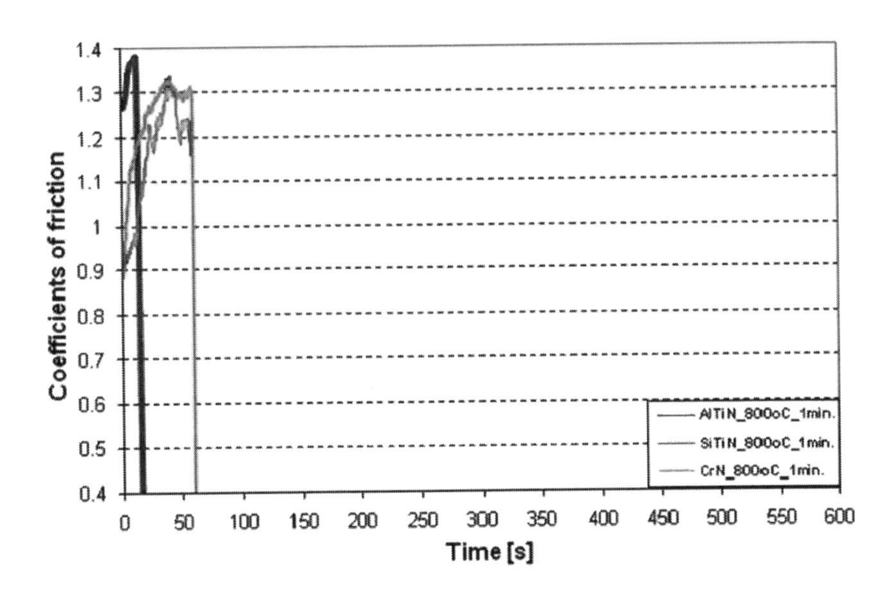

FIGURE 9.3 Friction coefficient of TiAlN, CrN and TiSiN coatings sliding against high-strength boron steel at 800°C for 1 minute.

experiments of TiAlN that showed that the friction coefficients of TiAlN increased over 1.7 during a very short time (less than 20 seconds), which is not shown in Figure 9.3. Increasing friction coefficients could be attributed to the fact that much force for plastic deformation is needed when the pin penetrates deep into the surface of the steel. In this case, larger areas of wear and more roughness were found.

9.5.2.2 Friction of Coatings during 10 Minutes

The sliding over a longer period of time leads to a reduction in friction. At 40°C, the friction coefficients as a function of time are shown in Figure 9.4. The observation pointed out that the friction coefficients increased rapidly at the beginning, as mentioned, and reduced and remained at stable values for TiAlN and CrN coatings, or increased slowly for TiSiN coating. Based on the curves recorded, the pair of TiAlN coating and boron steel shows the lowest friction (app. M = 0.89 ± 0.06) compared to CrN coating (app. μ = 1.05 ± 0.17), and TiSiN coating (app. μ = 1.1 ± 0.09). However, CrN coating was stable during sliding, and for TiAlN coating, it continued to increase.

The increase in friction coefficients could be explained by worn materials coming from the coatings and the boron steel disc during sliding, and the coatings were removed after a specific time when operating with very high hardness particles, and in some cases they act as cutting tools during the grinding processes. It was also observed that the higher roughness on the worn surfaces compared to their origins. At 800°C, the friction coefficients as a function of time were more complicated. The coefficients increased to 1.28 ± 0.2 for TiAlN, 1.29 ± 0.1 for CrN and 1.33 ± 0.13 for

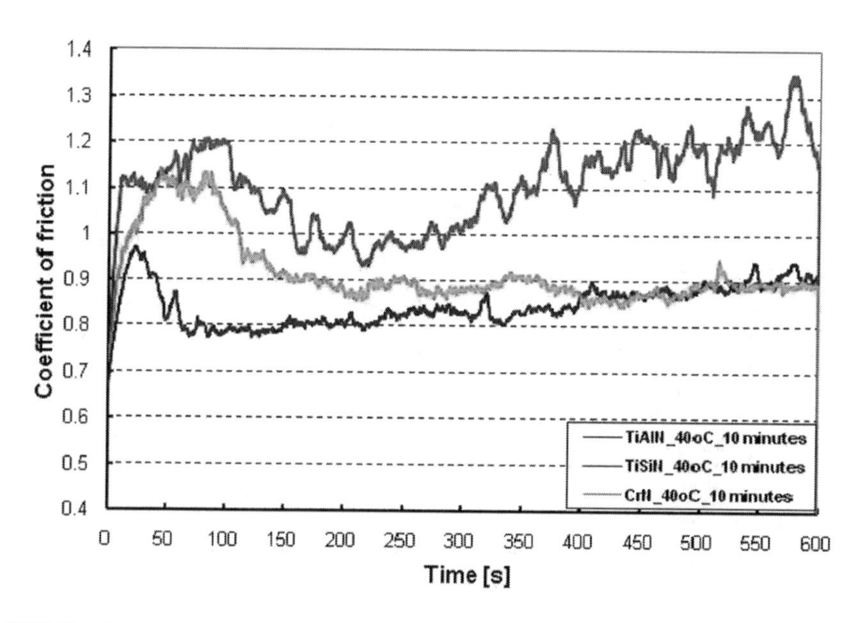

FIGURE 9.4 Friction coefficient of TiAlN, CrN and TiSiN coatings sliding against high-strength boron steel at 40°C for 10 minutes.

TiSiN, higher than operating at 40°C. The behaviours of friction coefficients with time are shown in Figure 9.5.

As mentioned above about the point contact at high temperature, the spherical pin penetrated deeply into the boron steel surface with large plastic deformation at elevated temperature. At 800°C, boron steel and substrate of tool steel were softer, while the coatings were still hard, and their deformation depended on mechanical properties, such as Young's modulus, hardness and the real contact conditions. Large plastic deformation of the spherical pin penetrating into the soft, flat disc caused plowing friction, and friction force is governed by plastic properties of penetrated surface and shear force at the interface [28-29].

Our experiments were conducted at high loading force and temperature. With these conditions, pins coated by TiAlN and CrN with higher hardness penetrated into the boron steel surfaces at the same level of indentation (δ = 3.61 E-3 mm) and depends strongly on the plastic property of boron steel. The friction of both TiAlN and CrN coatings was the same (Figure 9.5). Although these coatings have different physical and chemical properties, the coatings have been detached from substrates, and worn particles formed during sliding. The coefficients of friction seem to be constant during further sliding. Pins coated by TiSiN have lower hardness, and indentation was more complicated initially [30]. The coefficients of friction were of the same, but the experiments were not carried out further because of some extraordinary values above the machine's capacity. Clearly, the friction of TiSiN coatings was affected strongly by both the plastic properties of boron steel and substrate as well as shear at interfaces.

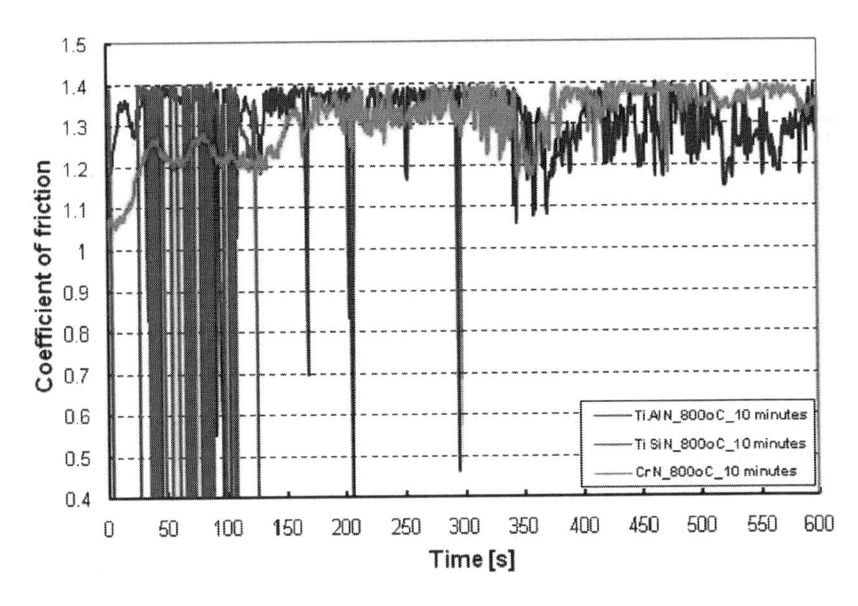

FIGURE 9.5 Friction coefficient of TiAlN, CrN and TiSiN coatings sliding against high-strength boron steel at 800°C for 10 minutes.

9.5.3 Wear Resistances

Wear resistance was studied by the characterization of wear scars after sliding the pin on the disc and combined with SEM/EDS analysis. There are some methods to study wear resistance, such as mass lost detecting, dimension change and volume lost, but in our study, wear scars have been detected using a surface topography analysis method.

9.5.3.1 Wear Resistance at 40°C

Wear scars at 40°C were detected and captured, which provided information about the worn surface. Figure 9.6 shows the surface profiles of the coatings after sliding. The profiles detected show that CrN coatings have good wear resistance, while TiSiN shows the least wear resistance. Even CrN coating was not very good initially. If the imperfection of CrN coating surface is not considered, the coating after sliding is very smooth and remained so, while the others looked rougher and the coatings were removed [31].

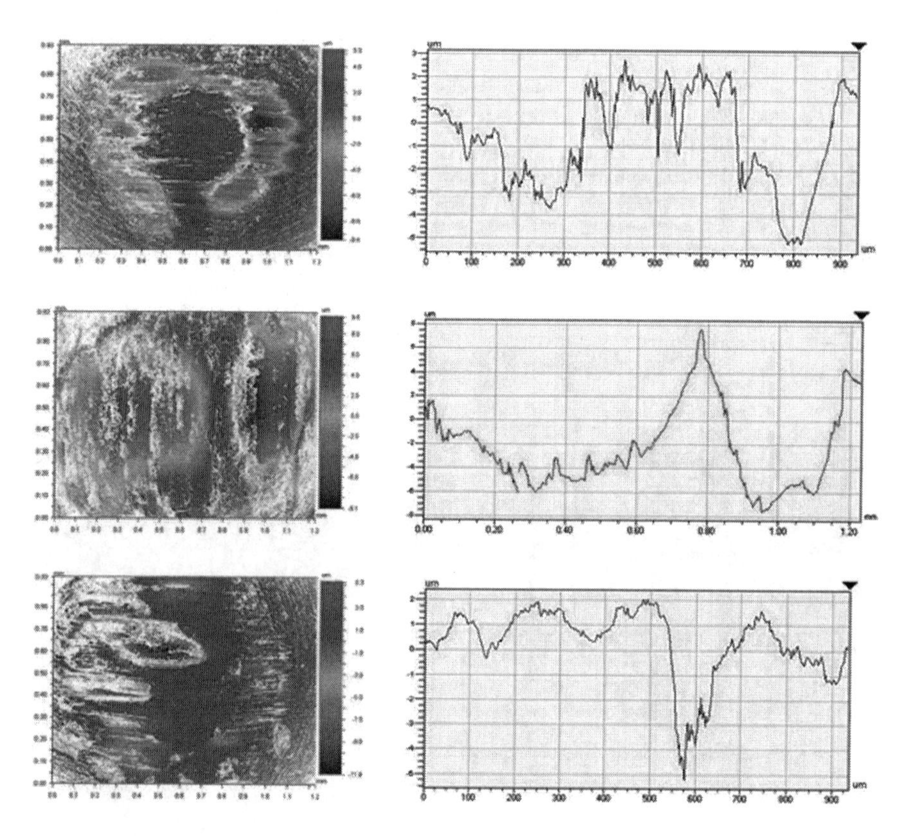

FIGURE 9.6 Profiles of wear scars of the coatings sliding on high-strength boron steel surface for 10 minutes at 40°C: (a) TiAlN coating profile, (b) TiSiN coating profile and (c) CrN coating profile.

Regarding the coefficients of friction, TiSiN coating had the highest friction and wear. Both coating and substrate wear were detected by SEM. The TiAlN coating had the lowest fiction initially, and the wear was also lower, but the coating did not survive for long [32]. The increasing friction coefficient of TiAlN explained this phenomenon. The CrN coating looked good during 10 minutes of sliding. SEM analysis showed that the coating remained after the test, and the friction coefficient was steady and resulted in a relatively smooth surface.

Regarding the hardness of the coatings, CrN coating had the highest hardness (app. 1600 HV), which showed the best wear resistance compared to others. TiSiN coating had the lowest hardness (app. 1200 HV) and showed the highest friction as well as wear scars. This shows that higher hardness results in higher wear resistance.

9.5.3.2 Wear Resistance at 800°C

Wear scars at 800°C were also studied based on imaged profiles. Investigations into wear resistance of the coatings at high temperature are quite complicated because of many factors, such as the steel becomes softer, lower hardness and oxidation of the surfaces. The profiles of coatings are shown in Figure 9.7.

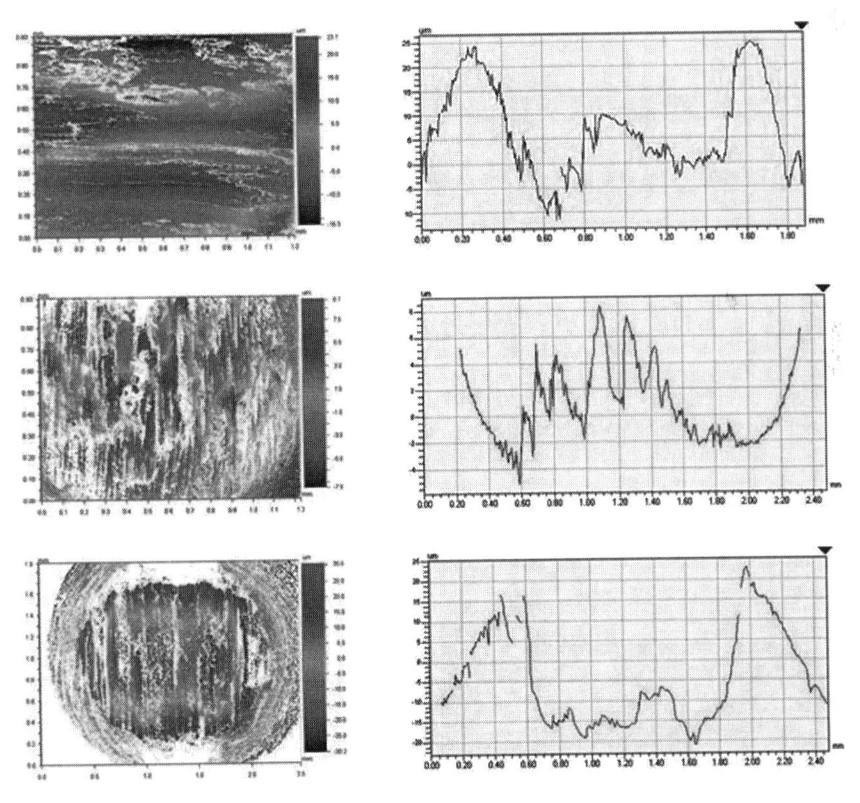

FIGURE 9.7 Profiles of wear scars of the coatings sliding on high-strength boron steel surface for 10 minutes at 800°C: (a) TiAlN coating profile, (b) TiSiN coating profile and (c) CrN coating profile.

Figure 9.7 indicates that the wear modes of the coatings are the same. TiSiN coating wear is less because the test ran over a shorter period of time (2 minutes compared to 10 minutes). For this case, the wear rate of TiSiN was estimated; however, we can compare the friction of those pairs. All experiments show that the very thin coatings were removed during the short time of sliding, and friction between the coated pin and the boron steel was as friction between substrate of tool steel and boron steel at high temperature [33]. The deep penetration of the tool to the surface leading to the abrasive wear was not the main effect. The same coefficients of friction indicate that tribological properties of the coatings were influenced by the plastic properties of substrates and boron steel, which produced deep indentation, thus worn materials and wear particles or oxidation films were less influencing.

9.5.4 Surface Damages

Wear scars were observed by SEM/EDS, and both cases of low and high temperature have been characterized to get more information about the damage generated by sliding friction.

9.5.4.1 Surface Damage at 40°C

During 1 minute of sliding, the coatings were not damaged, as can be seen from Figure 9.8. From Figure 9.8, in conjunction with Figure 9.6, the friction behaviour seems to be governed by coatings only. During this period, the coefficients of friction were steady, and the coating surfaces were polished by the breaking of sharp rough asperities. SEM/EDS analysis showed that there was no damage generated on the coating surface during this period. During 10 minutes of sliding, which corresponds to 60 metres, the TiSiN coating was completely removed, while the TiAlN and CrN coatings remained; however, SEM showed that TiAlN coating was removed and contact was remaining over half of the coating. Figure 9.9 shows detail about the wear on coating surfaces.

Referring to Figure 9.4, the friction coefficient of CrN coating was steady, implying that the sliding occurred between the coating and the boron steel. The TiAlN coating lasted for a short time, and it was removed because of the increasing friction and consequent contact of the substrate with the boron surface. The high friction In the case of TiSiN, the high friction removed the coating, leading to the substrate

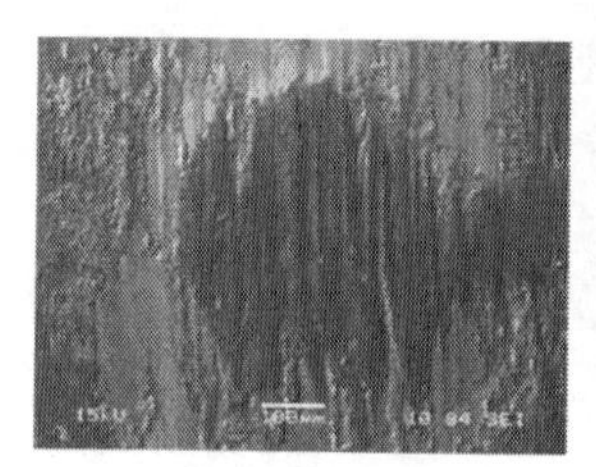

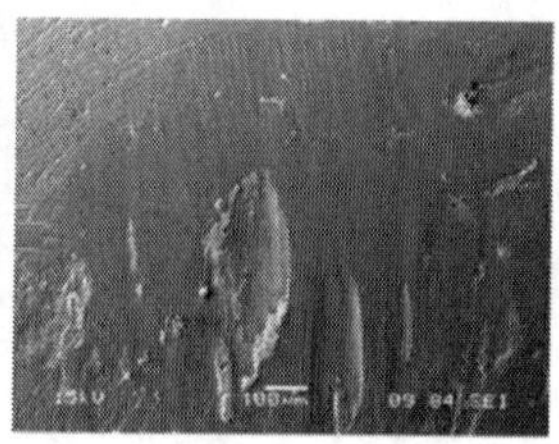

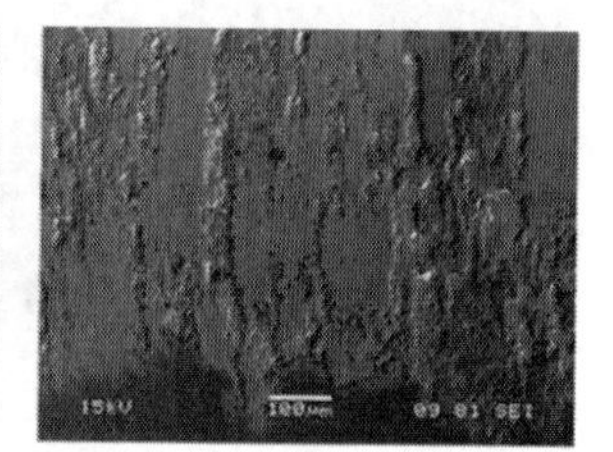

FIGURE 9.8 SEM photos of wear scars at 40°C for 1 minute of sliding, as shown in (a) TiAlN coating scar, (b) CrN coating scar and (c) TiSiN coating scar.

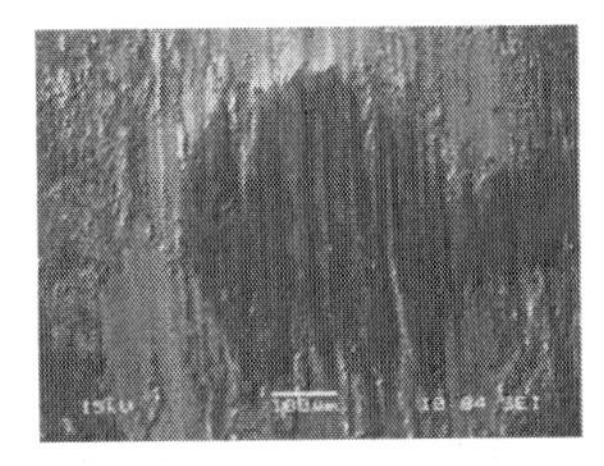
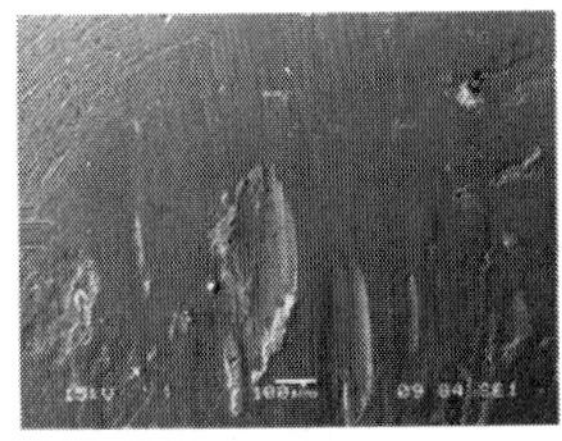
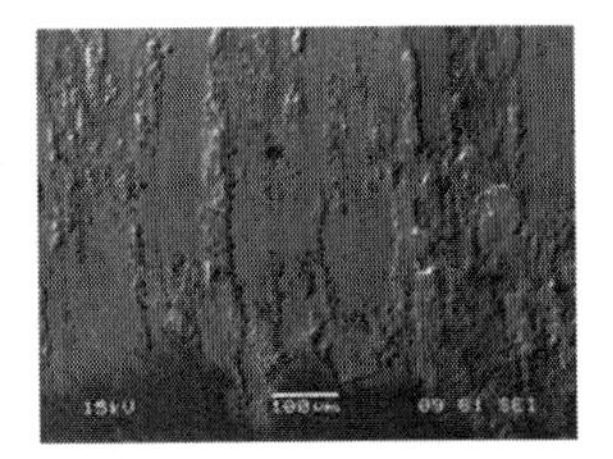

FIGURE 9.9 SEM photos of wear scars at 40°C for 10 minutes of sliding, as shown in (a) TiAlN coating scar, (b) CrN coating scar and (c) TiSiN coating scar.

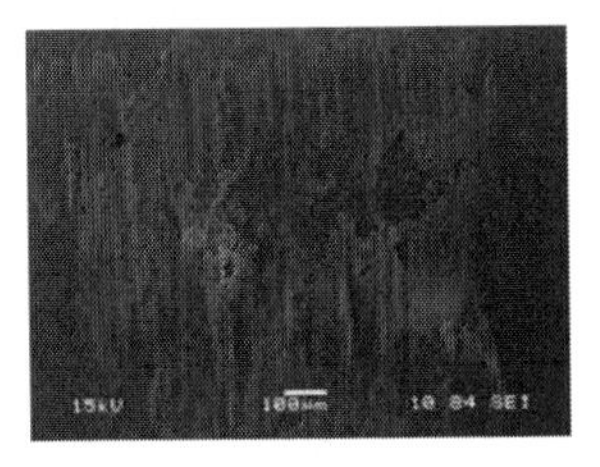
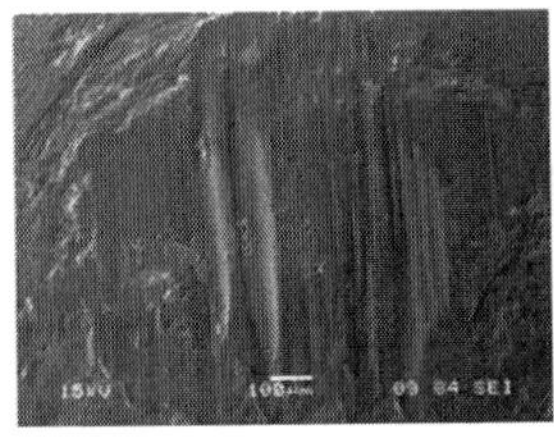
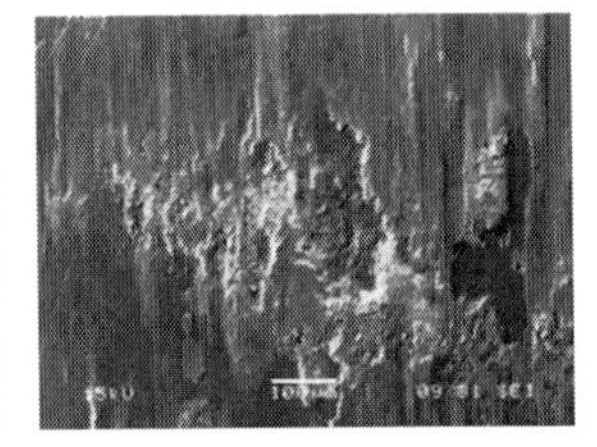

FIGURE 9.10 SEM photos of wear scars at 800°C for 1 minute of sliding, as shown in (a) TiAlN coating scar, (b) CrN coating scar and (c) TiSiN coating scar.

contacting the boron steel, thereby increasing the friction coefficient. The TiSiN coating has shown poorer tribological property compared to others.

The coating was removed totally and the friction occurred between the bulk materials [34]. The higher friction force and shorter survival time of the coating are shown in Figure 9.4. The lowest friction coefficient existed for a very short period and then increased quite rapidly. The wear scar damages were caused by abrasive wear from the pin coated by TiSiN and TiAlN. In these, the coatings were removed, and the substrates were exposed, which was not found in the pin coated with CrN [35].

9.5.4.2 Surface Damage at 800°C

During 1 min at 800°C, the friction increased rapidly. The worn surfaces of different coatings are shown in Figure 9.10. In observing both the coating surface and the boron steel surface, we saw that there was some adhesive wear on the disc surface, and the worn particles were attached on the surface of the coatings. The attachment could be explained by high pressure contact at high temperature (800°C) leading to localized bonding at the interface between contact surface and detached material from the boron surface during sliding (Figure 9.11).

The damage clearly showed that all coatings had been removed. TiSiN coating, with the highest friction coefficient, did not last long. The wear scar of TiAlN coating shows the occurrence of abrasive wear, and images of the worn surface also show some oxides and wear debris adhered to the worn surface of the coatings (Figure 9.12c). The wear scar of CrN coating, (Figure 9.12d) shows that both abrasive and adhesive wear can be seen. The surface was very rough, even the disc surface,

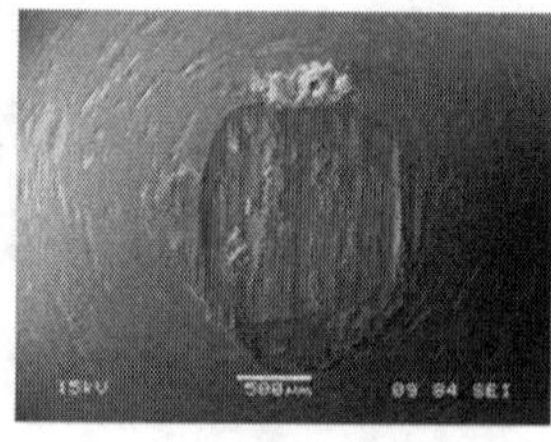
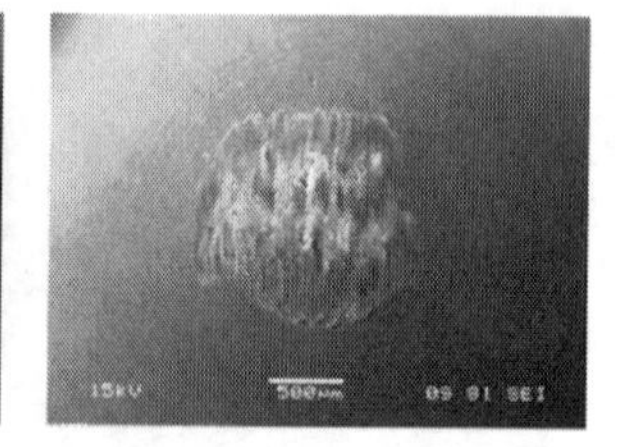

FIGURE 9.11 Wear scars for a longer time of sliding (10 minutes), as shown in (a) TiAlN_35X coating scar, (b) CrN coating scar and (c) TiSiN coating scar.

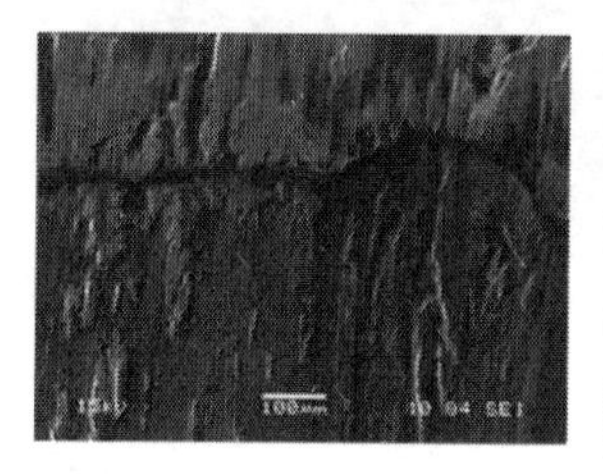
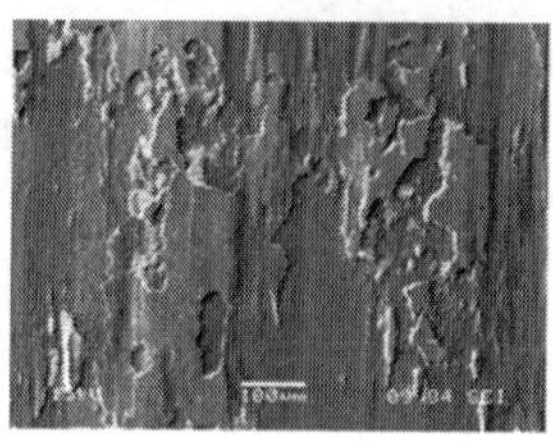

FIGURE 9.12 SEM photos of wear scars at 800°C for 10 minutes of sliding, as shown in (d) TiAlN coating scar, (e) CrN coating scar and (f) TiSiN coating scar.

with many detached areas. The bright areas show oxide layers, while dark areas show the substrate.

9.5.5 Oxidation of Sliding Surfaces

Throughout all the experiments, there was no oxidation on the surface of coatings, even at 800°C. Once the coating had been removed, the oxidation started, even at room temperature sliding. The EDS analysis found small amounts of oxygen within wear scars where the coatings had been removed, and it increased at high temperature. Oxide layers always exist on the surface of iron where protection is not provided, and the oxide layers, somehow, could be a protection layer against oxygen diffusing deeper. Figure 9.13 and Figure 9.14 show the presence of oxygen on the wear scar, in which, the bright areas appear as oxide films, while dark areas are substrates, as seen in Figure 9.9 and Figure 9.12.

Some researchers have pointed out that the oxide layers created a solid lubricant [36], which may reduce friction force if oxidation speed was fast enough. If the oxidation is severe then it could result in higher friction. As we know, oxidation depends on temperature, and by controlling temperature, we can control the friction force at the interface during sliding. In these experiments, at 40°C, oxidation had not occurred on the coating surface, but it could be done on the surface of its pair where the steel had no protection. At high temperature, the coatings had been removed, and the oxidation acted fast for both pins and disc; however, it did not affect the sliding friction much.

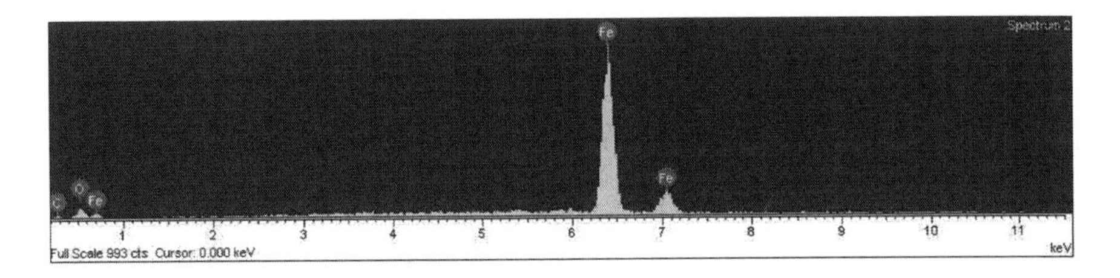

FIGURE 9.13 EDS spectrum of wear scar of tool steel coated TiSiN after sliding against high-strength boron steel at 40°C for 10 minutes.

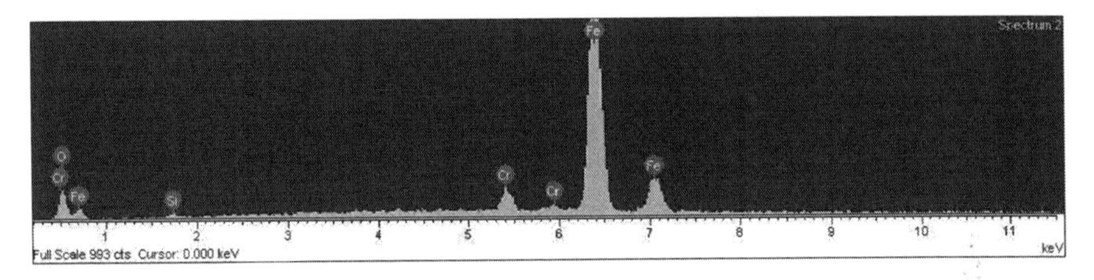

FIGURE 9.14 EDS spectrum of wear scar of tool steel coated CrN after sliding against high-strength boron steel at 800°C for 10 minutes.

9.6 CONCLUSION

In this work, the tribological properties of three different coatings of TiAlN, CrN and TiSiN with substrates of tool steel sliding against high-strength boron steel at low and high temperature were studied. The experiments were carried out in ambient air without any lubricants. The salient conclusions are as follows:

- At 40°C, friction had different effects on the three coatings. The CrN coating survived longer with a steady coefficient of friction. The TiAlN coatings showed the lowest friction, but its durability is shorter as compared to CrN coating, and the friction coefficient continued to increase as sliding progressed. The TiSiN coating showed the highest coefficient of friction with a short lifetime. Regarding the wear rate, CrN coating had very good wear resistance. TiSiN coating had the poorest wear resistance, as the coating was worn and removed after a short time. Considering the hardness of the coatings, the higher the hardness of coatings, the higher the wear resistance.
- At 800°C, the coatings were removed after a short time, and the tribological properties were not influenced strongly by coatings.
- The wear modes observed were abrasive and adhesive wear. At low temperature, abrasive wear occurred and became smoother. At high temperature and pressure, both abrasive and adhesive wear were detected with very rough surfaces.
- The coatings can resist oxidation, even at a very high temperature; however, oxidation occurred on steel even at low temperature during the sliding test. At high temperature, the oxidation is rapid and could be a cause of rapid wear.

REFERENCES

1. Bhushan, B. 2013. *Introduction to Tribology*. USA: John Wiley & Sons, Inc., 738p.
2. Hong Liang, David Craven 2005. *Tribology in Chemical-Mechanical Planarization*. USA: CRC Press, 200p.
3. Sahoo, P. 2006. *Engineering Tribology*. India: Prentice Hall of India, 336p.
4. Ludema, K. 2001. *History of Tribology and its Industrial Significance*. Dept. of Mechanical Engineering and Applied Mechanics – University of Michigan, 31 March, Ann Arbor, MI, 48109–2125, https://doi.org/10.1007/978-94-010-0736-8_1
5. Blau, P.J. 1997. Fifty years of research on the wear of metals. *Tribology International*, 30(5): 321–331.
6. Lubrication (Tribology) – Education and Research. 1966. *A Report on the Present Position and Industry Needs, (Jost Report), Department of Education and Science*. London, UK; Great Britain: HM Stationary Office, 79p.
7. ASME. 1977. *Strategy of Energy Conservation through Tribology*. USA: ASME, 125p.
8. Jost, L.S., Russell, J.A. and Debrodt, D.C. 1986. A review of DOE ETC tribology survey. *Transactions on ASME Journal of Tribology*, 108: 497–501.
9. Bhushan, B. 1977. *Principles and Applications of Tribology*. John Wiley & Sons, Inc.
10. Blau, Peter J. 2017. ASM Handbook, *Friction, Lubrication and Wear Technology*, 18, 1108p.
11. Burwell, J.T. and Rabinowicz, E. 1953. The nature of the coefficient of friction. *Journal of Applied Physics*, 24: 136–139.
12. Stachowiak, G.W. and Batchelor, A.W. 2000. *Engineering Tribology*. UK: Butterworth-Heinemann, 884p.
13. Hirst, W. and Hollander, A.E. 1974. Surface finish and damage in sliding. *Proceedings of the Royal Society of London Series A*, 337: 379–394.
14. Burakowski, T. and Wierzchoń, T. 1999. *Surface Engineering of Metals – Principles, Equipments, Technologies*. UK: CRC Press, 608p.
15. Nealeasm, M.J. 2001. *The Tribology Handbook*, 2nd edition. UK: Butterworth-Heinemann, 640p.
16. Murakamia, T., Umedaa, K., Sasakia, S. and Ouyang, J. 2006. High-temperature tribological properties of strontium sulfate films formed on zirconia-alumina, alumina and silicon nitride substrates. *Tribology International*, 39: 1576–1583.
17. Bhushan, B., Kato, K. and Adachi, K. 2001. Modern Tribology Handbook, *Metals and Ceramics. CRC* 21, 1760p.
18. Bhushan, B. 2005. *Nanotribology and Nanomechanism – An Introduction*. Germany: Springer, 930p.
19. Trent, E.M. and Wright, P.K. 2000. *Metal Cutting*, 4th edition. Netherlands: Elsevier, 464.
20. Atik, E., Yunker, U. and Menc, C. 2003. The effects of conventional heat treatment and boronizing on abrasive wear and corrosion of SAE 1010, SAE 1040, D2 and 304 steels. *Tribology International*, 36: 155–161.
21. Prabhudev, K.H 1999. *Handbook of Heat Treatment of Steels*. USA: McGraw-Hill, 762p.
22. Ramachandran, V.S. and Beaudoin, J.J. 2001. *Handbook or Analytical Techniques in Concrete Science and Technology*. Netherlands: William Andrew Publishing/Noyes, 1003.
23. Goodhew, P.J. 2000. *Electron Microscopy and Analysis*. London, UK: Taylor & Francis, Limited.
24. Birkholz, M. 2006. *Thin Film Analysis by X-Ray Scattering*. United States: Wiley-VCH, 378p.

25. Ewald, P. 1962. *Fifty Years of X-Ray Diffraction, Publish for the International Union of Crystallography*. The Netherlands: N.V.A Oosthoek's Uitgeversmaatschappij Utrecht.
26. Shackelford, J.F. 2005. *Introduction to Materials Science for Engineers*, 6th edition. United States: Pearson Prentice Hall, 896p.
27. Guinier 1994. *Elements of X-ray Diffraction*. Bernard Dennis Cullity, Stuart R. Stock and United States: Courier Dover Publications, 664p.
28. Mizutani, U. 2001. *Introduction to the Electron Theory of Metals*. Cambridge University Press, 163.
29. Stachowiak G.W., Batchelor, A.W. and Stachowiak, G.B. 2004. *Experimental Methods in Tribology*. Netherlands: Elsevier, 372.
30. Panjan, M., Sturm, S., Panjan, P. and Cekada, M. 2007. TEM investigation of TiAlN/CrN multilayer coatings prepared by magnetron sputtering. *Surface and Coatings Technology*, 202: 815–819.
31. Yang, S.-M., Yin-Yu, C., Lin, D.-Y, Wang, C. and Wu, W. 2007. Mechanical and tribological properties of multilayer TiSiN/CrN coatings synthesized by cathodic arc deposition process, *Surface and Coatings Technology* 202(10): 2176–2181.
32. Staia, M.H., D'Alessandria, M., Quibto, D.T., Roudet, F. and Marsal Astort, M. 2006. High-temperature tribological characterization of commercial TiAlN coatings. *Journal of Physics – Condensed Matter*, 18: S1727–S1736.
33. Fisher-Cripps, C. 1999. The Hertz contact surface. *Journal of Materials Science*, 34: 129–137.
34. Cheng, H.-Y., Tsai, C.-J. and Lu, F.-H. 2004. The Young's modulus of chromium nitride films. *Surface and Coatings Technology*, 184: 69–73.
35. Walter, C., Antretter, T., Daniel, R. and Mitterer, C. 2007. Finite element simulation of the effect of surface roughness on nanoindentation of thin films with sphere indenters. *Surface and Coatings Technology*, 202: 1103–1107.
36. Kamminga, J.-D. and Janssen, G.C.A.M. 2007. Experimental discrimination of plowing and shear friction. *Tribology Letters*, 25(2), 149–152.

10 Process Modelling and Optimization of Hardness in Laser Cladding of Inconel® 625 Powder on AISI 304 Stainless Steel

S. Sivamani, M. Vijayanand, A. Umesh Bala and R. Varahamoorthi

CONTENTS

10.1 INTRODUCTION

A large number of components used in transportation, mining and aerospace sectors are subjected to wear failures due to fatigue and overloaded operating conditions (Shariff et al. 2010; Kaysser 2001). In order to improve the properties of mechanical parts, several surface treatment methods are available. Laser surface alloying is considered to be a potential surface modification method, especially in wear-resistant applications (Song et al. 2016).

Laser cladding, also known as hardfacing or surfacing, is one of the welding-related techniques used to cover the original surface or to refurbish the worn out surfaces by depositing a welded overlay that provides abrasion, hardness, galling, erosion, corrosion or heat resistance so that it performs better in their working environments. Laser cladding uses a laser beam with a selected spot size to fuse materials onto a substrate. The powder coating material is carried by an inert gas such as argon or helium through a powder nozzle into the melt pool. The laser optics and powder nozzle are moved across the workpiece surface to deposit single tracks, complete layers or even high-volume build-ups (Chryssolouris et al. 2002; Singh et al. 2020). Compared with conventional hardfacing techniques, such as arc welding and thermal spraying, laser cladding can produce much better coatings, such as dense microstructure, high-wear resistance, low dilution and good metallurgical bonding to substrate (Xu et al. 2014).

Stainless steel is used in motor vehicle applications because it is resistant to corrosion and high-temperature oxidation, offers energy absorption properties and maintains its mechanical properties over a wide temperature range. For applications in which corrosion resistance is of primary importance, austenitic stainless steel grade 1.4301 (AISI 304) and its derivatives are suitable for mild environments, such as interior components, fuel tanks, etc. (International Stainless Steel Forum 2020).

Inconel® 625 is a nickel-based super alloy widely used for various cladding applications in turbines, industrial boilers, aerospace, chemical, petrochemical and marine applications because of its good mechanical properties, ease of weldability and resistance to high-temperature corrosion in prolonged exposure to aggressive environments (Sandhu and Shahi 2016). Inconel® is the trademark of the Special Metals Corporation group. Nominal chemical composition of Inconel® 625 provided by Special Metals Corporation is the following (in wt.%): Ni (58.00 min), Cr (20.00–23.00), Mo (8.00–10.00), Nb and Ta (3.15–4.15), Fe ($\leq$5.00), Ti ($\leq$0.40), Al ($\leq$0.40), Co ($\leq$1.0), C ($\leq$0.10), Si and Mn (each $\leq$ 0.50) and S and P (each $\leq$ 0.015) (Gupta et al. 2015; Special Metals Corporation 2013; Dubiel and Sieniawski 2019).

Abioye et al. (2017) compared powder and wire deposited Inconel® 625 on AISI 304 substrate by laser cladding, and reported that the powder laser track demonstrated higher hardness compared to the corresponding wire laser track. The laser-cladded Inconel® 625 coating is preferred when compared to shielded metal arc welding because of its better mechanical performance, such as hardness and wear resistance, at both room and elevated temperature (Feng et al. 2017). Abioye et al. (2013) investigated the individual effects of laser power, traverse speed and wire feed rate on the track characteristics and found that, with increasing the wire feed rate, the contact angle increases, whereas dilution ratio and aspect ratio decreases. Mathematical models were developed by Lian et al. (2019) to predict the microhardness and wear resistance of the cladding layer by controlling the laser cladding processing parameters and reported a higher microhardness in the cladding layer can be obtained by increasing the laser power and decreasing scanning speed.

However, all researchers developed a mathematical model for maximizing the hardness in laser cladding and compared the experimental values with the predicted values, but not compared with the different optimization techniques. Hence, in this research, the AISI 304 stainless steel was hard faced with nickel-based Inconel® 625 using a CO_2 laser. A model was developed relating the input conditions, laser power, powder-feed rate, laser-scanning speed and focal position of the laser beam with the hardness as an output using curve fitting and artificial neural network (ANN) models.

The developed model using a curve fitting method was compared with the first derivative test, grey relational analysis (GRA) and generalized reduced gradient (GRG).

10.2 EXPERIMENTAL METHODS

Laser hardfacing was achieved using a coaxial powder-feed method with a maximum laser power of 4000 W and a frequency of 20 KHz. After the deposition, the substrate was machined into the required size using a wire electrical discharge machine (WEDM). A Vickers microhardness testing machine (Make: Shimadzu, Japan; Model: HMV – 2T) was used to measure the hardness with a load of 0.5 kg and dwell time of 15 s. The process parameters that influence the finished product by laser cladding are laser power, powder-feed rate, laser-scanning speed, focal position of laser beam, shield gas type, shield gas flow rate, laser pulse shape, powder-feed angle and powder-feed position.

10.3 PROCESS MODEL DEVELOPMENT BY CURVE FITTING METHOD

Curve fitting is the process of specifying the model that provides the best fit to the specific curves in the dataset (Juliano 2001). In this study, Microsoft Excel 16.0 (2016) from Microsoft Corporation was used to perform the curve fitting method to develop a mathematical model to relate independent and dependent variables. Equations 10.1 and 10.2 were used to develop the model.

$$\text{Linear:} \quad Y = \alpha_0 + \sum_{i=1}^{4} \alpha_i X_i \tag{10.1}$$

$$\text{Quadratic:} \quad Y = \alpha_0 + \sum_{i=1}^{4} \alpha_i X_i + \sum_{i,j=1}^{4} \alpha_{ij} X_i X_j + \sum_{i=1}^{4} \alpha_{ii} X_i^2 \tag{10.2}$$

The model coefficients were evaluated by fitting the experimental data to the above models. The model consistency should be checked by comparing the number of equations and the number of coefficients. If there are fewer equations than coefficients, then a unique solution is not possible or infinitesimal solutions are possible. Similarly, a unique solution is possible only if the number of equations is equal to or greater than the number of coefficients.

If the number of equations is equal to the number of coefficients, then the model coefficients can be evaluated as follows:

If the equation is written in the form of a matrix, then the equation will be $X.\alpha = Y$ where X is the independent variable matrix, α is the coefficient matrix and Y is the dependent variable matrix.

$$X \cdot \alpha = Y \tag{10.3}$$

Multiply by X^{-1} on both sides,

$$X^{-1} \cdot X \cdot \alpha = X^{-1} \cdot Y \tag{10.4}$$

$$\alpha = X^{-1} \cdot Y \tag{10.5}$$

If the number of equations is greater than the number of coefficients, then the model coefficients can be evaluated by the following steps:

The principle of least squares (PLS) is used to find the coefficient matrix in Equation 10.3. The principle of least squares states that the sum of the square of residuals is zero. Residual is defined as the difference between the experimental value and the predicted value, i.e. $\varepsilon = Y_e - Y_p$

$$\therefore S = \sum_{i=1}^{n} \varepsilon_i^2 = 0 \tag{10.6}$$

$$S = \sum_{i=1}^{n} \left(Y_{e,i} - Y_{p,i}\right)^2 = 0 \tag{10.7}$$

Partially differentiate S with respect to ε and let it to zero,

$$\frac{\partial S}{\partial \varepsilon} = 2\sum_{i,j=1}^{n} \varepsilon_i \frac{\partial \varepsilon_i}{\partial \alpha_j} = 0 \tag{10.8}$$

The coefficient matrix α is given by,

$$\alpha = \left(X^T \cdot X\right)^{-1} \cdot \left(X^T \cdot Y\right) \tag{10.9}$$

In this study, laser power, powder-feed rate, laser-scanning speed and focal position of the laser beam were considered as independent variables, and hardness number as a dependent variable. Table 10.1 shows the independent variables and their levels used in the laser cladding of Inconel® 625 powder on AISI 304 SS substrate. Table 10.2 shows the mean, standard deviation and range of variables used in this experimental system.

Table 10.3 shows 25 designed experiments with hardness number. The extreme star points are at some distance from the centre based on the properties desired for the design and the number of factors in the design. The experimental design has circular or spherical symmetry and requires five levels for each factor. Augmenting a resolution *V* fractional factorial design with star points can produce this design.

Fitting the experimental data to the linear equation yields the following model:

$$Y = -63.505 + 0.046 * P + 0.075 * V + 1.858 * D + 2.225 * F \tag{10.10}$$

TABLE 10.1
Laser Cladding Parameters and Their Working Ranges

Parameter	Unit	Notation	Levels				
			−2	−1	0	1	2
Laser power	W	P	2200	2350	2500	2650	2800
Laser-scanning speed	mm/min	V	800	900	1000	1100	1200
Focal position of laser beam	mm	D	15	20	25	30	35
Powder-feed rate	g/min	F	30	35	40	45	50

TABLE 10.2
Mean, Standard Deviation and Range of Variables Used in this Experiment

Variables	Mean	Standard Deviation	Range
Laser power	2500	237.17	600
Laser-scanning speed	1000	158.11	400
Focal position of laser beam	25	7.91	20
Powder-feed rate	40	7.91	20
Hardness	261	29.29	112

Figure 10.1(a) shows the experimental vs. predicted values of hardness for the linear model. The value of $R^2 > 0.9$ shows the good fit. Because $R^2 < 0$, the model does not fit well to the experimental data, so the quadratic equation was tested for the fitness of the model.

Fitting the experimental data to the quadratic equation yields the following model:

$$\begin{aligned} Y = & -10157.79 + 5.14 * P + 3.48 * V + 24.30 * D + 92.50 * F + 0.0002 * P * V \\ & -0.0032 * P * D - 0.0034 * P * F + 0.0054 * V * D - 0.0039 * V * F \\ & -0.2575 * D * F - 0.0010 * P^2 - 0.0020 * V^2 - 0.1879 * D^2 - 0.8929 * F^2 \end{aligned} \quad (10.11)$$

Figure 10.1(b) shows the experimental vs. predicted values of hardness for the quadratic model. Because $R^2 > 0.9$, the model fits well to the experimental data. So, it was concluded that the quadratic equation is the suitable model for the experimental data. The analysis of regression shows that the values of R, R^2, adjusted R^2 and standard error were 0.997, 0.994, 0.994 and 2.25 respectively. The terms in the quadratic equation reveal that the positive sign of linear terms and negative sign of quadratic terms tend to increase the response.

TABLE 10.3
Hardness Results for Cladded Specimens Using Inconel® 625 Powder

S. No.	Laser power (W)	Laser-scanning speed (mm/min)	Focal position of laser beam (mm)	Powder-feed rate (g/min)	Experimental hardness number
1	2350	900	20	35	216
2	2650	900	20	35	233
3	2350	1100	20	35	222
4	2650	1100	20	35	255
5	2350	900	30	35	247
6	2650	900	30	35	254
7	2350	1100	30	35	264
8	2650	1100	30	35	287
9	2350	900	20	45	262
10	2650	900	20	45	268
11	2350	1100	20	45	260
12	2650	1100	20	45	283
13	2350	900	30	45	264
14	2650	900	30	45	267
15	2350	1100	30	45	279
16	2650	1100	30	45	286
17	2200	1000	25	40	223
18	2800	1000	25	40	246
19	2500	800	25	40	234
20	2500	1200	25	40	261
21	2500	1000	15	40	289
22	2500	1000	35	40	326
23	2500	1000	25	30	218
24	2500	1000	25	50	256
25	2500	1000	25	40	328

Table 10.4 shows the analysis of variance (ANOVA) for the quadratic model. The ANOVA shows the significance of the model from F- and significance F-values. The model is significant if the F-value is high with a significance F-value of <0.05, which was revealed from the ANOVA.

Figures 10.2(a)–(d) show the linear effect of laser power, laser-scanning speed, focal position of the laser beam and powder-feed rate on the hardness number. The linear effect was studied by varying one parameter at a time, keeping all others at a constant value. Laser power, laser-scanning speed, focal position of the laser beam and powder-feed rate were varied in the range of 2200–2800 W, 800–1200 mm/min, 15–35 mm and 30–50 g/min, respectively.

The linear effect of laser power on the hardness number was studied by varying from 2200 to 2800 W in the step size of 50 W at constant laser-scanning speed, focal

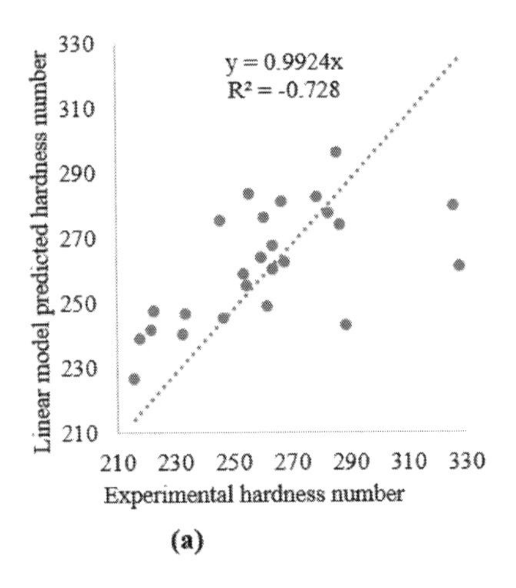

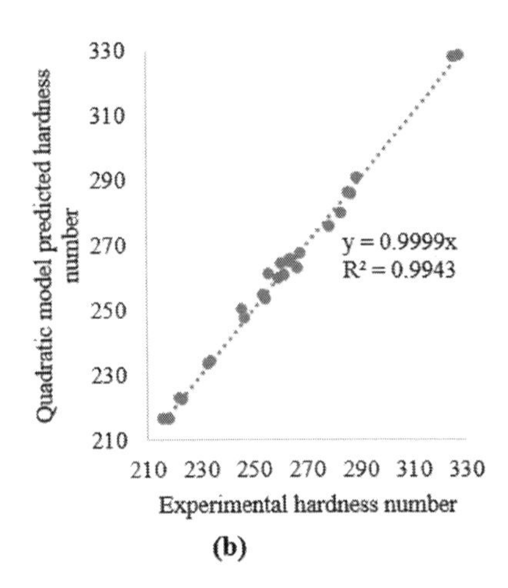

FIGURE 10.1 Experimental vs. predicted values of hardness for (a) linear and (b) quadratic models.

TABLE 10.4
ANOVA for the Quadratic Model

Source	df	Sum of squares	Mean square	F-value	Significance F
Model	1	20477.89	20477.89	4034.188	< 0.001
Residual	23	116.75	5.076087		
Total	24	20594.64			

position of the laser beam and a powder-feed rate of 1000 mm/min, 25 mm and 40 g/min, respectively. The variation in laser power exhibited a bell-shaped curve. As the laser power increased from 2200 to 2550 and then further to 2800 W, the hardness number increased from 238 to 369 and then decreased to 312, respectively. The linear effect showed that the hardness was at its maximum at the laser power of 2550 W.

The linear effect of laser-scanning speed on the hardness number was studied by varying from 800 to 1200 mm/min in the step size of 50 mm/min at constant laser power, focal position of the laser beam and a powder-feed rate of 2550 W, 25 mm and 40 g/min, respectively. The variation in laser-scanning speed exhibited a bell-shaped curve. As the laser-scanning speed increased from 800 to 1000 and then further to 1200 mm/min, the hardness number increased from 295.03 to 369 and then decreased to 284, respectively. The linear effect showed that the hardness was at its maximum at the laser-scanning speed of 1000 mm/min.

The linear effect of the focal position of the laser beam on the hardness number was studied by varying from 15 to 35 mm in the step size of 5 mm at constant laser power, laser-scanning speed and a powder-feed rate of 2550 W, 1000 mm/min and 40 g/min, respectively. The variation in the focal position of the laser beam exhibited

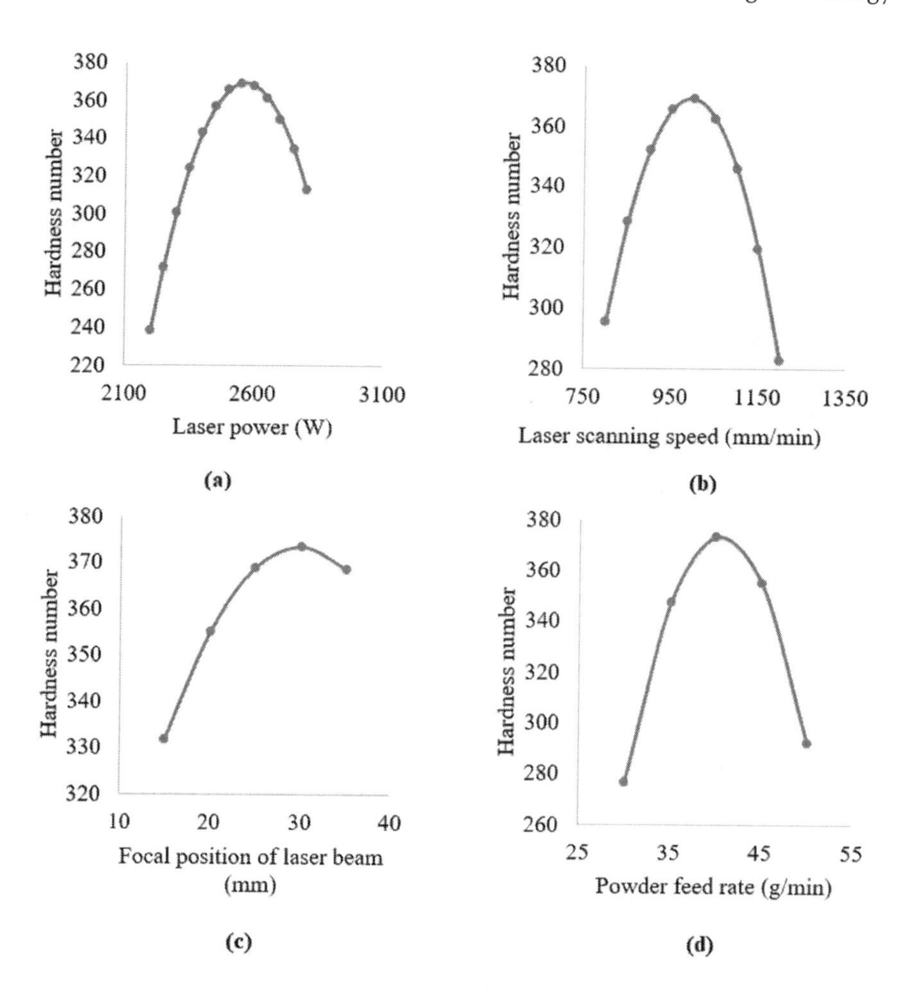

FIGURE 10.2 Linear effect of (a) P, (b) V, (c) D and (d) F on hardness number.

a bell-shaped curve. As the focal position of the laser beam increased from 15 to 30 and then further to 35 mm, the hardness number increased from 332 to 373 and then decreased to 368, respectively. The linear effect showed that the hardness was at its maximum at the focal position of the laser beam of 30 mm.

The linear effect of powder-feed rate on the hardness number was studied by varying from 30 to 50 g/min in the step size of 5 g/min at constant laser power, laser-scanning speed and focal position of the laser beam of 2550 W, 1000 mm/min and 30 mm, respectively. The variation in powder-feed rate exhibited a bell-shaped curve. As the powder-feed rate increased from 30 to 40 and then further to 50 g/min, the hardness number increased from 276 to 373 and then decreased to 292, respectively. The linear effect showed that the hardness was at its maximum at the powder-feed rate of 40 g/min. From the linear effect of laser power, laser-scanning speed, focal position of the laser beam and powder-feed rate on the hardness number, it was concluded that the maximum hardness of 373 was achieved at the optimum values of 2550 W, 1000 mm/min, 30 mm and 40 g/min, respectively.

Next, the interactive effect between laser power (*P*) and scanning speed (*V*), laser power (*P*) and focal position of laser beam (*D*), laser power (*P*) and powder-feed rate (*F*), laser-scanning speed (*V*) and focal position of the laser beam (*D*), laser-scanning speed (*V*) and powder-feed rate (*F*) and focal position of the laser beam (*D*) and powder-feed rate (*F*) on hardness are explained.

Figures 10.3(a)–(f) show the three-dimensional (3D) interactive effect of independent variables on a dependent variable. If the curves intersect at a certain point, then the interaction effect is significant. The interactive effect between laser power and scanning speed was studied by varying laser power from 2200 to 2800 W and laser-scanning speed from 800 to 1200 mm/min at a constant focal point of the laser beam and powder-feed rate of 30 mm and 40 g/min, respectively. The hardness of 182 was obtained at a low laser power and scanning speed of 2200 W and 800 mm/min, respectively. At this point, when the laser power increased to 2550 W, the hardness enhanced to 294; further increment of laser power to 2800 W led to a decrease in hardness to 224.

Similarly, the hardness increased to 248 at the laser power and scanning speed of 2200 W and 1000 mm/min, respectively. The hardness decreased to 153 when the laser-scanning speed increased to 1200 mm/min at the same laser power of 2200 W. Also, the hardness values of 373 and 293 were obtained at laser-scanning speed of 1000 and 1200 mm/min at the laser power of 2550 W. Finally, the hardness values of 313 and 242 were observed at laser-scanning speed of 1000 and 1200 mm/min at the laser power of 2800 W. From the interactive effect between laser power and scanning speed at a constant focal point of the laser beam and powder-feed rate of 30 mm and 40 g/min, respectively, the optimal conditions were found to be laser power and scanning speed of 2550 W and 1000 mm/min, respectively.

The interactive effect between laser power and focal position of the laser beam was studied by varying laser power from 2200 to 2800 W and focal position of the laser beam from 15 to 35 mm at a constant laser-scanning speed and powder-feed rate of 1000 mm/min and 40 g/min, respectively. The hardness of 189 was obtained at a low laser power and focal position of the laser beam of 2200 W and 15 mm, respectively. At this point, when the laser power increased to 2550 W, the hardness enhanced to 332; further increment of laser power to 2800 W led to a decrease in hardness to 283.

Similarly, the hardness increased to 248 at the laser power and focal position of the laser beam of 2200 W and 30 mm, respectively. The hardness almost remained at 249 when the focal position of the laser beam increased to 35 mm at the same laser power of 2200 W. Also, the hardness values of 373 and 368 were obtained at focal position of the laser beam of 30 and 35 mm at the laser power of 2550 W. Finally, the hardness values of 313 and 304 were observed at focal position of the laser beam of 30 and 35 mm at the laser power of 2800 W. From the interactive effect between laser power and focal position of laser beam at a constant laser-scanning speed and powder-feed rate of 1000 mm/min and 40 g/min, respectively, the optimal conditions were found to be laser power and focal position of the laser beam of 2550 W and 30 mm, respectively.

The interactive effect between laser power and powder-feed rate was studied by varying laser power from 2200 to 2800 W and powder-feed rate from 30 to 50 g/min

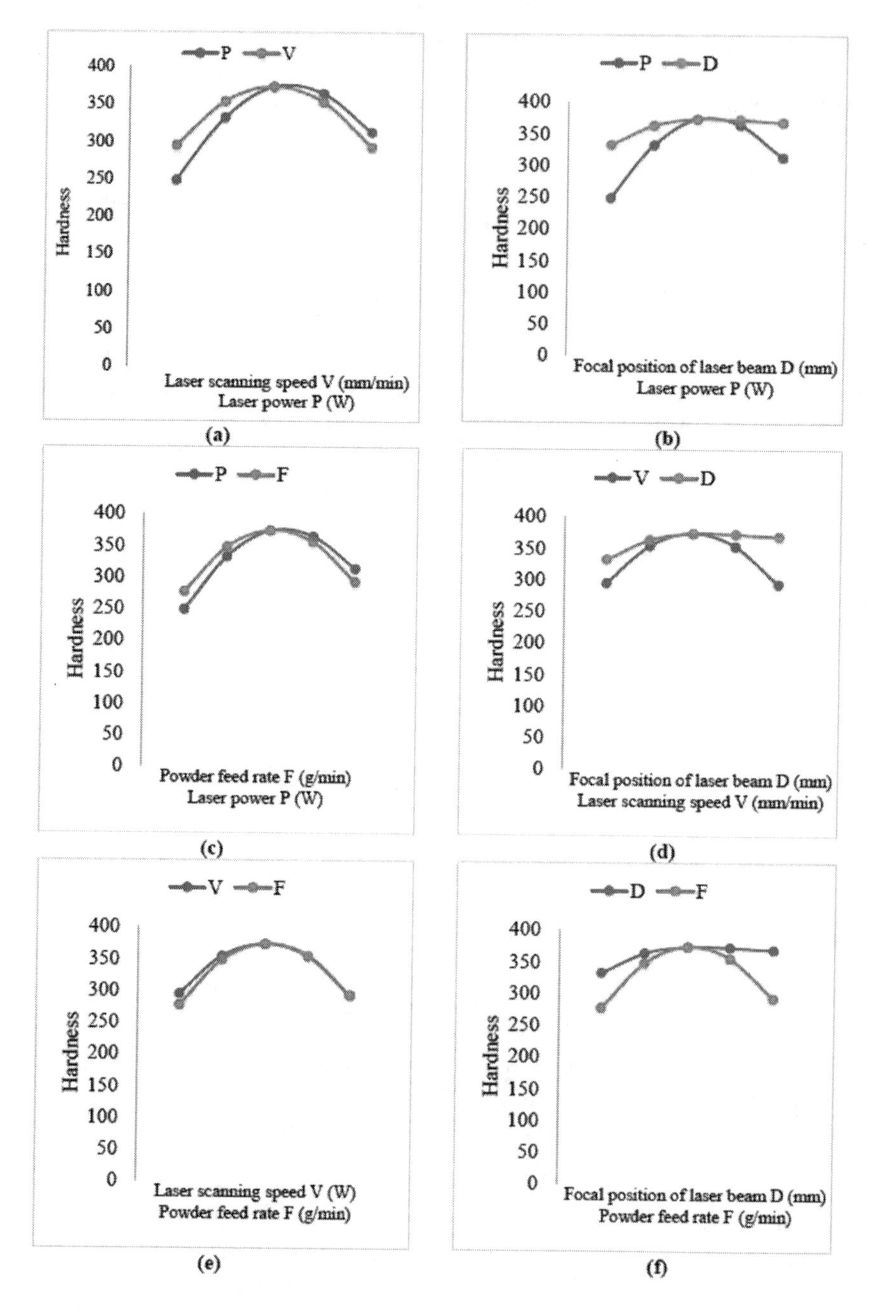

FIGURE 10.3 Effect of (a) P and V, (b) P and D, (c) P and F, (d) V and D, (e) V and F and (f) D and F on hardness.

at a constant laser-scanning speed and focal position of the laser beam of 1000 mm/min and 30 mm, respectively. The hardness of 139 was obtained at a low laser power and powder-feed rate of 2200 W and 30 g/min, respectively. At this point, when the laser power increased to 2550 W, the hardness enhanced to 276; further increment of laser power to 2800 W led to a decrease in hardness to 224.

Similarly, the hardness increased to 248 at the laser power and powder-feed rate of 2200 W and 40 g/min, respectively. The hardness decreased to 178 when the powder-feed rate increased to 50 g/min at the same laser power of 2200 W. Also, the hardness values of 373 and 292 were obtained at powder-feed rate of 40 and 50 g/min at the laser power of 2550 W. Finally, the hardness values of 313 and 223 were observed at powder-feed rate of 40 and 50 g/min at the laser power of 2800 W. From the interactive effect between laser power and powder-feed rate at a constant laser-scanning speed and focal position of the laser beam of 1000 mm/min and 30 mm, respectively, the optimal conditions were found to be laser power and powder-feed rate of 2550 W and 40 g/min, respectively.

The interactive effect between laser-scanning speed and focal position of the laser beam was studied by varying laser-scanning speed from 800 to 1200 mm/min and focal position of the laser beam from 15 to 35 mm at a constant laser power and powder-feed rate of 2550 W and 40 g/min, respectively. The hardness of 269 was obtained at a low laser-scanning speed and focal position of the laser beam of 800 mm/min and 15 mm, respectively. At this point, when the laser-scanning speed increased to 1000 mm/min, the hardness enhanced to 332; further increment of laser-scanning speed to 1200 mm/min led to a decrease in hardness to 236.

Similarly, the hardness increased to 294 at the laser-scanning speed and focal position of the laser beam of 800 mm/min and 30 mm, respectively. The hardness decreased to 284 when the focal position of the laser beam increased to 35 mm at the same laser-scanning speed of 800 mm/min. Also, the hardness values of 373 and 368 were obtained at focal position of the laser beam of 30 and 35 mm at the laser-scanning speed of 1000 mm/min. Finally, the hardness values of 293 and 294 were observed at focal position of the laser beam of 30 and 35 mm at the laser-scanning speed of 1200 mm/min. From the interactive effect between laser-scanning speed and focal position of the laser beam at a constant laser power and powder-feed rate of 2550 W and 40 g/min, respectively, the optimal conditions were found to be laser-scanning speed and focal position of the laser beam of 1000 mm/min and 30 mm, respectively.

The interactive effect between laser-scanning speed and powder-feed rate was studied by varying laser-scanning speed from 800 to 1200 mm/min and powder-feed rate from 30 to 50 g/min at a constant laser power and focal position of the laser beam of 2550 W and 30 mm, respectively. The hardness of 189 was obtained at a low laser-scanning speed and a powder-feed rate of 800 mm/min and 30 g/min, respectively. At this point, when the laser-scanning speed increased to 1000 mm/min, the hardness enhanced to 274; further increment of laser-scanning speed to 1200 mm/min led to a decrease in hardness to 203.

Similarly, the hardness increased to 294 at the laser-scanning speed and powder-feed rate of 800 mm/min and 40 g/min, respectively. The hardness decreased to 220 when the powder-feed rate increased to 50 g/min at the same laser-scanning speed of 800 mm/min. Also, the hardness values of 373 and 292 were obtained at powder-feed rate of 40 and 50 g/min at the laser-scanning speed of 1000 mm/min. Finally, the hardness values of 293 and 203 were observed at powder-feed rate of 30 and 35 g/min at the laser-scanning speed of 1200 mm/min. From the interactive effect between laser-scanning speed and powder-feed rate at a constant laser power and focal position of the laser beam of 2550 W and 30 mm, respectively, the optimal conditions were found to be laser-scanning speed and powder-feed rate of 1000 mm/min and 40 g/min, respectively.

The interactive effect between focal position of laser beam and powder-feed rate was studied by varying focal position of laser beam from 800 to 1200 mm and powder-feed rate from 30 to 50 g/min at a constant laser power and focal position of the laser beam of 2550 W and 30 mm, respectively. The hardness of 189 was obtained at a low focal position of laser beam and a powder-feed rate of 800 mm and 30 g/min, respectively. At this point, when the focal position of laser beam increased to 1000 mm, the hardness enhanced to 276; further increment of focal position of the laser beam to 1200 mm led to a decrease in hardness to 203.

Similarly, the hardness increased to 294 at the focal position of the laser beam and powder-feed rate of 800 mm and 40 g/min, respectively. The hardness decreased to 220 when the powder-feed rate increased to 50 g/min at the same focal position of the laser beam of 800 mm. Also, the hardness values of 373 and 292 were obtained at a powder-feed rate of 40 and 50 g/min at the focal position of the laser beam of 1000 mm. Finally, the hardness values of 293 and 203 were observed at a powder-feed rate of 30 and 35 g/min at the focal position of the laser beam of 1200 mm. From the interactive effect between focal position of the laser beam and powder-feed rate at a constant laser power and laser-scanning speed of 2550 W and 1000 mm/min, respectively, the optimal conditions were found to be focal position of the laser beam and a powder-feed rate of 30 mm and 40 g/min, respectively.

10.4 ANN FOR MODEL DEVELOPMENT

MATLAB R2019a from MathWorks was used for model development by nntool command. The ANN is a component of artificial intelligence that is meant to simulate the functioning of a human brain. Processing units make up ANNs, which in turn consist of inputs and outputs. The inputs are what the ANN learns from to produce the desired output. Backpropagation is the set of learning rules used to guide ANNs (Daniel 2013).

The ANN involves training, testing and validation of data. Normally, training, testing and validation are performed in the ratio of 70:15:15 (Kalogirou 2000). The following network settings were used for ANN in MATLAB:

Network type: Feed-forward backpropagation

Training function: *Levenberg–Marquardt (TRAINLM)*

Adaptation learning function: Gradient descent with momentum weight and bias (LEARNGDM)

Performance function: Mean squared error (MSE)

Number of layers: 2

Properties for Layer 1:
Number of neurons: 10
Transfer function: Tangent sigmoid (TANSIG)
Properties for Layer 2:
Number of neurons: Automatically taken by the system
Transfer function: Tangent sigmoid (TANSIG)
The network was trained using the following parameters:
Epochs to train (epochs): 100
Performance goal (goal): 0 (Zero)
Maximum validation failures (max_fail): 5
Membrane/speed tradeoff (mem_reduc): 1
Minimum gradient (min_grad): 1×10^{-10}
Initial training gain (mu): 0.001
Training gain decrease factor (mu_dec): 0.1
Training gain increase factor (mu_inc): 10
Maximum training gain (mu_max): 1×10^{10}
Epochs between displays (show): 25

Figure 10.4 shows the ANN architecture for laser cladding of Inconel® powder on AISI 304 SS substrate. The network flows from input factor to neuron and then to output factor. Let X_i represent the input factor, w_{ij} represent the weight from i^{th} input factor of the input layer to j^{th} neuron of the hidden layer, b_{1j} represent bias from the Layer 1 to j^{th} neuron of the hidden layer.

When the signal flows from the input factor to the neuron, the function is represented as

$$\sum_{j=1}^{10} N_j = \sum_{i=1,j=1}^{i=4,j=10} w_{ij} X_i \tag{10.12}$$

When bias is added to the neuron, the input function to the neuron from the input factor is written as,

$$\sum_{j=1}^{10} n_j = \sum_{j=1}^{10} (N_j + b_{1j}) \tag{10.13}$$

The output function from the neuron is

$$\sum_{j=1}^{10} y_{j2} = \text{tansig}\left(\sum_{j=1}^{10} n_j\right) = \frac{2}{1 + e^{\sum_{j=1}^{10} n_j}} - 1 \tag{10.14}$$

The output function from the neuron to the output factor is

$$\sum_{j=1}^{10} Y_{j2} = \sum_{j=1}^{10} (y_{j2} + w_j) + b_2 \tag{10.15}$$

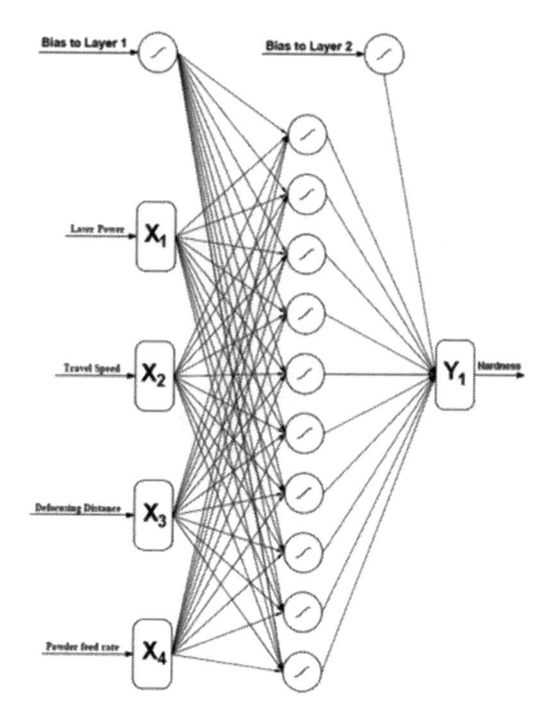

FIGURE 10.4 ANN architecture for laser cladding of Inconel® powder on AISI 304 SS substrate.

The output function from the output factor is

$$F_o = \text{tansig}\left(\sum_{j=1}^{10} Y_{j2}\right) = \frac{2}{1+e^{\sum_{j=1}^{10} Y_{j2}}} - 1 \tag{10.16}$$

The experimental data was fitted to the model, and the following coefficients were determined. Weights from input factor to neurons w_{ij} is

$$\begin{pmatrix}
-1.2466 & 0.68366 & -0.11444 & -3.6369 \\
1.9145 & -1.9357 & 0.95102 & -2.6124 \\
-0.61341 & 1.8961 & -0.36938 & -0.53356 \\
1.7628 & -0.956 & 1.9106 & 2.0771 \\
1.7008 & 2.0192 & 2.0273 & 0.067656 \\
1.7519 & 1.5161 & 0.62612 & -2.0113 \\
2.0357 & -1.3777 & -1.4769 & 0.13436 \\
0.26945 & -0.64557 & 2.041 & -0.19565 \\
-1.0926 & -1.5107 & -0.81874 & 1.0869 \\
0.4868 & 0.23717 & 2.089 & -0.20336
\end{pmatrix}$$

Bias to neurons in first layer b_{1j} is

$$\begin{pmatrix} 1.3122 \\ -1.5278 \\ -2.128 \\ 1.5249 \\ 0.49342 \\ 1.5942 \\ 1.7848 \\ -1.9391 \\ -2.4198 \\ 3.0075 \end{pmatrix}$$

Weights from neurons to output factor w_j is

$$\begin{pmatrix} 0.7224 \\ -1.2882 \\ -1.2383 \\ 1.2674 \\ 0.13598 \\ 0.57522 \\ 0.40404 \\ 0.37206 \\ 0.1297 \\ -1.9553 \end{pmatrix}$$

Bias to output factor in second layer b_2 is

$$(-1.1878)$$

Table 10.4 shows the experimental values, quadratic model and ANN predicted values and the corresponding percentage residual. The fractional residual is evaluated as the ratio between residual and experimental value (Sivamani 2015). Percentage residual is fractional residual multiplied by 100. Residual is difference between the experimental value and the predicted value. i.e.

$$\text{Fractional residual} = \text{Residual/Experimental value} \tag{10.17}$$

$$\text{Percentage residual} = \text{Fractional residual} \times 100 \tag{10.18}$$

$$\text{Residual} = \text{Experimental value} - \text{Predicted value} \tag{10.19}$$

TABLE 10.5
Experimental, Quadratic Model and ANN Predicted Values of Hardness for Cladded Specimens Using Inconel® 625 Powder

		Predicted hardness			Percentage residual		
S. No.	Experimental hardness	Linear model	Quadratic model	ANN model	Linear model	Quadratic model	ANN model
1	216	226	216	227	–4.80	–0.10	–5.22
2	233	240	233	231	–3.06	–0.14	1.01
3	222	241	223	221	–8.69	–0.45	0.36
4	255	255	253	264	–0.01	0.64	–3.64
5	247	245	247	261	0.83	–0.07	–5.60
6	254	259	255	252	–1.85	–0.21	0.75
7	264	260	265	256	1.56	–0.27	3.18
8	287	274	285	284	4.66	0.58	1.09
9	262	249	260	264	5.11	0.64	–0.91
10	268	262	267	271	2.10	0.30	–1.15
11	260	264	259	254	–1.36	0.24	2.16
12	283	277	280	305	2.02	1.24	–3.65
13	264	267	267	268	–1.21	–0.58	–1.67
14	267	281	263	265	–5.23	1.62	0.63
15	279	282	275	300	–1.12	1.31	–7.64
16	286	296	286	306	–3.45	0.10	–7.01
17	223	247	222	228	–10.93	0.24	–2.16
18	246	275	250	249	–11.74	–1.61	–1.14
19	234	246	234	234	–5.22	–0.13	–0.01
20	261	276	264	280	–5.76	–1.20	–7.30
21	289	243	291	284	16.08	–0.56	1.56
22	326	260	328	321	14.20	–0.55	1.59
23	218	238	216	217	–9.57	0.71	0.51
24	256	283	261	256	–10.69	–1.94	0.19
25	328	261	328	323	20.39	0	1.59

From Table 10.5, the percentage residual ranges from -11.74 to 20.39%, -1.94 to 1.62% and -3.65 to 3.18% for linear, quadratic and ANN models, respectively. The percentage residual should be within ±5%; this is high for a linear model, but it is within the limit for quadratic and ANN models. Hence, it can be concluded that quadratic and ANN models fit well to the experimental data.

10.5 PARAMETRIC OPTIMIZATION BY FIRST DERIVATIVE TEST

The first derivative test is used to find local extreme points, either maximum or minimum. A local maximum is where the function starts to decrease and a local minimum is where the function starts to increase. Critical points are where a function can have a local maximum or minimum, and are the only places where the first derivative can change sign (Stewart 2012). Let *c* be a critical point for a continuous function f(x):

- If f′(x) changes from positive to negative at *c*, then f(*c*) is a local maximum.
- If f′(x) changes from negative to positive at *c*, then f(*c*) is a local minimum.
- If f′(x) does not change sign at *c*, then f(*c*) is neither a local maximum or minimum.

The model that fits the experimental data is given by Equation 10.11. Equation 10.11 should be differentiated with respect to each variable and equate them to zero to obtain critical points.

$$\frac{\partial Y}{\partial P} = 5.14 + 0.0002 * V - 0.0032 * D - 0.0034 * F - 0.0020 * P \tag{10.20}$$

$$\frac{\partial Y}{\partial V} = 3.48 + 0.0002 * P + 0.0054 * D - 0.0039 * F - 0.0040 * V \tag{10.21}$$

$$\frac{\partial Y}{\partial D} = 24.30 - 0.0032 * P + 0.0054 * V - 0.2575 * F - 0.3758 * D \tag{10.22}$$

$$\frac{\partial Y}{\partial F} = 92.50 + 0.0034 * P - 0.0039 * V - 0.2575 * D - 1.7858 * F \tag{10.23}$$

Equating all the first derivatives to zero,

$$5.14 + 0.0002 * V - 0.0032 * D - 0.0034 * F - 0.0020 * P = 0 \tag{10.24}$$

$$3.48 + 0.0002 * P + 0.0054 * D - 0.0039 * F - 0.0040 * V = 0 \tag{10.25}$$

$$24.30 - 0.0032 * P + 0.0054 * V - 0.2575 * F - 0.3758 * D = 0 \tag{10.26}$$

$$92.50 + 0.0034 * P - 0.0039 * V - 0.2575 * D - 1.7858 * F = 0 \tag{10.27}$$

Solving the above equations, the critical points of P, V, D and F are calculated as 2554 W, 999 mm/min, 30 mm and 40.5 g/min, respectively. At the critical point, the hardness was found to be 373.54.

10.6 GENERALIZED REDUCED GRADIENT FOR PROCESS OPTIMIZATION

Solver, an add-in program for Microsoft Excel 16.0 from Microsoft Corporation, was used to find maximum hardness in laser cladding of Inconel® 625 powder on AISI 304 SS substrate. Solver is used to find the maximum or minimum value for a model in one cell called the objective cell either subject to or not subject to constraints. Solver works with a group of cells, called decision variables or simply variable cells, that are used in computing the functions in the objective and constraint cells. Solver adjusts the values in the decision variable cells to satisfy the constraints and produce the desired result. In a simpler way, Solver can be used to determine the maximum or minimum value of one cell by changing other cells (Nenov and Fylstra 2003).

The GRG method is an extension of the reduced gradient method that was presented originally for solving problems with linear constraints only. The GRG method is based on the idea of eliminating variables using the constraints. Thus, theoretically, one variable can be reduced from the set of input factors for each of the constraints (Bharathiraja et al. 2017). Solver uses the following procedure to find the optimal solution (Rao 2019).

Step 1: Specify the design and state variables.

Start with an initial trial vector Z. Identify the design and state variables (X and Y) for the problem using the following guidelines.

- The state variables are to be selected to avoid singularity of the matrix, [D].
- Because the state variables are adjusted during the iterative process to maintain feasibility, any component of Z that is equal to its lower or upper bound initially is to be designated a design variable.
- Because the slack variables appear as linear terms in the inequality constraints, they should be designated as state variables; however, if the initial value of any state variable is zero (its lower bound value), it should be designated a design variable.

It is convenient to divide the design variables arbitrarily into two sets as

$$Z = \begin{pmatrix} X \\ Y \end{pmatrix} \tag{10.28}$$

where Z is initial value matrix, X is input or independent or design variable matrix, Y is output or dependent or state variable matrix.

Step 2. Compute the GRG.

The GRG is determined using the equation below. The derivatives involved in the equation can be evaluated numerically, if necessary.

$$G_R = \nabla_Y f - \left([D]^{-1} [C] \right)^T \nabla_Z f \tag{10.29}$$

Step 3. Test for convergence.

If all the components of the GRG are close to zero, the method can be considered to have converged, and the current vector Z can be taken as the optimum solution of the problem. For this, the following test can be used:

$$\left\| G_R \right\| \leq \varepsilon \tag{10.30}$$

where ε is a number close to zero. If this relation is not satisfied, go to Step 4.

Step 4. Determine the search direction.

The GRG can be used similarly to a gradient of an unconstrained objective function to generate a suitable search direction, S. The techniques, such as steepest descent, Fletcher–Reeves, Davidon–Fletcher–Powell or Broydon–Fletcher–Goldfarb–Shanno methods can be used for this purpose. For example, if a steepest descent method is used, the vector S is determined as

$$S = -G_R \tag{10.31}$$

Step 5. Find the minimum along the search direction.

To find a local minimum of function along the search direction S, the following procedure can be used conveniently.

- Find an estimate for λ as the distance to the nearest side constraint. When design variables are considered, then

$$\lambda = \begin{pmatrix} \dfrac{x_i^u - x_i^{old}}{s_i} & \text{if } s_i > 0 \\ \dfrac{x_i^l - x_i^{old}}{s_i} & \text{if } s < 0 \end{pmatrix} \tag{10.32}$$

  where s_i is the ith component of S. Similarly, when state variables are considered, then

$$dY = -\left[D\right]^{-1} \cdot \left[C\right] \cdot dX \tag{10.33}$$

  Using $dY = \lambda S$, the above equation gives the search direction for the variables Y as

$$T = -\left[D\right]^{-1} \cdot \left[C\right] \cdot S \tag{10.34}$$

  Thus,

$$\lambda = \begin{pmatrix} \dfrac{y_i^u - y_i^{old}}{t_i} & \text{if } t_i > 0 \\ \dfrac{y_i^l - y_i^{old}}{t_i} & \text{if } t_i < 0 \end{pmatrix} \tag{10.35}$$

  where t_i is the ith component of T.
- The minimum value of λ, given by Equation 10.32, λ_1, makes some design variable attain its lower or upper bound. Similarly, the minimum value of λ given by Equation 10.35, λ_2, will make some state variable attain its lower or upper bound. The smaller of λ_1 or λ_2 can be used as an upper bound on the value of λ for initializing a suitable

one-dimensional minimization procedure. The quadratic interpolation method can be used conveniently for finding the optimal step length λ*.

- Find the new vector Z_{new}:

$$Z_{new} = \begin{pmatrix} X_{old} + dX \\ Y_{old} + dY \end{pmatrix} = \begin{pmatrix} X_{old} + \lambda^{*} S \\ Y_{old} + \lambda^{*} T \end{pmatrix} \tag{10.36}$$

If the vector Z_{new} corresponding to λ^* is found infeasible, then X_{new} is held constant and Y_{new} is modified with $dY = Y_{new} - Y_{old}$. Finally, when convergence is achieved, then

$$Z_{new} = \begin{pmatrix} X_{old} + \Delta X \\ Y_{old} + \Delta Y \end{pmatrix} \tag{10.37}$$

Repeat the iteration from Step 1.

The optimal values of laser power, laser-scanning speed, the focal position of the laser beam and powder-feed rate were calculated as 2517 W, 1026 mm/min, 29.9 mm and 40.5 g/min, respectively. At the optimal point, the hardness was found to be 334.

10.7 GREY RELATIONAL ANALYSIS FOR PARAMETRIC OPTIMIZATION

Grey theory is an effective model to acquire uncertain knowledge, proposed by Deng (Deng, 1989). It mainly focuses on small sample and limited information, which is only known partially. GRA is an important task of grey theory to scale the similar or different level of development trends among various factors. It has drawn more and more researchers' attention in recent years and achieved many research results (Singh et al. 2004).

The GRA procedure can be followed by normalizing either experimental value or signal to noise ratio (S/N) of experimental value. The sequential procedure for GRA is explained as follows (Singh and Singh 2015):

Step 1: Collect the experimental data.

Step 2: In GRA, experimental data of each trial should be normalized from 0 to 1. This process is known as relational generation. In relational generation, normalized results follow either smaller-the-better (STB) or larger-the-better (LTB) criteria, which can be expressed as follows:
For STB,

$$\text{Normalized value } x = \frac{\max(y) - y}{\max(y) - \min(y)} \tag{10.38}$$

For LTB,

$$\text{Normalized value } x = \frac{y - \min(y)}{\max(y) - \min(y)} \tag{10.39}$$

where y is the experimental data,

Step 3: Based on normalized experimental data, the grey relational coefficient (GRC) is calculated to represent the correlation between the desired and the actual experimental data. GRC is calculated using the following equation:

$$\text{GRC } \xi_o = \frac{\min(x) + \zeta.\max(x)}{x + \zeta.\max(x)} \tag{10.40}$$

where ζ is taken as 0.5 to provide equal weightage to all factors.

TABLE 10.6
GRA for Laser-Cladded Specimens Using Inconel® 625 Powder to Obtain Maximum Hardness

S. No.	Experimental value	Normalized value	GRC	Rank
1	216	0	1	25
2	233	0.151786	0.767123	21
3	222	0.053571	0.903226	23
4	255	0.348214	0.589474	16
5	247	0.276786	0.643678	18
6	254	0.339286	0.595745	17
7	264	0.428571	0.538462	10
8	287	0.633929	0.440945	4
9	262	0.410714	0.54902	12
10	268	0.464286	0.518519	8
11	260	0.392857	0.56	14
12	283	0.598214	0.455285	6
13	264	0.428571	0.538462	10
14	267	0.455357	0.523364	9
15	279	0.5625	0.470588	7
16	286	0.625	0.444444	5
17	223	0.0625	0.888889	22
18	246	0.267857	0.651163	19
19	234	0.160714	0.756757	20
20	261	0.401786	0.554455	13
21	289	0.651786	0.434109	3
22	326	0.982143	0.337349	2
23	218	0.017857	0.965517	24
24	256	0.357143	0.583333	15
25	328	1	0.333333	1

Step 4: Grey relational degree (GRD) is calculated by averaging the GRC values for each trial as:

$$r_i = \frac{1}{m}\sum_{j=1}^{n} \xi_{0,j} \tag{10.41}$$

Thus by applying Equation 10.41, all GRD can be computed, which provides ranking for each trial. Here, the highest GRD value represents that the corresponding S/N is closer to the ideal normalized value. The GRD obtained for trial and the ranking order of the trial is shown in Table 10.6. It is seen that Trial 25 has the best performance characteristic among 25 performed trials and has the highest value of GRC and is optimal.

The optimal values of laser power, laser-scanning speed, the focal position of the laser beam and powder-feed rate were calculated as 2500 W, 1000 mm/min, 25 mm and 40 g/min, respectively. At the optimal point, the hardness was found to be 328.

REFERENCES

Abioye, T.E., Farayibi, P.K. and Clare, A.T. 2017. A comparative study of Inconel 625 laser cladding by wire and powder feedstock. *Materials and Manufacturing Processes*, 32(14): 1653–1659.

Abioye, T.E., Folkes, J. and Clare, A.T. 2013. A parametric study of Inconel 625 wire laser deposition. *Journal of Materials Processing Technology*, 213(12): 2145–2151.

Bharathiraja, G., Jayabal, S. and Kalyana Sundaram, S. 2017. Gradient-based intuitive search intelligence for the optimization of mechanical behaviors in hybrid bioparticle-impregnated coir-polyester composites. *Journal of Vinyl and Additive Technology*, 23(4): 275–283.

Chryssolouris, G., Zannis, S., Tsirbas, K. and Lalas, C. 2002. An experimental investigation of laser cladding. *CIRP Annals*, 51(1): 145–148.

Daniel, G. 2013. *Principles of Artificial Neural Networks*. Vol. 7. Singapore: World Scientific Publishing Company.

Deng, J. 1989. Introduction to grey theory system. *The Journal of Grey System*, 1(1), 1–24.

Dubiel, B. and Sieniawski, J. 2019. Precipitates in additively manufactured Inconel 625 superalloy. *Materials*, 12(7): 1144.

Feng, K., Chen, Y., Deng, P., Li, Y., Zhao, H., Lu, F., … and Li, Z. 2017. Improved high-temperature hardness and wear resistance of Inconel 625 coatings fabricated by laser cladding. *Journal of Materials Processing Technology*, 243: 82–91.

Gupta, R. K., Anil Kumar, V., Gururaja, U. V., Shivaram, B. R. N. V., Maruti Prasad, Y., Ramkumar, P., … and Sarkar, P. 2015. Processing and characterization of Inconel 625 Nickel base superalloy. In *Materials Science Forum*, 830: 38–40. Switzerland: Trans Tech Publications Ltd.

International Stainless Steel Forum. 2020 https://www.worldstainless.org/Files/issf/non-image-files/PDF/Automotiveapplications.pdf. (Accessed on 18 January 2020).

Juliano, S.A. 2001. Nonlinear curve fitting: predation and functional response curves. *Design and Analysis of Ecological Experiments*, 2: 178–196.

Kalogirou, S.A. 2000. Applications of artificial neural-networks for energy systems. *Applied Energy*, 67(1–2): 17–35.

Kaysser, W. 2001. Surface modifications in aerospace applications. *Surface Engineering*, 17(4): 305–312.

Lian, G., Zhang, H., Zhang, Y., Yao, M., Huang, X. and Chen, C. 2019. Computational and experimental investigation of micro-hardness and wear resistance of Ni-Based alloy and TiC composite coating obtained by laser cladding. *Materials*, 12(5): 793.

Nenov, I.P. and Fylstra, D.H. 2003. Interval methods for accelerated global search in the Microsoft Excel Solver. *Reliable Computing*, 9(2): 143–159.

Rao, S.S. 2019. *Engineering Optimization: Theory and Practice.* USA: John Wiley & Sons.

Sandhu, S.S. and Shahi, A.S. 2016. Metallurgical, wear and fatigue performance of Inconel 625 weld claddings. *Journal of Materials Processing Technology*, 233: 1–8.

Shariff, S.M., Pal, T.K., Padmanabham, G. and Joshi, S.V. 2010. Sliding wear behaviour of laser surface modified pearlitic rail steel. *Surface Engineering*, 26(3): 199–208.

Singh, P.N., Raghukandan, K. and Pai, B.C. 2004. Optimization by grey relational analysis of EDM parameters on machining Al–10% SiCP composites. *Journal of Materials Processing Technology*, 155: 1658–1661.

Singh, S., Goyal, D.K., Kumar, P. and Bansal, A. 2020. Laser cladding technique for erosive wear applications: A review. *Materials Research Express*, 7(1): 012007.

Singh, V.K. and Singh, S. 2015. Multi-objective optimization using Taguchi based grey relational analysis for wire EDM of Inconel 625. *Journal of Materials Science and Mechanical Engineering*, 2(11): 38–42.

Sivamani, S. 2015. Studies on bioethanol production from Cassava industrial residues Statistical optimization kinetics and thermodynamics, PhD Thesis, Anna University, Chennai.

Song, L., Zeng, G., Xiao, H., Xiao, X. and Li, S. 2016. Repair of 304 stainless steel by laser cladding with 316L stainless steel powders followed by laser surface alloying with WC powders. *Journal of Manufacturing Processes*, 24: 116–124.

Special Metals Corporation. 2013. Inconel® alloy 625. New Hartford, NY: Special Metals Corporation.

Stewart, J. 2012. *Essential calculus: Early transcendentals.* USA: Cengage Learning

Xu, P., Lin, C., Zhou, C. and Yi, X. 2014. Wear and corrosion resistance of laser cladding $AISI_30_4$ stainless steel/Al_2O_3 composite coatings. *Surface and Coatings Technology*, 238: 9–14.

11 Effect of Wire Electric Discharge Machining Process Parameters on Surface Roughness of Monel 400 Alloy

M. Mahalingam, A. Umesh Bala and R. Varahamoorthi

CONTENTS

11.1 INTRODUCTION

Wire electric discharge machining (WEDM) plays a major role in manufacturing industries because of its superior hardness and ease of use. The use of nickel-based material is increasing, because of its ability to retain high mechanical and chemical properties at elevated temperature. Using conventional machining methods, nickel-based alloy shows poor machinability because of its high hardness toughness.

WEDM is the best nonconventional machining process to machine complex geometries in high strength, high hardness with high precision.

Surface integrity is a composite property that describes the deviation of surface characteristics from the substrate. Its constituent parameters include change in surface metallurgy, residual stresses on surface, depth of surface affected by metallurgy and residual stresses, geometrical accuracy, including accuracy of size and form and surface roughness. In other words, the surface characteristics may be aptly considered to define the surface integrity of the machined component. Keeping this in view, this chapter chose important surface characteristics to define surface integrity.

A large numbers of experiments are required to obtain the adequate model related to surface roughness. The machining parameters include pulse on time, pulse off time, wire tension and wire feed. Moreover, it requires different experiments for each and every combination of electrode and workpiece material. In this work, response surface methodology (RSM) is used for modelling. RSM has been used to plan and analyze the experiments. It is a collection of mathematical and statistical techniques that are useful for the modelling and analysis of problems in which a response of interest is influenced by several variables, and the objective is to optimize the response (Montgomery 1997). It is a sequential experimentation strategy for empirical model building and optimization. By conducting experiments and applying regression analysis, a model of the response to some independent input variables can be obtained. Based on the model of the response, a near optimal point can be deduced. RSM is often applied in the characterization and optimization of processes (El Baradie 1981). The objective of using RSM is not only to investigate the response over the entire space (Sundaram and Lambert 1981), but also to locate the region of interest where the response reaches its optimum or near optimum value. By carefully studying the response surface model, the combination of factors, which gives the best response, can then be established.

11.2 EXPERIMENTAL METHOD

11.2.1 Wire Electrical Discharge Machining

The experimental investigation was done on a computer numerical control WEDM (Brand: Maxi cut WEDM machine, Electronica Machine Tools Ltd.). WEDM is a thermoelectrically process in which material is eroded by a series of sparks between the workpiece and the wire electrode (tool). The part and wire are immersed in a dielectric (electrically nonconducting) fluid that also acts as a coolant and flushes away debris. In this process, there is no contact between workpiece and electrode, thus materials of any hardness can be cut as long as they can conduct electricity.

11.2.2 Material Used: Monel 400

Monel 400 alloy of 5 mm thickness was selected as the workpiece for this experiment. Monel is widely used because it can be hardened only by cold working. It has high strength and toughness over a wide temperature range and excellent resistance to many corrosive environments (Shoemaker and Smith 2006).

11.2.3 Tool Electrode Material: Brass

In this experiment, brass wire is used as a tool electrode material. Brass EDM wire is a combination of copper and zinc, typically alloyed in the range of 63–65% Cu and 35–37% Zn. Additionally, zinc provides significantly higher tensile strength, a lower melting point and higher vapours pressure rating, which more than offsets the relative losses in conductivity. Brass quickly became the most widely used electrode material for general-purpose WEDM. Brass wire of 0.25 mm is used.

11.2.4 Dielectric Fluid: Deionized Water

Deionized water is the dielectric fluid that is generally used for WEDM. Water was used because it flows better into the small slots than other dielectrics and provides good cooling. The major advantage of water is its good cooling qualities, which are needed for the energy transmission during the wire-cutting process.

11.2.5 Measuring Apparatus

The surface roughness of the machined workpieces was measured using a surface roughness gauge. The surface roughness gauge used was Perth meter, MahrSurf PS 1. A set of three readings were taken on each workpiece at different places. The average surface roughness values for different workpiece were considered.

11.2.6 Parameters and Its Range

In total, four experimental parameters were considered. The pulse on time, pulse off time, wire tension and wire feed are the input process parameters of the WEDM process. The levels of these parameters were decided based on published information regarding previous studies (Baig and Venkaiah 2001). The experiments were conducted based on four factors with five levels of central composite designs. Table 11.1 shows the details of the parameters.

TABLE 11.1
Parameters with Their Ranges

PARAMETER	-2	-1	0	1	2
Pulse on time(μs)	2	4	6	8	10
Pulse off time(μs)	2	4	6	8	10
Wire tension(g)	700	850	1000	1150	1300
Wire feed (m/min)	3	4	5	6	7

11.2.7 Design of Experiments

A designed experiment is an approach to symmetrically vary the controllable input factors in the process and determine whether these changes affect the product parameters. It is extremely helpful to discover the key variable and quality characteristic of input that influences the process (Ross 1988). The experimental work is carried out as per the central composite design using RSM methodology. The design is prepared with the help of Design-Expert software MINI TAB, which is used to create experimental designs. The experimental investigation is carried out on a CNC WEDM. A total of 30 experiments were conducted as per the design (Muthu Kumar et al. 2010).

11.3 RESULTS AND DISCUSSION

Table 11.2 shows the experiments input values and their response (surface roughness (R_a) in μm). The experiments were conducted as per the central composite design in WEDM with a brass electrode for all 30 experiments.

The final response equation for surface roughness is given in equation below.

$$\begin{aligned}\mathrm{Ra} = {} & 2.85 + 0.14A - 0.2B + 0.41C - 0.45D + 0.35AB + 0.20AC \\ & - 0.34AD - 0.031BC - 0.17BD - 0.41CD - 9.167E - 003A^2 \\ & + 0.41B^2 + 0.17C^2 + 0.18D^2\end{aligned} \tag{11.1}$$

11.3.1 Effect of Pulse On Time

Figure 11.1 shows the main effect of pulse on time versus response characteristics. The pulse on time is varied from 2 to 10 μs in steps of 2 μs. The value of the other parameters is kept constant. The surface roughness increased as the pulse on time increased. The greater discharge energy produces a larger crater, causing a larger surface roughness value on the workpiece. Surface roughness value is mainly influenced by pulse on time (Shayan et al. 2013). The brass wire of the cutting tool accelerates depletion, which generates a built-up layer. This increased built-up layer results in a rougher surface (Yang et al. 2011).

11.3.2 Effect of Pulse Off Time

Figure 11.2 shows the main effect of pulse off time versus response characteristics. The pulse off time is varied from 2 to 10 μs in steps of 2 μs. The values of the other parameters are kept constant. Pulse off time is the time duration in between two simultaneous sparks. The voltage is absent during this part of cycle. Higher pulse off time removes the formation of ionized bridges across the gap and results in higher ignition and decreased surface roughness (Umanath and Devika 2018).

TABLE 11.2
Experimental Design Matrix

Sl. No	Pulse on Time (µs)	Pulse off Time (µs)	Wire Tension (g)	Wire Feed (m/min)	Surface Roughness R_a (µm)
1	4	4	850	4	2.84
2	8	4	850	4	4.1
3	4	8	850	4	3.13
4	8	8	850	4	4.17
5	4	4	1150	4	6.8
6	8	4	1150	4	4.55
7	4	8	1150	4	2.6
8	8	8	1150	4	8.5
9	4	4	850	6	3.2
10	8	4	850	6	3.38
11	4	8	850	6	3.03
12	8	8	850	6	2.17
13	4	4	1150	6	3.0
14	8	4	1150	6	4.34
15	4	8	1150	6	3.12
16	8	8	1150	6	2.95
17	2	6	1000	5	3.05
18	10	6	1000	5	1.49
19	6	2	1000	5	4.5
20	6	10	1000	5	3.4
21	6	6	700	5	3.0
22	6	6	1300	5	2.96
23	6	6	1000	3	2.83
24	6	6	1000	7	3.19
25	6	6	1000	5	2.85
26	6	6	1000	5	2.85
27	6	6	1000	5	2.85
28	6	6	1000	5	2.85
29	6	6	1000	5	2.85
30	6	6	1000	5	2.85

11.3.3 Effect of Wire Tension

Figure 11.3 shows the main effect of wire tension versus response characteristics. The wire tension is varied from 700 to 1300 grams in steps of 150 grams. The values of the other parameters are kept constant. It was observed that surface roughness decreased as the wire tension increased. This can be attributed to minimized wire bending because of the increased wire tension, which leads to a dynamic, stable condition of the wire.

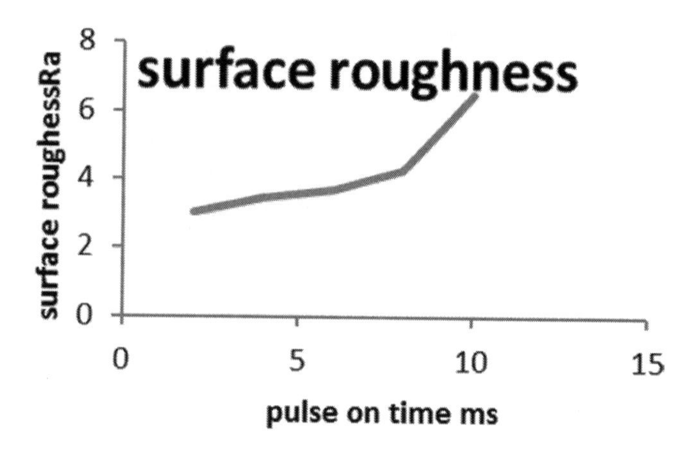

FIGURE 11.1 Effect of pulse on time on surface roughness.

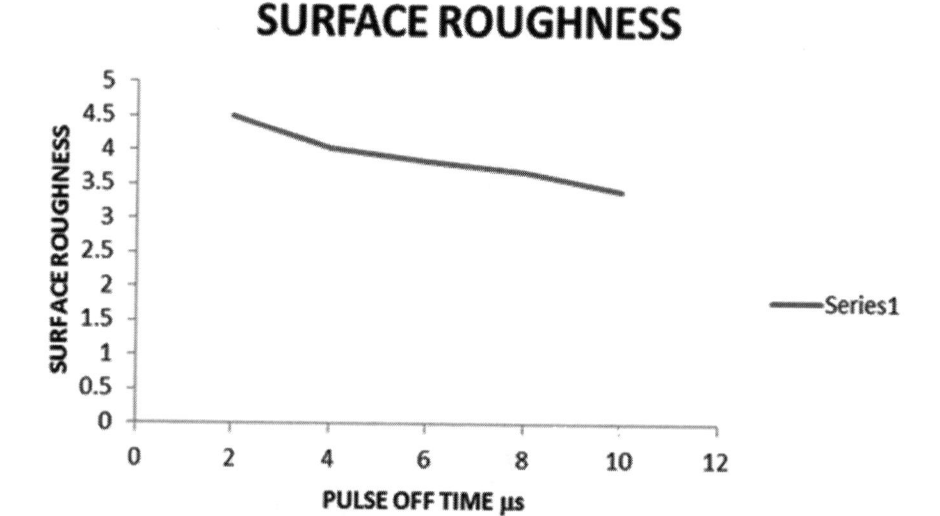

FIGURE 11.2 Effect of pulse off time on surface roughness.

11.3.4 Effect of Wire Feed

The effect of wire feed rate is represented in Figure 11.4. The wire feed is varied from 3 to 7 m/min in steps of 1 m/min. The other parameters are kept constant. When the wire feed rate is increased, the surface roughness is decreased. The spark produced between the tool and the workpiece is stable, thereby generating constant discharge conditions, which resulted in lower surface roughness. At the optimum feed rate of 4.6 m/min with these peak parameters, eroded debris is easily cleared from the spark gap (Ali et al. 2019).

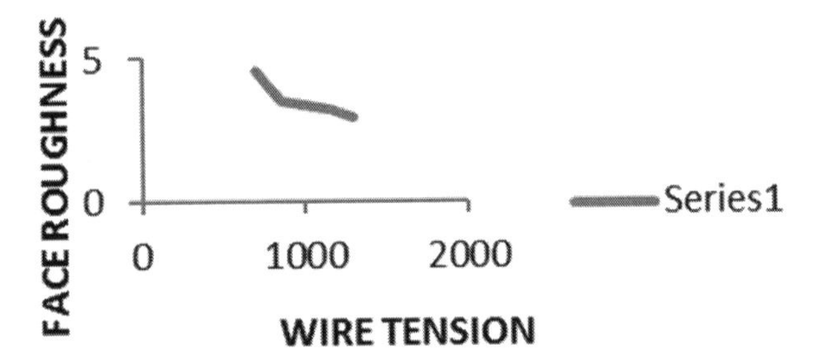

FIGURE 11.3 Effect of wire tension on surface roughness.

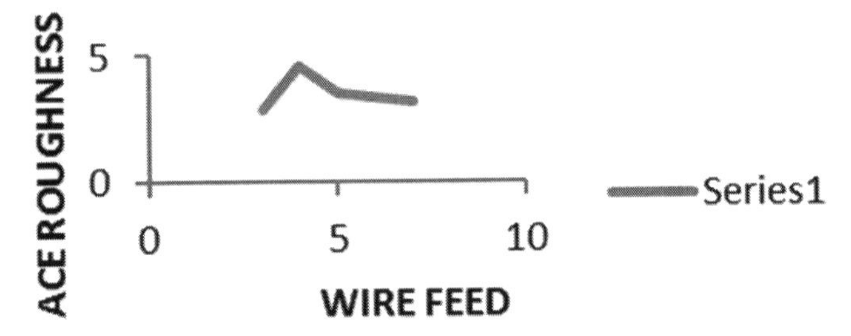

FIGURE 11.4 Effect of wire feed on surface roughness.

11.3.5 Effect of Surface Roughness on Machining Parameters

Figure 11.5 (a–d) shows the micrographs of the machined surface. In this study, an attempt has been made to investigate surface roughness during machining of Monel 400 alloy by using the WEDM process. The thermal stresses developed at the interface zone between the tool and work resulted in microstructure changes ([Tosun et al. 2003). Hence, the surface characteristics of surface roughness were investigated. From the examination of the surface characteristics, it was observed that WEDM process causes damage to the surface, such as globules of debris, pockmarks, melted drops, varying size craters and cracks. Thus, the surface texture is responsible for the uneven surface profile.

Figure 11.6 (a–d) shows the combined effect of input parameters on surface roughness. As the pulse on value (Ton) increases, the value of surface roughness increases. High discharge energy due to the high value of pulse on time results in melting and evaporation of work material that causes formation of gas bubbles that explode when the discharge ceases (Li et al. 2013). Explosion of gas bubbles causes high-pressure energy that creates large-size craters on the work surface. The diameter and depth of the crater increases as the pulse on time increases, thereby increasing the surface roughness (Hewidy et al. 2005).

High discharge energy results in the formation of deep, overlapping craters with large diameters. The high-density melted globules accumulate at the machined surface, resulting in a poorer surface finish.

In addition to that, residual stresses are created at the surface during the machining process and when the stress in the surface exceeds the material's ultimate tensile strength, cracks are formed (Demuth and Beale 2006).

Increase in the value of pulse off time (Toff) decreases the spark frequency, thereby reducing the discharge energy, resulting in a smooth surface with small crater sizes is generated. Increasing the pulse off time (Toff) results in quick and easy flushing of melted debris that results in minimal accumulation of melted globules, decreasing surface roughness.

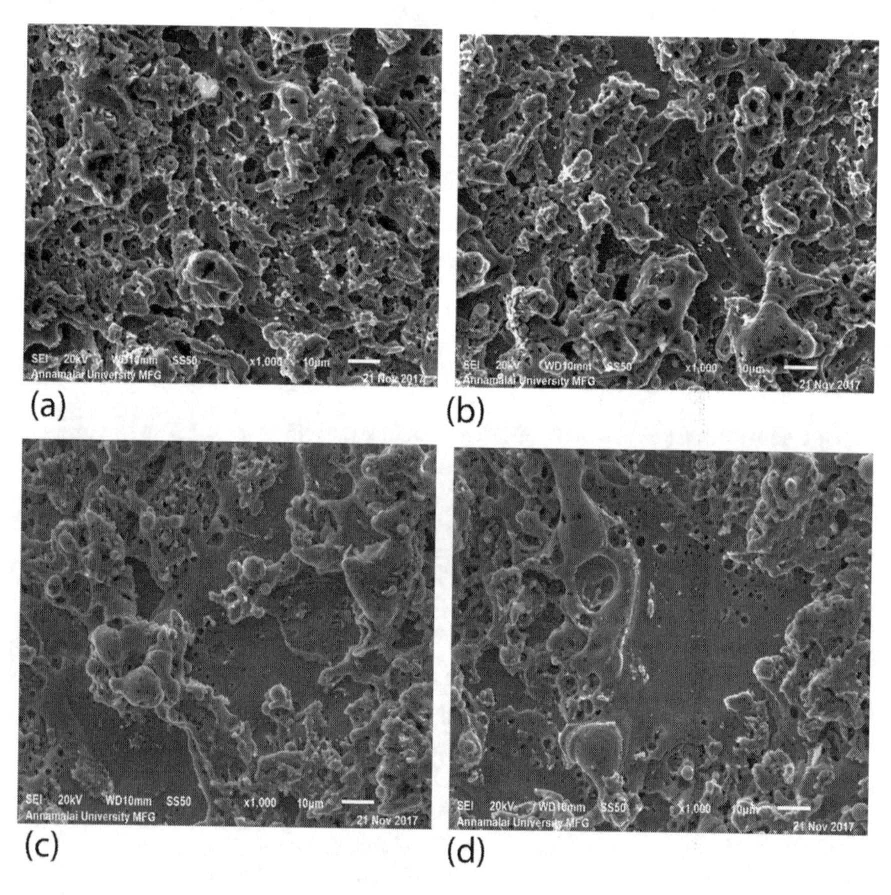

FIGURE 11.5 (a–b) Crater and micro cracks during WEDM of Monel 400 alloy. (c–d) Spherical nodule debris and irregular debris of Monel 400 alloy.

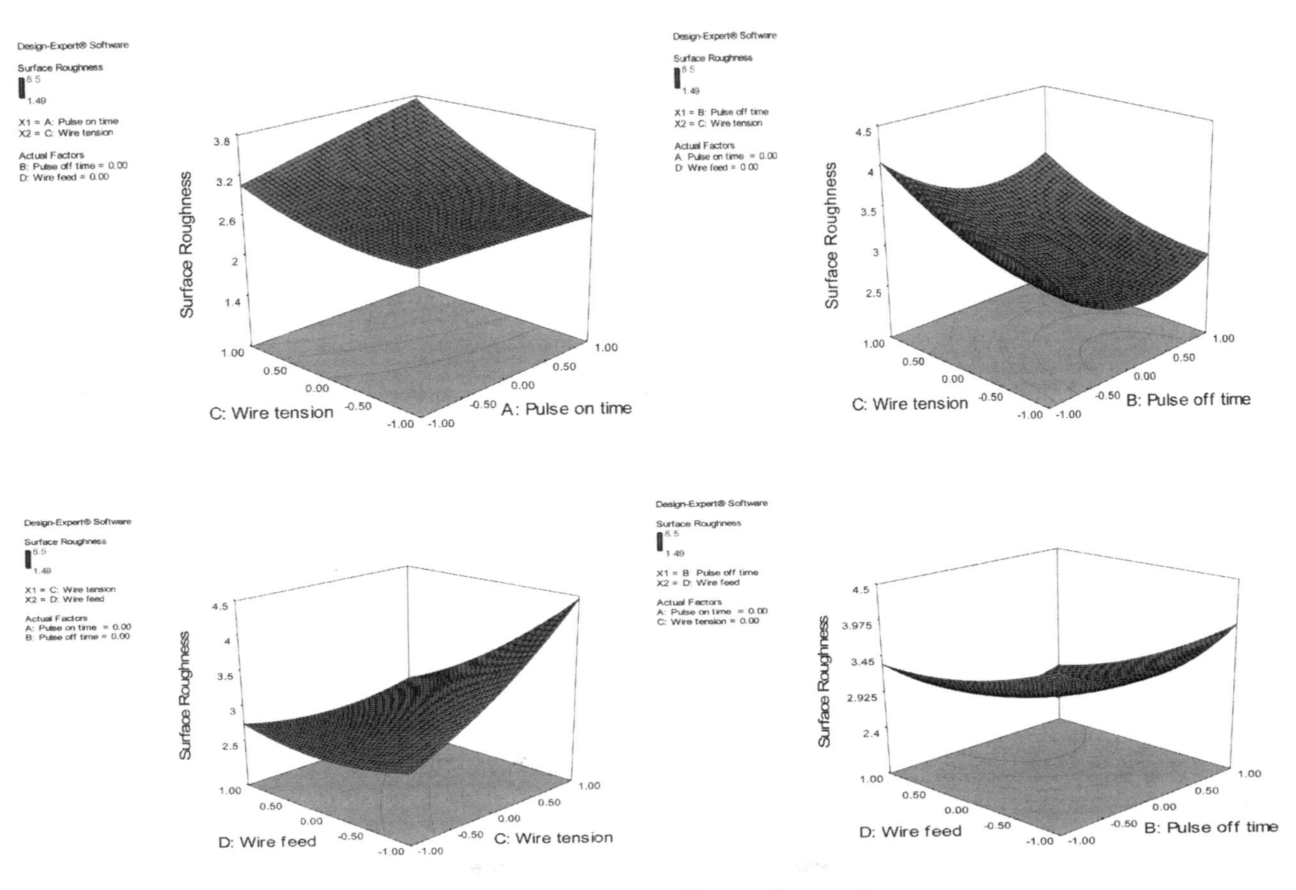

FIGURE 11.6 Effect of (a) A and C, (b) B and C (c), C and D and (d) B and D on surface roughness.

11.4 CONCLUSION

In the present work, the Monel 400 alloy was machined using the WEDM process with different machining conditions, and brass as electrode materials. Based on the experimental results and subsequent analysis, the following conclusions are drawn.

- The effects of the process parameters viz. pulse on time, pulse off time, wire tension and wire feed on surface roughness were studied.
- RSM was applied to develop the mathematical models in the form of equations correlating the dependent parameters with the independent parameters in EDM of Monel 400 alloy. Using the model equations, the response surfaces were plotted to study the effects of the process parameters on the performance characteristics.
- EDM is an adequate process to machine Monel 400 alloy with minimal surface roughness. The predicted values match the experimental values for surface roughness.
- The surface roughness increases as pulse on time increases. Surface roughness decreases as pulse off time increases, whereas surface roughness increases as the wire feed increases.
- The minimum surface roughness of 1.49 μm was obtained for the process parameter combination of pulse on time = 10 μs, pulse off time = 6 μs, wire tension = 1000 g and wire feed = 5 m/min.
- The surface integrity of WEDM in Monel 400 alloy includes surface roughening because of surface micro cracks, debris and melted drops.

REFERENCES

Ali, L.F., Kuppuswamy, N., Narayanan, B., Hari, S. and Vellingiri, S. 2019. Optimization of process parameters in electrical discharge machining of EN 8 steel by using Taguchi technique. *International Journal of Innovative Technology and Exploring Engineering*, 8(7): 2507–2511.

Baig, M.D.K. and Venkaiah, N. 2001. Parametric optimization of WEDM for Hastelloy C276, using GRA method. *International Journal of Engineering Development and Research*, 1: 1–7.

Demuth, H. and Beale, M. 2006. *Neural Network Toolbox User's Guide*, Version 7.0. Natick, MA: Math Works Inc.

El Baradie, M.A. 1981. A surface roughness model for turning gray cast iron (154BHN). *Proceedings of the Institution of Mechanical Engineers, Part B: Journal of Engineering Manufacture*, 207: 43–54.

Hewidy, M.S., El-Taweel, T.A. and El-Safty, M.F. 2005. Modeling the machining parameters of wire electrical discharge machining of Inconel 601 using RSM. *Journal of Materials Processing Technology*, 169: 328–336.

Li, L., Guo, Y.B., Wei, X.T. and Li, W. 2013.Surface integrity characteristics in wire-EDM of Inconel 718 at different discharge energy. *Procedia CIRP*, 6: 221–226.

Montgomery, D.C. 1997. *Design and Analysis of Experiments*, 5th edition. Singapore: John Wiley Publications.

Muthu Kumar, V., Babu, A.S., Venkatasamy, R. and Raajenthiren, M. 2010. Optimization of the WEDM parameters on machining Incoloy 800 super alloy with multiple quality characteristics. *International Journal of Engineering, Science and Technology*, 2(6): 1538–1547.

Ross, P.J. 1988. *Taguchi Techniques for Quality Engineering*. New York: McGraw-Hill Book Company.

Shayan, A.V., Afza, R.A. and Teimouri, R. 2013. Parametric study along with selection of optimal solutions in dry wire cut machining of cemented tungsten carbide (WC-Co). *Journal of Manufacturing Processes*. 35 (4): 644–658.

Shoemaker, L.E. and Smith, G.D. 2006. A century of monel metal: 1906–2006. *Journal of the Minerals Metals and Materials*, 58(9): 22–26.

Sundaram, R.M. and Lambert, B.K. 1981. Mathematical models to predict surface finish in fine turning of steel. Part I. *International Journal of Production Research*, 19: 547–556.

Tosun, N., Cogun, C. and Inan, A. 2003. The effect of cutting parameters on work piece surface roughness in wire EDM. *Machining Science and Technology*, 7(2): 209–219.

Umanath, K. and Devika, D. 2018. Optimization of electric discharge machining parameters on titanium alloy (ti-6al-4v) using Taguchi parametric design and genetic algorithm. *MATEC Web of Conferences*, 172(04007): 1–9.

Yang, R.T., Tzeng, C.J., Yang, Y.K. and Hsieh, M.H. 2011. Optimization of wire electrical discharge machining process parameters for cutting tungsten. *International Journal of Advanced Manufacturing Technology*. 60: 135–147.

12 Experimental Investigation of the Effect of Various Chemical Treatments on *Agave angustifolia* Marginata Fibre

T. Ramakrishnan, R. Suresh Kumar, S. Balasubramani, R. Jeyakumar, Samson Jerold Samuel Chelladurai and R. Ramamoorthi

CONTENTS

12.1 INTRODUCTION

New products have been introduced to replace materials such as metal, cement, etc., that are heavy, corrosive and not environmentally friendly. The usages of artificial materials will certainly affect the environment, and they are dangerous to all living things; therefore, alternative materials are needed to overcome this problem. Natural fibre-reinforced composites are a potential solution.

Fibres deliver strength and stiffness and provide support in fibre-reinforced composite materials. Synthetic fibres, such as glass, carbon and aramid are used as reinforcement in the majority of polymer composite materials; however, ecological concerns identified with manufactured filaments have led to support for the use of natural fibres because of their biodegradable property. The utilization of normal plant filaments as a fortification in polymer composites used to create materials has drawn much attention and discussion lately. In addition to their biodegradability, natural fibres have other benefits over conventional glass fibres, including good specific strength and modulus, economic viability, low density and reduced tool wear (Dhakal et al. 2007).

Furthermore, the uses of natural fibre in fibre-reinforced plastics (FRPs) will significantly reduce the cost, making FRPs even more competitive. Natural fibres are renewable and obtained from natural resources that present several advantages, including low density, acceptable specific strength properties, good sound abatement capability, low abrasiveness, low cost, high biodegradability and the existence of vast resources. Several studies have shown that natural fibre composites combined good mechanical properties with low specific mass to offer an alternative material to glass fibre-reinforced composites (Rao et al. 2010; Thakur et al. 2014; Brahmakumar et al. 2005). Similarly, automobile manufacturers in developed countries have embraced plant-fibre composites for various applications, such as interior trim parts, instrument panels and bumpers (Megiatto et al. 2007; Riedel and Nickel 1999). Normal filaments are commonly incongruent with the hydrophobic polymer lattice and tend to frame totals, these are hydrophilic strands and therefore show helpless protection from dampness. This change of characteristic filaments was endeavouring to make the strands hydrophobic and improve the interfacial attachment between the fibre and the framework polymer. In recent years, natural fiber made composites were evolving area in the polymer matrix composites. Basically, natural fiber was very low cost, easily available nature with low density and good strength to weight ratio. The important thing in the natural fibers was decompose in nature and non-abrasive. The natural fiber reinforced composites had better specific properties when compared with other conventional or synthetic fibers. This review paper studied that the type of fibers used, polymer, chemical treatment on fiber (Saheb and Jog 1999). The utilization of normal strands as modern segments improves the natural maintainability of the parts being built, particularly for car parts (Muller and Krobjilowski 2003). Various researchers have investigated the properties and applications of natural fibres for various applications.

Boobathi et al. (2012) extracted Borassus fruit fibres and experimentally studied the physical and chemical properties of the fibre. In this study, they found that the NaOH surface treatment significantly improved the mechanical properties. A Fourier-transform infrared spectroscopy (FTIR) and scanning electron microscopy (SEM) study was also conducted to identify the surface morphology of the fibre. Mylsamy and Rajendran (2010), conducted experiments on *Agave americana* natural fibre and analyzed the tensile strength, elongation and water absorption ability of the fibre. They proved that the NaOH treatment improved the surface properties and removed the surface impurities. Mohan and Kanny (2012) directed the surface chemical treatment on sisal fibre utilizing both clay and NaOH to remove the surface impurities. The result proved that the 20 wt.% of muds in NaOH improved the reduction of moisture content in the sisal fibre and improved mechanical and moisture absorption properties.

Kabir et al. (2012) stated that chemical treatment was essential to reduce the hydrophilic nature of the fibres and thus improves adhesion with the matrix. Pretreating fibre made significant changes to the fibre structure and surface morphology. Significant improvements were found in the mechanical properties of the composites by using different chemical treatments on the reinforcing fibre. Sgriccia et al. (2008) observed that the NaOH treatment eliminated lignin and hemicellulose contents from the surface of natural fibres. Sghaier et al. (2009) investigated the mechanical behaviour of treated/untreated palm fibre and found that the treated natural fibres had noteworthy advantages over the glass fibre. Brígida et al. (2010) treated the green coconut fibre in three chemical treatment methods NaOCl, NaOCl/NaOH, or H_2O_2. Various testing processes, such as SEM, FTIR, x-ray photoelectron spectroscopy (XPS) and thermogravimetric analysis (TGA), were used, and while the H_2O_2 treated fibre showed that fatty acids and waxy contents were removed, it did not improve the surface of the fibre; the hemicelluloses and cellulose content in the fibre were reduced by the NaOCl/NaOH surface treatment. Edeerozey et al. (2007) chemically modified the surface of kenaf fibres and found that the NaOH treatment improved the mechanical properties. Khanam and Reddy (2007) treated the surface of sisal fibre with NaOH and discovered that the substance treatment vastly improved sisal/silk composites' mechanical properties, such as tensile, flexural and compressive.

A study of the effect of sodium hydroxide base treatment on *Prosopis juliflora* fibres (PJFs) was conducted in an attempt to change their physical-synthetic properties (Saravanakumar et al. 2014a). They found that a soaking the fibres in a 5% (w/v)

NaOH solution for 60 min produced optimal results and cleaned the surface of the fibres of pollutants (2014a). The Prosopis juliflora fibers (PJFs) were subjected to 5% (w/v) of NaOH solution concentration with 60 min soaking time. Synthetic adjustments positively changed the physicochemical properties of PJFs and clearly lessened the waxy substance from the fibre (Saravanakumar et al. 2014b). The particular properties of Areca fruit husk fibre (AFHF) polymer composite are much higher than those of the widely used E-glass fibre composite, which positions AFHF composite as an optional auxiliary material (Binoj et al. 2016). The bunch of natural fiber materials were Focused by FTIR and XRD test. The free substance bunches on it were concentrated on by FTIR and XRD. It had a rigidity of 558 ± 13.4 MPa with a normal strain rate of 1.77 ± 0.04% and a microfibril edge of 10.64° ± 0.45°. High temperature examinations conducted using thermogravimetry (TG) and differential thermal analysis (DTG) and it was demonstrated that the fibres began debasing at 217°C with representation vitality of 76.72 kJ/mol [2121].

In the current study, investigational results on the physical, chemical and mechanical properties of chemically treated *Agave angustifolia* 'Marginata' (AAM) fibres are presented. The effect of NaClO, NaOH and H_2O_2 (2%, 5%, 10% and 15%) on the above properties was also studied in the detail. The statistical analysis was carried out after the tests were conducted on the fibre to reveal the applications of the AAM natural fibre.

12.2 MATERIALS AND EXPERIMENTS

12.2.1 Natural Fibre Material

Agave angustifolia 'Marginata' (AAM) fibres were collected from Coimbatore, Tamilnadu, India. Agave angustifolia 'Marginata' (variegata) is a plant with tall, rigid, sword-shaped leaves with cream-coloured edges. The fibres were taken extracted from the leaves using a decorticator machine.

12.2.2 Extraction of Fibre From Plants

A mechanical process known as decortication was used to extract the fibre from the collected AAM leaves. In this method, the collected fibres were fed into a machine consisting of the three rollers used to crush and beat the leaves. The fibres are collected on one side and the leaves' pulp is collected on the other side. The fibres were dried in the sunlight up in temperatures up to 32°C for 24 hours to remove the moisture content and humidity.

12.3 CHEMICAL TREATMENT OF FIBRE

12.3.1 NaOH Treatment

The raw, dried AAM fibres were immersed in stainless steel vessels containing 2%, 5%, 10% and 15% solutions of NaOH with water for up to 1 hour with subsequent stirring. The fibres were then washed thoroughly with water to remove the excess NaOH, and were rinsed again distilled water. The washed fibres were kept in the sunlight for about 48 hours. The dried fibre was subjected to various physical and chemical properties testing.

12.3.2 NaClO Treatment

The raw, dried AAM fibres were immersed in stainless steel vessels containing 2%, 5%, 10% and 15% solutions of NaClO with water for up to 1 hour with subsequent stirring. To maintain pH11 of the NaClO solution, 0.5 M NaOH was used. The fibres were then washed thoroughly with water to remove the excess NaClO, and were rinsed again with distilled water. The washed fibres were kept in the sunlight for about 48 hours. The dried fibre was subjected to various physical and chemical properties testing.

12.3.3 H_2O_2 Treated

The raw, dried AAM fibres were immersed in stainless steel vessels containing 2%, 5%, 10% and 15% solutions of H_2O_2 with water for up to 1 hour at pH11 with subsequent stirring. To maintain pH11 of the H_2O_2 solution, 0.5 M NaOH was used. The fibres were then washed thoroughly with water to remove the excess H_2O_2, and were rinsed again with distilled water. The washed fibres were kept in the sunlight for about 48 hours. The dried fibre was subjected to various physical and chemical properties testing.

12.4 PHYSICAL PROPERTIES OF FIBRE

12.4.1 Fibre Density

To determine the density of the AAM fibre, the water displacement method was used. The weighed quantity of fibre was immersed in water, and the volumetric displacement was observed. Finally, based on the weight to volume ratio, the density value was determined.

12.4.2 Fibre Diameter

An image analyzer was used to calculate the diameter of the raw, NaOH-treated fibres. To get accurate results, the image analyzer was used at 20 different locations in the fibres and average mean values were taken.

12.4.3 Chemical Properties of Fibre

The effectiveness of the reinforcing method of natural fibre depends on the nature of cellulose and its crystallinity.

12.4.4 Wax Content

The extraction or removal of the naturally adhered waxes on the natural fibre was done by organic solutions. These waxy constituents consist of diverse kinds of alcohol, which are indecipherable in water, and several acids, such as palmitic acid, oleaginous acid and stearic acid. The wax content was measured with the aid of the soxhlet apparatus. Fuel benzene liquid was heated to 70°C, and one gram of fibre was placed in the liquid. The fibre was then desiccated for 1 hour before it was weighed. The mass difference provides the amount of wax content.

12.4.5 Moisture Content

The AAM fibre was positioned in the rotisserie at the temperatures between 105 ± 3°C for up to 4 hrs. After this process was completed, the weight of the fibre was measured. The difference in weight accounts for the moisture content present in the fibre.

12.4.6 Cellulose Content

A weighed quantity AAM fibre was immersed in solutions of 1.72% NaOH, NaClO and H_2O_2 and some drops of sulfuric acid mixed in the water for up to 1 hr. Using the suction process, the excess fluids were removed and ammonia was added. The remaining fibre was washed with distilled water a maximum of five times, and it was allowed dry at room temperature. The percentage of cellulose weight loss was noted, and the difference was identified.

12.4.7 Lignin Content

A weighed amount of AAM fibre was submerged in a mixture of sulfuric acid 12.5 ml and 300 ml water at room temperature for 2 hrs. The solvents were then removed by rinsing the fibre with distilled water. Finally, the residue weight was noted as the lignin content.

12.4.8 Hemicellulose Content

A weighed quantity of dry AAM fibre was dipped in a mixture of 2%, 5%, 10% and 15% of NaOH, NaClO and H_2O_2 solution at room temperature for 30 mins., and then it was deactivated with HCI. The fibre was dried in a hot air oven and weighed. The weight transformation interprets the presence of hemicellulose content.

12.4.9 Fourier-Transform Infrared Spectrometry (FTIR)

The FTIR was executed by a thermoscientific Nicolet iS10 spectrometer at room temperature. The infrared (IR) rays were passed across the sample of AAM fibre. Absorption takes place when the IR frequency is the same as the vibrational frequency. The interferogram was accounted for by the FTIR spectrometer, and it was executed by Fourier transform on this interferogram to complete the spectrum. The preoccupation band was produced and depends on the inspection spectrum, the useful compounds of the fibre were owed. The already existing band and inspection spectrum were used based on the type of fiber involved for inspection. The FTIR band was completed for raw, 5%, 10% and 15% NaOH preserved AAM fibres. The FTIR test was conducted only for NaOH-treated fibre because of favourable results observed in the physical and mechanical behaviour of the AAM natural fibre.

12.4.10 Tensile Strength of Agave Angustifolia 'Marginata' Fibre

The AAM fibre tensile properties were identified by the Instron 5500R-60211 machine with a gauge length of 50 mm and a crosshead speed of 2 mm/min. Thirty samples of raw, 5%, 10% and 15% NaOH, NaClO and H_2O_2-treated fibres were subjected to a tensile test, and the average values were noted.

12.4.11 Thermal Analysis

Thermal stability of the various chemically treated AAM fibres was identified with the help of TGA. Thermal stability was obtained using Perkin-Elmer Pyris 1. The experiment was conducted with a temperature range of 30–650°C with a heat range of 10 °C/min. A range of 5–6 mg of fibre mass was maintained during the test. During the TGA analysis, weight loss and its derivative was the main function of temperature.

12.5 RESULTS AND DISCUSSION

12.5.1 Physical Properties of AAM Fibre

12.5.1.1 Fibre Density

The density values were improved more using the NaOH treatment than the other chemical treatments, such as NaClO and H_2O_2, and the results are shown in Table 12.1. It can be understood that the density values are proportionally reduced based on the percentage of NaOH present in the solution used. Also, the density values are less as compared to other synthetic fibres. This property predicted that the AAM fibres can be used as lightweight composite structure reinforcements. Additionally, the supplementary feature of the AAM fibre is biodegradability. Table 12.1 shows the detailed study of density and tensile strength of the chemically treated AAM fibre.

The density values for the AAM fibre treated with NaClO and H_2O_2 increased somewhat when compared with the NaOH-treated fibre. The tensile strength of the AAM fibre improved in 5% NaOH solution than the others; however, it was observed that all of the 5% chemical treatments improved the tensile strength more than the other weight percentage.

TABLE 12.1
The Density and Tensile Strength Values of Raw and Various Chemically Treated AAM Fibres

S. No	Type of Fibre	Density (g/cc)	Tensile Strength (MPa)
1.	Raw fibre (RF) (untreated fibre)	1.608	512
	Sodium Hypochlorite (NaClO) Treated		
2	2% NaClO	1.721	515
3	5% NaClO	1.786	586
4	10% NaClO	1.742	541
5	15% NaClO	1.772	520
	NaOH Treated		
6	2% NaOH (2T)	1.617	522
7	5% NaOH (5T)	1.639	626
8	10% NaOH (10T)	1.621	587
9	15% NaOH (15T)	1.67	540
	H_2O_2 Treated		
10	2% H_2O_2	1.799	519
11	5% H_2O_2	1.832	602
12	10% H_2O_2	1.802	545
13	15% H_2O_2	1.867	532

12.5.1.2 Fibre Diameter

Generally, natural fibre diameter will not be uniform. There were 100 samples of AAM fibres used for testing, and the fibre diameters were measured in various locations such as middle and end of the fibres and the average values were determined. The diameters of raw, 2%, 5%, 10% and 15% NaOH, NaClO and H_2O_2-treated fibres were measured, and the values are given in Table 12.2. The results clearly show that when the NaOH weight percentage increases, the fibre diameters decrease gradually. This is because the impurities of the fibre surface are being removed by the NaOH treatment.

The diameter did not reduce as much in the NaClO and H_2O_2-treated fibre as it did in the NaOH. Because of the high concentration of the chemical on the fibre surface, a large number of surface impurities were removed. The NaOH surface treatment was considered for bonding or making of composite purposes over the NaClO and H_2O_2 surface chemical treatments. Because of the safest and considerable amount of surface removal of NaOH treatment, it considered for useful applications. The exact concentration of the NaOH used on the fiber surface become the safest one and it will remove the surface impurities in considerable range. Figure 12.1 presents the analysis of the various fibre diameters for the different chemical treatments.

TABLE 12.2
Maximum Diameter, Minimum Diameter and Average Diameter (μm) Values of Raw, NaClO-, NaOH- and H_2O_2-Treated AAM Natural Fibres

S. No	Type of Fibre	Maximum Diameter (μm)	Minimum Diameter (μm)	Average Fibre Diameter (μm)
1.	Raw fibre (RF)	232	153	186
		Sodium Hypochlorite (Nacol) Treated		
2	2% NaClO	181	176	178
3	5% NaClO	163	123	143
4	10% NaClO	146	113	129
5	15% NaClO	136	100	118
		NaOH Treated		
6	2% NaOH	196	189	152
7	5% NaOH	181	112	134
8	10% NaOH	154	106	123
9	15% NaOH	146	101	114
		H_2O_2 Treated		
10	2% H_2O_2	193	173	183
11	5% H_2O_2	179	165	172
12	10% H_2O_2	143	129	136
13	15% H_2O_2	141	99	120

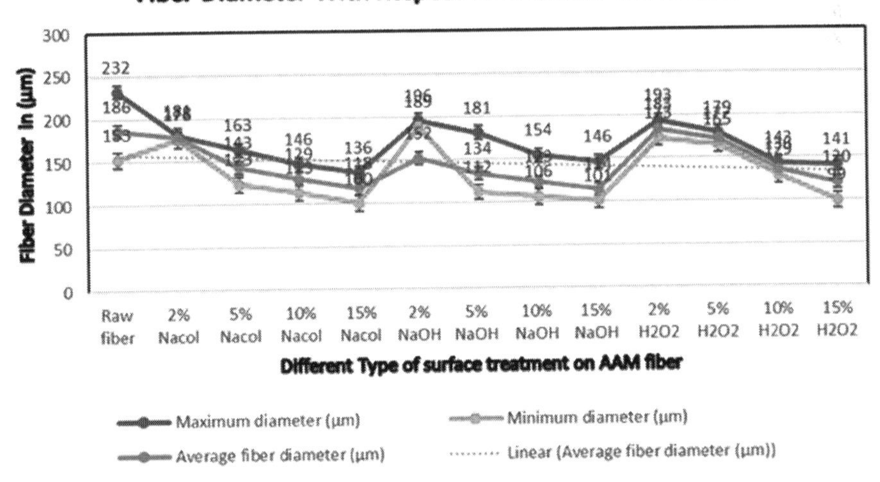

FIGURE 12.1 Fibre diameter concerning the chemical treatments. The maximum, minimum, and average diameters of the different chemical treated AAM fibres brief in the graph.

12.5.1.3 Chemical Properties of Fibre

Table 12.3 shows the chemical properties of the raw and treated fibres. The raw AAM fibre consists of cellulose (72.60%) hemicelluloses (16.03%), lignin (8.09%), wax (0.34%) and moisture (8.77%). The NaOH treatment stimulated the cellulosic fibre to swell. Additionally, the hemicellulose and other impurities were removed from the fibre surface. The microfibrils of cellulose remained unaffected because of the NaOH treatment. The removal of the surface impurities helped to attain greater mechanical properties, fibre wetting features and fibre–matrix surface bonding in composite applications.

The NaClO treatment stimulated the cellulosic fibre to swell. Additionally, the hemicellulose and other impurities were partially removed from the fibre surface. The microfibrils of cellulose remained unaffected by the NaClO treatment. The removal of the surface impurities helped to attain reasonable mechanical properties but not as much as the NaOH treatment. In this treatment, the 5% solution of the NaClO-treated fibre improved the tensile strength more than the other weight percentage of the chemical treatments.

Table 12.3 shows the H_2O_2 treatment also stimulated the cellulosic fibre to swell. Additionally, the hemicellulose and other impurities were partially removed from the fibre surface. The microfibrils of cellulose remained unaffected by the H_2O_2 treatment. The removal of the surface impurities helped to attain reasonable mechanical properties but not as much as the NaOH treatment. In this treatment, the 5% solution of the H_2O_2-treated fibre improved the tensile strength more than the other weight percentage of the chemical treatments.

TABLE 12.3
Chemical Properties of Raw and Treated Fibres

Fibre Category	Wax Content (wt %)	Moisture Content (Wt. %)	Cellulose Content (Wt. %)	Lignin Content (Wt. %)	Hemi cellulose Content (Wt. %)	Ash Content (Wt. %)
Raw Fibre (RF)	0.34	8.77	72.60	8.09	16.03	4.35
2% NaClO	0.22	8.73	73.32	8.01	10.03	4.10
5% NaClO	0.13	8.71	76.86	7.89	8.93	4.01
10% NaClO	0.11	8.69	79.33	6.59	6.32	3.86
15% NaClO	0.09	8.63	74.42	5.583	4.21	3.12
2% NaOH (2T)	0.10	8.64	78.50	7.09	8.03	3.12
5%NaOH (5T)	0.08	8.62	81.95	6.82	4.06	2.58
10% NaOH (10T)	0.06	8.49	80.11	5.50	Nil	2.48
15% NaOH (15T)	0.04	9.16	76.06	4.63	Nil	2.36
2% H_2O_2	0.20	8.31	72.31	7.99	9.01	4.01
5% H_2O_2	0.11	8.20	74.80	7.64	8.23	3.91
10% H_2O_2	0.10	8.36	78.32	6.42	5.95	3.36
15% H_2O_2	0.08	8.53	73.41	5.52	4.11	2.11

12.5.1.4 Fourier-Transform Infrared Spectrometry (FTIR)

The results attained from the tensile strength and physical properties tests showed that the NaOH-treated fibre improved the properties more than the chemical treatments. The FTIR test was conducted only for NaOH-treated fibre because of the results attained previously. FTIR has been mostly successful in accurately analyzing both major (cellulose, hemicellulose and lignin) and minor (mineral, pectin and waxes) constituents of natural fibres. FTIR could also effectively identify changes in chemical compositions, interface and properties of natural fibres and composites. The difference between the waves was identified using the superimposing techniques with the help of an interferometer. Figures 12.2–12.7 show the FTIR spectra images of untreated and all the NaOH-treated (2%, 5%, 10% and 15%) fibres.

The raw fiber and 2 wt.%,5 wt.%,10 wt.% and 15 wt.% agave angustifolia marginata fiber were undergone for the FTIR and chemical composition test. Figure 12.2 shows the raw fibre FTIR absorbance vs. wavenumbers (cm^{-1}). The peak at 1026.96 cm^{-1} was comprised of Si-O cellulose. This peak signals the existence of sulfur and lignin (Brigida et al. 2010). The band about 1464.84 cm^{-1} and 1490.41 cm^{-1} directed the occurrence of the C=O stretch of the acetyl group of hemicellulose (Edeerozey et al. 2007). The band at about 2359.83 cm^{-1} was O-H extending, yet the expansion of the oxygen particles was insufficient to frame carbonate molecules (Noorunnisa Khanam et al. 2007). The peak at 2920.98 cm^{-1} signals the presence of C-H methods of methyl and methylene groups (Saravanakumar et al. 2014). The band 3331.76 and 3697.46 cm^{-1} indicate the presence of N-H and surface impurities presence on the fibre. The wide band up to 3911.06 cm^{-1} was indication of O-H and presence of the hydrogen bonding in the AAM fiber surface.

Figure 12.3 indicates the FTIR spectrum of 2% NaOH-treated fibres. The peak at 1027.29 cm-1 was comprised Si-O cellulose, and it shows the presence of the sulfur and lignin. The slight changes shown in the 2% NaOH-treated filaments contrasted

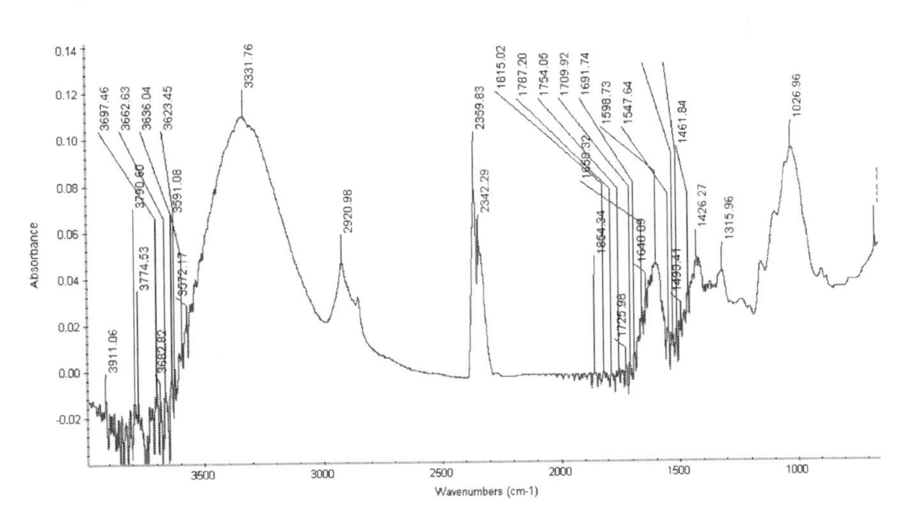

FIGURE 12.2 Raw(untreated) AAM fibre or untreated fibre FTIR results show that the presence of the surface impurities on the fibre surface. Before the start of the chemical treatment, surface impurities were present.

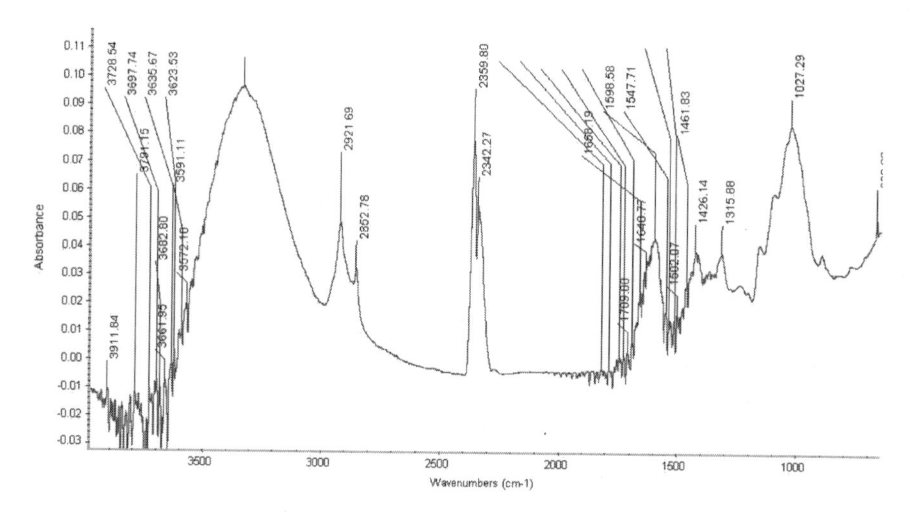

FIGURE 12.3 shows that the 2% NaOH treated AAM Fibre FTIR result and surface impurities slightly start to remove on the natural AAM fibre surface.

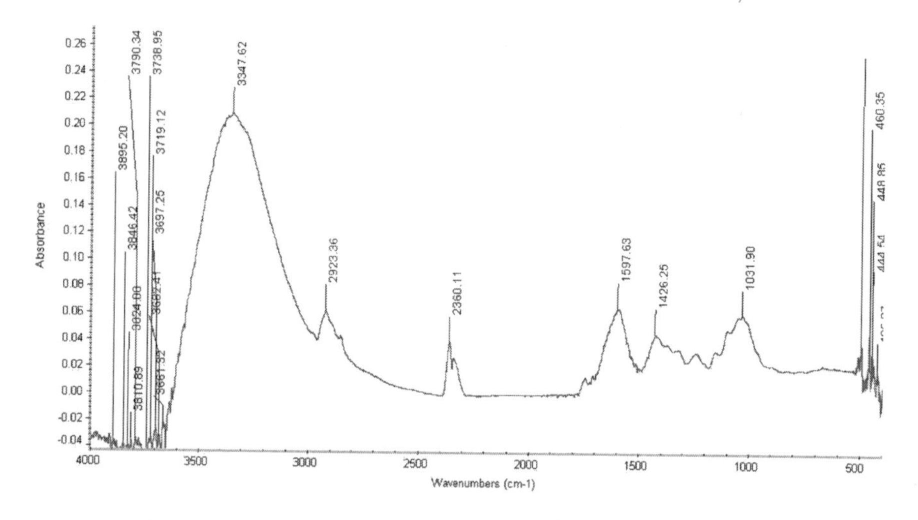

FIGURE 12.4 shows that the 5% NaOH treated AAM Fibre FTIR result and surface impurities such as lignin, hemicellulose and wax content removed when compared with the other NaOH treated AAM fibre.

with the untreated fibre. The band 1426.14 cm^{-1} and 1502.07 cm^{-1} indicates the presence of the C=O stretch and it confirmed that the hemicellulose not removed for this 2wt.% of chemical treatment.

Figure 12.4 demonstrates the FTIR range for 5% NaOH-treated filaments. The Si-O bowing was allocated to the peaks between 444.54 cm^{-1} and 460.35 cm^{-1}. The result exhibited that the peak 1026.96 cm^{-1} not observed in this 5wt.% of NaOH treatment. This is because of the way that NaOH responded to the C-O bonds present in the fragrant rings of lignin. The treated fibre was not expected to contain sulfur. Lignin content was decreased because of the alkali treatment [4]. It was observed that the peak at 1490.41 cm^{-1} was decreased here when compared to the raw fibres.

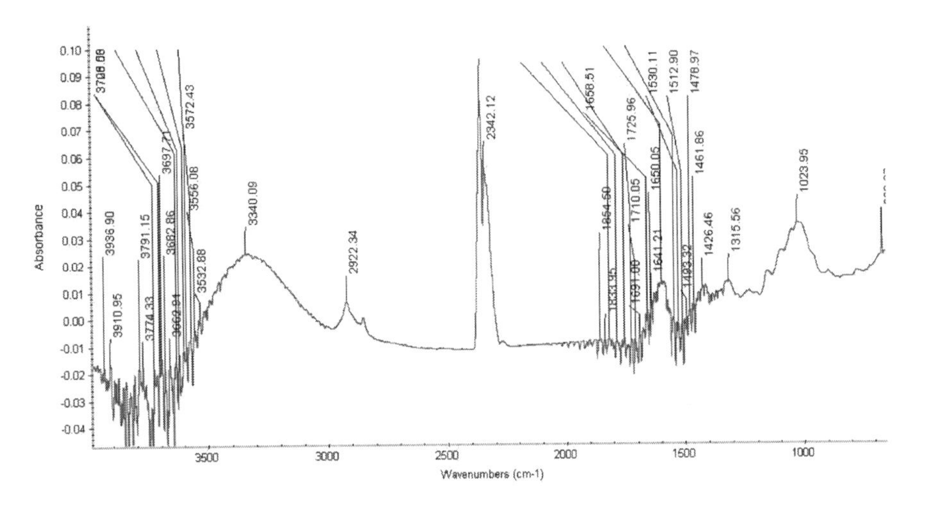

FIGURE 12.5 shows that the 10% NaOH treated AAM Fibre FTIR result. Due to huge NaOH concentration on the fibre surface, cellulose content starts to damage, and weight loss of the fibre was increased.

This showed the expulsion of hemicellulose from the fibre surface of NaOH-treated fibres. The expulsion of hemicellulose in the alkali base treatment is recorded in Table 12.3. The band at about 2360.11 cm^{-1} was higher than that of the untreated fibres. It was the sign of carboxylic acids with O-H extending with the expansion of oxygen molecules. The peaks after 3661.32 cm^{-1} seemed higher than those of the raw strands because of reinforcing hydrogen bonds and O-H extending.

Figure 12.5 shows the FTIR range of 10% NaOH-treated fibres. The expulsion of sulfur and the diminishment of lignin was proved by the absence of peaks around 1023.95 cm^{-1}. The most extreme expulsion of hemicellulose was demonstrated by the lower band around the peak at 1691 cm^{-1}. The band around 2342.12 cm^{-1} was radically raised and showed the expansion of oxygen molecules and O-H extending further. The distinctive band between 3522.88 cm^{-1} and 3936.90 cm^{-1} demonstrated further fortification hydrogen bonds because of the expansion of oxygen particles and O-H stretching.

Figure 12.6 shows the 15% NaOH-treated FTIR spectra. No highest was seen around 1026.96 cm^{-1} and meant that the most extreme expulsion of sulfur and lignin. The absence of hemicellulose was observed because of the lower band around the peak at 1547.67 cm^{-1}. The higher peak at 2922.22 cm^{-1} occurred because of the further increase of oxygen molecules and O-H extending. During the reinforcement of the fiber hydrogen bonds were important, but O-H outspreading bond were exceeded the limit between 3623.85 cm^{-1} and 3895.43 cm^{-1}.

12.6 TENSILE PROPERTIES OF AAM FIBRE

The tensile properties of AAM fibre are listed in Table 12.1. The tensile strength was higher in the 5% NaOH-treated fibre than those treated in the 2%, 10% and 15% NaOH and all the NaClO and H_2O_2-treated fibres. The tensile strength of the

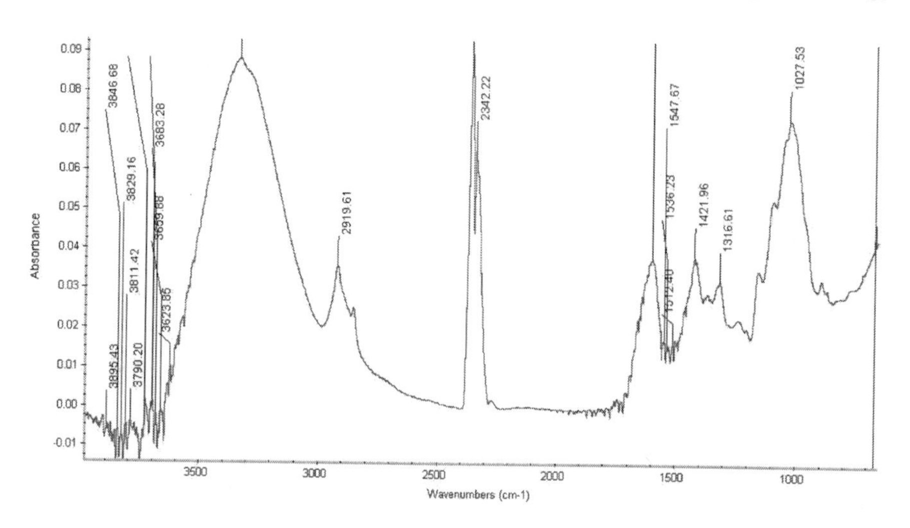

FIGURE 12.6 shows that the 15% NaOH treated AAM Fibre FTIR result. Due to high NaOH concentration on the fibre surface, cellulose content on the fibre surface was damaged and weight loss of the fibre was increased drastically.

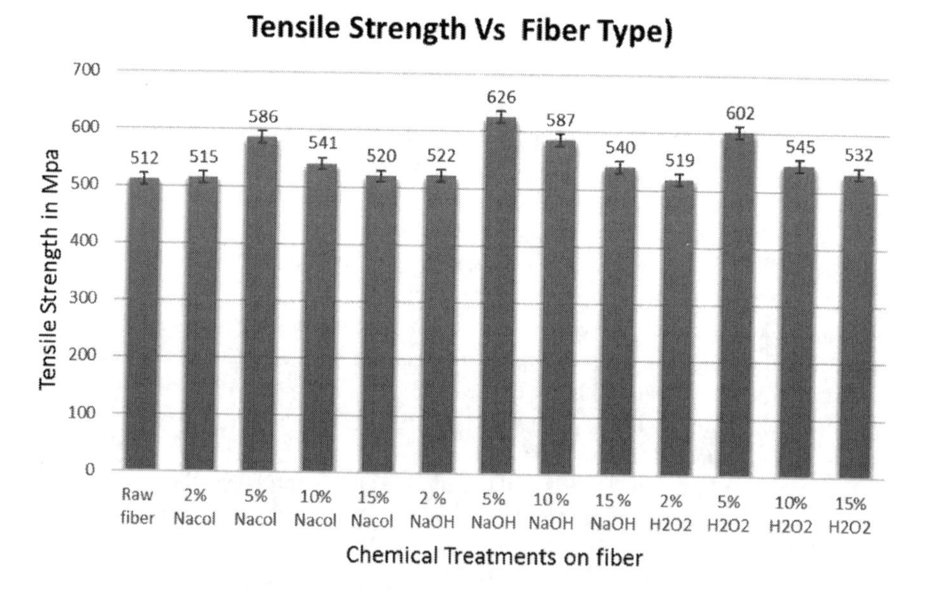

FIGURE 12.7 Illustrates Fibre Type Vs Tensile Strength. The different types of chemical treatments conducted on the fibre surface to identify better mechanical properties. Out of three different chemical treatments, NaOH treated AAM fiber showed better mechanical properties.

5% alkali-treated fibre was 626 MPa. It was higher than the raw AAM fibre and other percentages of NaOH-treated fibres. The tensile test was also conducted on the untreated fibre and NaOH-treated AAM fibres, and the values are shown in Figure 12.7. It is evident from the results that the 5% NaOH-treated fibre has the highest tensile strength.

12.7 THERMAL ANALYSIS

The summarized TGA data of raw and chemically treated AAM fibre is shown in Table 12.4. The raw and treated AAM fibre went through four different stages of the thermograms and degradation. TG curves were obtained from the various degradation steps. All the testing samples' weight loss started at the peak temperature of 50°C when the moisture loss initiated. The NaOH-treated fibre somewhat increased the TGA properties when compared to the NaClO surface treatment on the AAM fibres. The H_2O_2 surface treatment on the AAM natural fibre also increased the thermal stability when compared with the other two surface chemical treatments. It was observed that the weight loss increased in the TGA tested AAM fibre affected the cellulose contents present in the fibre. Based on the experimental study on thermal stability, it proved that the 5% NaOH-treated AAM fibre improved the thermal stability more than the other chemical treatments.

12.8 CONCLUSIONS

The physical, chemical and mechanical properties of AAM fiber were examined and the results included:

- The fibre diameter and weight ratios were analyzed in the experiments. The diameters of fibre and weight of the fibres were reduced when the NaOH treatment exceeded more than 5%.
- Compared to the synthetic fibre, the AAM fibre was less dense, but density was improved in the 5% NaOH treatment over the raw fibre. This fibre could be used to make lightweight composites.

TABLE 12.4
Thermogravimetric Results of Raw or Untreated and Various Chemically Treated AAM Fibres

Fibre	Raw Fibre (RF)	NaClO	NaOH	H_2O_2
Transition Temperature Range (°C)	25–95 210–330 330–450	25–90 190–310 310–450	35–100 235–540	30–90 210–330 330–510
Transition Peak (°C)	45 315 360	40 280 360	55 360	40 310 410
Onset (°C)	- 258.8 300	- 245.8 300.3	- 298.4	- 264.1 310.4
Weight Loss (%)	4.29 27.72 46.32	3.16 22.55 48.73	5.39 66.42	3.64 26.85 44.37

- The NaOH-treated fibres were detached from surface layers, such as lignin, hemicellulose and other impurities, which made the surface of the fibre smoother and better able to bond between fibre surface matrices. FTIR spectrum showed a strong hydrogen bond in treated fibres, showing that surface impurities on the fibre had been removed.
- The thermal stability of the AAM fibre improved in the 5% NaoH treatment more than all other surface treatments such as NaClO and H_2O_2.

REFERENCES

Binoj, J.S., Raj, R.E., Daniel, B.S. and Saravanakumar, S.S. 2016. Optimization of short Indian Areca fruit husk fiber (*Areca catechu* L.)-reinforced polymer composites for maximizing mechanical properties. *International Journal of Polymer Analysis and Characterization*, 21(2): 112–122.

Boopathi, L., Sampath, P.S. and Mylsamy, K. 2012. Investigation of physical, chemical and mechanical properties of raw and NaOH treated Borassus fruit fiber. *Composites Part B: Engineering*, 43(8): 3044–3052.

Brahmakumar, M., Pavithran, C. and Pillai, R.M. 2005. Coconut fiber-reinforced polyethylene composites: Effect of natural waxy surface layer of the fiber on fiber/matrix interfacial bonding and strength of composites. *Composites Science and Technology*, 65: 563–569.

Brígida, A.I.S., Calado, V.M., Gonçalves, L.R. and Coelho, M.A. 2010. Effect of chemical treatments on properties of green coconut fiber. *Carbohydrate Polymers*, 79(4): 832–838.

Dhakal, H.N., Zhang, Z.Y. and Richardson, M.O.W. 2007. Effect of water absorption on the mechanical properties of hemp fiber reinforced unsaturated polyester composites. *Composites Science and Technology*, 67: 1674–1683.

Edeerozey, A.M.M., Akil, H.M., Azhar, A.B. and Ariffin, M.I.Z. 2007. Chemical modification of kenaf fibers. *Materials Letters*, 61: 2023–2025.

Kabir, M.M., Wang, H., Lau, K.T. and Cardona, F. 2012. Chemical treatments on plant-based natural fiber reinforced polymer composites: An overview. *Composites Part B: Engineering*, 43(7): 2883–2892.

Mohan, T.P. and Kanny, K. 2012. Chemical treatment of sisal fiber using NaOH and clay method. *Composites Part A: Applied Science and Manufacturing*, 43(11): 1989–1998.

Megiatto, J.D. Jr, Oliveira, F.B., Rosa, D.S., Gardrat, C., Castellan, A. and Frollini, E. 2007. Renewable resources as reinforcement of polymeric matrices: Composites based on phenolic thermosets and chemically modified sisal fibers. *Macromolecular Bioscience*, 7: 1121–1131.

Mueller, D.H. and Krobjilowski, A. 2003. New discovery in the properties of composites reinforced with natural fibers. *Journal of Industrial Textiles*, 33(2): 111–130.

Mylsamy, K. and Rajendran, I. 2010. Investigation on physio-chemical and mechanical properties of raw and NaOH-treated Agave americana fiber. *Journal of Reinforced Plastics and composites*, 29(19): 2925–2935.

Noorunnisa Khanam, P., Reddy, M.M., Raghu, K., John, K. and Venkata Naidu, S. 2007. Tensile, flexural and compressive properties of sisal/silk hybrid composites. *Journal of Reinforced Plastics and Composites*, 26: 1065.

Rao, K.M., Rao, K.M. and Prasad, A.R. 2010. Fabrication and testing of natural fibre composites: Vakka, sisal, bamboo and banana. *Materials and Design*, 31: 508–513.

Riedel, U. and Nickel, J. 1999. Natural fibre-reinforced biopolymers as construction materials – New discoveries. *Angewandte makromolekulare Chemie*, 272: 34–40.

Saheb, D.N. and Jog, J.P. 1999. Natural fiber polymer composites: A Review. *Advances in Polymer Technology*, 18(4): 351–363.

Saravanakumar, S.S., Kumaravel, A., Nagarajan, T. and Moorthy, I.G. 2014a. Investigation of physico-chemical properties of alkali-treated *Prosopis juliflora* fibers. *International Journal of Polymer Analysis and Characterization*, 19(4): 309–317.

Saravanakumar, S.S., Kumaravel, A., Nagarajan, T. and Moorthy, I.G. 2014b. Effect of chemical treatments on physicochemical properties of *Prosopis juliflora* fibers. *International Journal of Polymer Analysis and Characterization*, 19(5): 383–390.

Saravanakumar, S.S., Kumaravel, A., Nagarajan, T., Sudhakar, P. and Baskaran, R. 2013. Characterization of a novel natural cellulosic fiber from *Prosopis juliflora* bark. *Carbohydrate polymers*, 92(2): 1928–1933.

Sghaier, S., Zbidi, F. and Zidi, M. 2009. Characterization of Doum palm fibers after chemical treatment. *Textile Research Journal*, 79(12): 1108–1114.

Sgriccia, N., Hawley, M.C. and Misra, M. 2008. Characterization of natural fiber surfaces and natural fiber composites, *Composites Part A: Applied Science and Manufacturing*, 39(10): 1632–1637.

Thakur, V.K., Thakur, M.K. and Gupta, R.K. 2014. Review: Raw natural fiber-based polymer composites. *International Journal of Polymer Analysis and Characterization*, 19(3): 256–271.

13 Defect Identification in Casting Surface Using Image Processing Techniques

S. Balasubramani, N. Balaji, T. Ramakrishnan, R. Sureshkumar and Samson Jerold Samuel Chelladurai

CONTENTS

13.1 INTRODUCTION TO SURFACE INSPECTION

Metals such as aluminium, gold, copper, silver and iron naturally have a strong reflection capacity. Today, these metals are widely used in different industries as raw material for producing various finished products. These industries include power electronics, automobile industries, shipping, domestic electronics, communications and construction industries. The surface quality of these materials inevitably affects the quality and durability of the finished products. On the other hand, defect analysis of these reflective materials is strongly recommended in large-scale industries. Though the traditional methods such as visual detection and flashlight detection are applied in defect analysis. The traditional method gives less accuracy due to low random inspection, poor inspection atmosphere. The above method does not meet the industrial requirements. As a solution, the development of a newly enhanced vision-based method to detect defects in these materials is encouraged.

Dhenge et al. (2013) studied image recognition of bolts and nuts. The objective of this study was to develop the image processing algorithm using principal component analysis to get the normalized resize images that would be suitable inputs for processing and detection.

Holba et al. (2015) discussed the image processing in the analysis of product quality. The goal is to design and implement a camera system that will work on the principles of industrial microscopy to measure and compare the porosity of metal casting cuts. In conclusion, the authors found that clustering and threshold imaging methods appear to be optimal.

The enhanced quality dimensions are described below:

a). Process control is improved immensely.
b). The efficiency of the algorithm to detect defected region is enhanced.
c). It reflects in smooth relationships with the customer.

Most of the presently applied inspection systems perform defect detection; however, defect classification remains a research subject.

13.2 CHALLENGES AND ISSUES IN DEFECT DETECTION

There are many defect identification systems that can perform surface defect detection in strong reflective metals, but these systems are commercially available and are expensive (Song and Yan 2013). Though problem defect identification is addressed by various systems, the technique of defect classification, i.e. defect region identification, prediction of affected feature, and the optimal classification, still needs to be addressed.

The strong reflective metal has a high coefficient and which makes the designing of lighting system for defect enrichment. As the metal surface moves under the defect inspection system, it results in a higher volume of data, i.e. approximately metal is 1–2 m wide and they move in an average speed of 2–3 m/s. Accordingly, the huge volume of data for cen-percent inspection is tedious. There are variety of defects in metal surface. Though it is classified as a single class, the defect can be entirely different in its size, texture or depth, which would result in a faulty finished product. Here, even efficient classification procedures experience difficulty in identifying the defects in strongly reflective metals because most defects in metals are tiny in nature, and the holes, stains, scratches, pilling, indentations, burrs and other ill-defined defects are deeply rooted in the metal. As these defects are refractive in light, it makes the automatic defect detection system faulty. In a recent scenario, advancements in image processing, computer vision techniques, artificial intelligence (AI) and similar fields have considerably enhanced the ability of visual inspection techniques (Yamana et al. 2005). To date, several types of inspection systems have been proposed, nearly all of which include image processing and pattern recognition techniques to an extent.

13.3 PROPOSED METHOD

Noise is a part of the image acquisition process. The noise would extremely affect the captured image, resulting in blurred vision and difficulty in edge detection. Hence, it is recommended to denoise or weaken the noise present in the image. Noise can be removed by applying appropriate image filtering techniques. In the present work, authors have implemented the Wiener filter to denoise the image. Application of Wiener filters resulted in analyzing the known signal to detect the unknown signals by relating the known signals as input. The Weiner filter uses a statistical approach to filter the trained dataset.

Phase I also includes the process of image resize, cropping, contrast enhancement and colour conversion to obtain the desired result. Further, the captured image is resized to standard resolution in order to reduce the amount of computation capacity of the model, which, in turn, increases the computation speed. The image is converted into greyscale to identify the defect region (Song and Yan 2013). Phase II involves image segmentation by applying k-means clustering algorithm to identify the defected region. The segmented region is further passed as input to Phase III, where the affected features are extracted from the defected region of the image. Finally, in Phase IV the defected region is classified from the input image (Figure 13.1 to Figure 13.6).

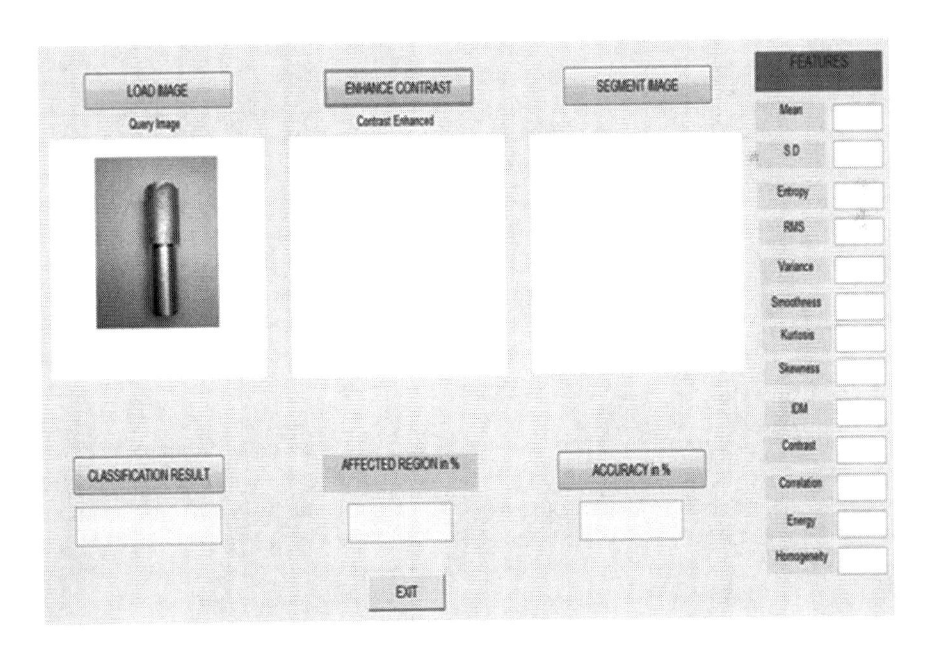

FIGURE 13.1 Original image.

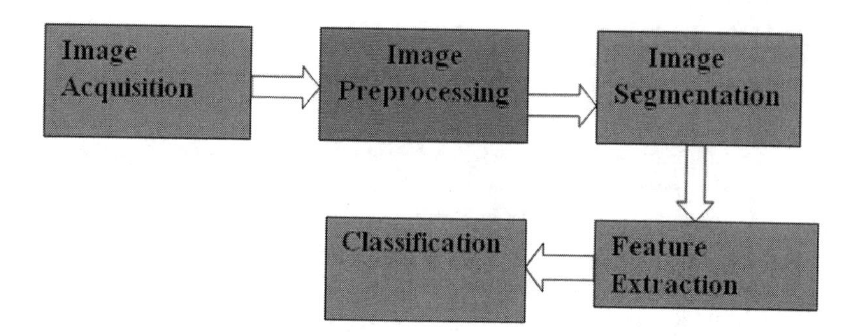

FIGURE 13.2 Flow of defect identification in metal surface.

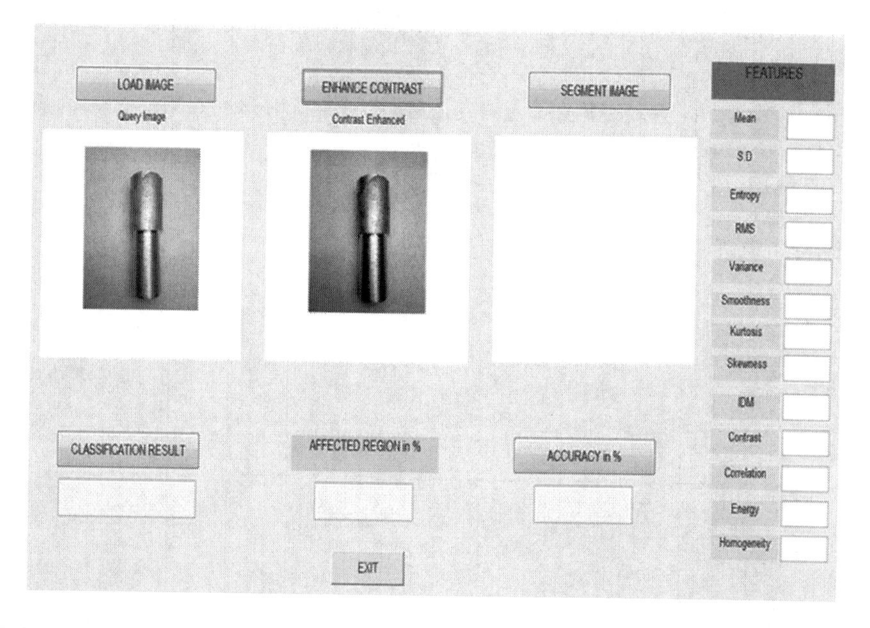

FIGURE 13.3 Contrast enhancement.

13.4 CONTRAST ENHANCEMENT

From Phase I the denoised image is further included for contrast enhancement by converting the image into greyscale. The greyscale images support the application of defect identification techniques as the intensity differs consistently in the image. Tonal enhancements results are showed as highlight, midtone, and shadow. The highlight, midtone, and shadow are mentioned as bright, grey, and dark regions in the image. (Figure 13.3).

13.5 IMAGE SEGMENTATION

Image segmentation is attempted to extract a part of an image or a segment to make it easier to analyze. It also ensures the task of restoring the more meaningful and

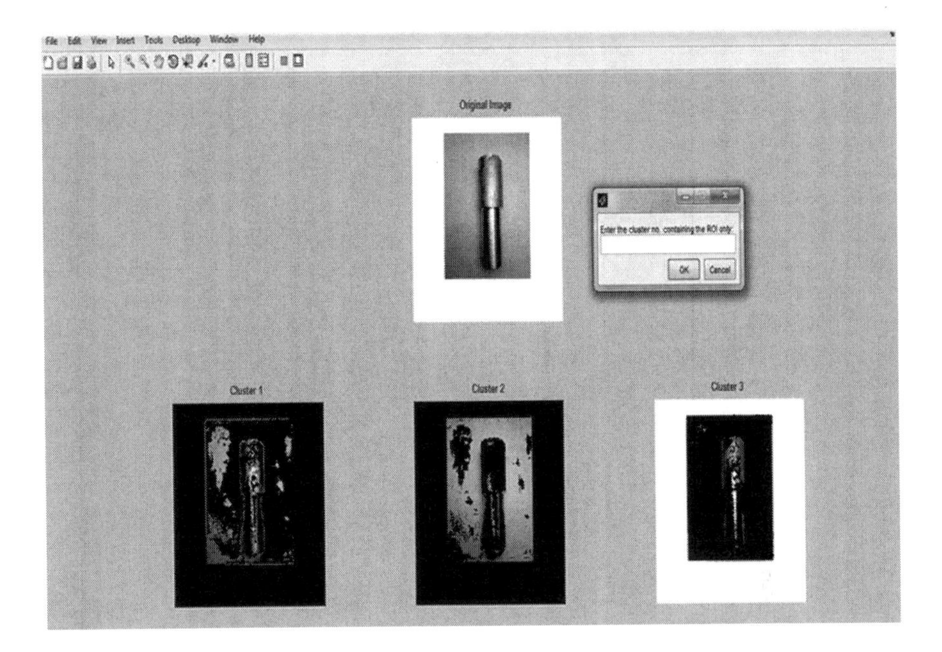

FIGURE 13.4 Segmentation of defected region.

related data. The process of segmentation has to be carried out to a point where the required part of the image is isolated from the core image. The current work concentrates on segmentation to identify the region to be classified as a defected area. There are various techniques for image segmentation, such as clustering methods, compression-based methods, histogram-based methods and region-growing methods. K-means clustering is the technique used in the present work to proceed with the segmentation of images. K-means clustering is a technique of analyzing the clusters with the aim of partitioning *n* observations into *k* mutually exclusive clusters that belong to the clusters' nearest mean (Figure 13.4).

13.6 FEATURE EXTRACTION

Feature extraction technique is used to feature the denoised image. Feature extraction is carried out by withholding the maximum possible information from the given large dateset. Though feature extraction is a required process in image processing, its efficiency and effectiveness of feature selection and extraction present challenges. In this study, features such as colour, texture and shape are extracted for segmentation process (Table 13.1).

13.7 CLASSIFICATION OF DEFECTS

In the final phase, the segmented image is passed to classify the defective region in the processed image. Support vector machine (SVM) tends to classify the classes into binary, i.e. (L = 2), where the problems, in reality, may require multiple

TABLE 13.1
Extraction Features of Cracks

S.No	Features	Value
1	Mean	153.828
2	S.D.	123.255
3	Entropy	1.49592
4	RMS	12.1146
5	Variance	7543.67
6	Smoothness	1
7	Kurtosis	1.19436
8	Skewness	−0.420249
9	IDM	255
10	Contrast	0.540533
11	Correlation	0.97655
12	Energy	0.481375
13	Homogeneity	0.980098

categories. Here, multiclass SVM technique is used to classify the defective region in various categories. Multiclass SVM is widely applied in various fields because of its efficiency of classifying the problems into ($L > 2$) category (Figure 13.5 and Figure 13.6).

13.8 CONCLUSION

Multiclass SVM algorithm is implemented to classify the multiple categories of defects in the reflective metal surface. The image is passed into different phases after denoising the original image. The image is resized into a fixed resolution, and the data is converted into greyscale to enhance the contrast in the image. From there, segmentation of the defective region is identified by applying k-means cluster algorithm. Therefore, the desired result is obtained with greater efficiency.

13.9 APPLICATIONS

- It can be effectively implemented in the manufacturing industry for detection of cylindrical casted products.
- It is also applicable for other products with alterations of algorithm to identify the defect detections.
- This method of image processing can be implemented in the textile industry for defect detection.

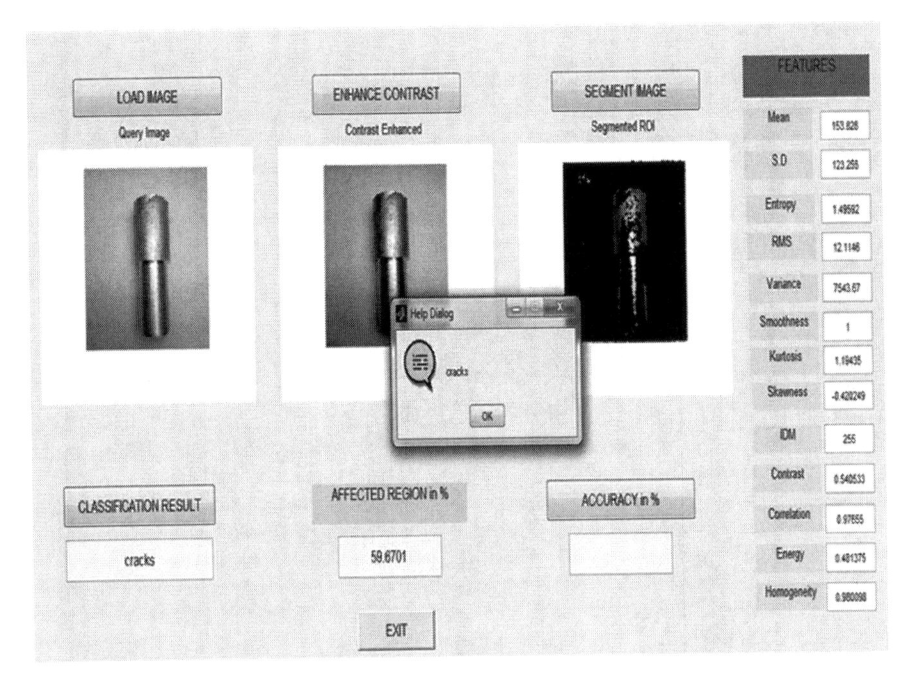

FIGURE 13.5 Classification using multiclass support vector machine.

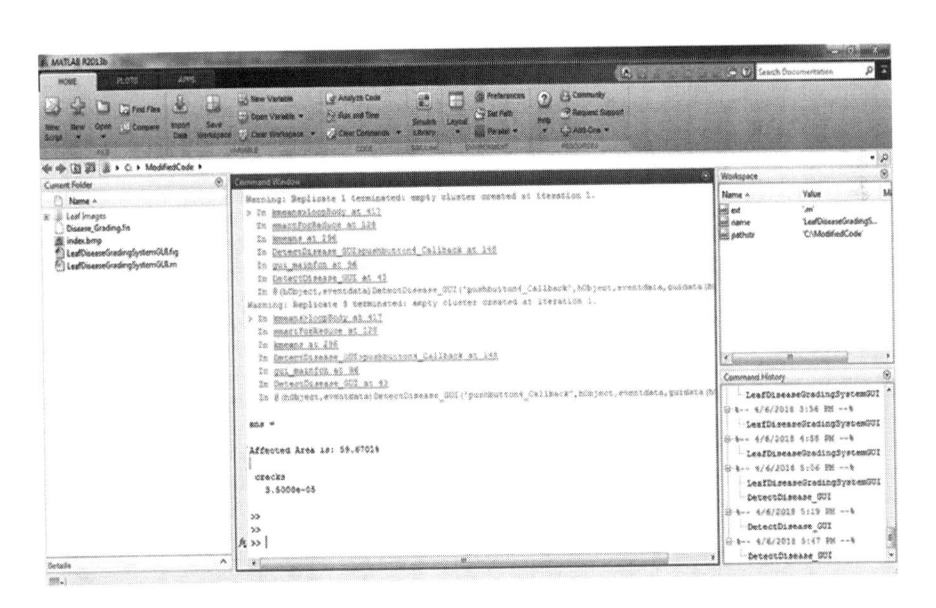

FIGURE 13.6 Area of cracks.

BIBLIOGRAPHY

Bento, M.P., de Medeiros, F.N., de Paula Jr., I.C. and Ramalho, G.L. 2009. Image processing techniques applied for corrosion damage analysis. In *Proceedings of the XXII Brazilian Symposium on Computer Graphics and Image Processing*, Rio de Janeiro, Brazil.

Dhenge, A., Khobragade, A.S. and Salodkar, A. 2013.Mechanical nut-bolt sorting using principal component analysis and artificial neural network. *International Journal of Applied Information Systems (IJAIS),* 2nd National Conference on Innovative Paradigms in Engineering and Technology, 1: 30–32.

Dupont, F., Odet, C. and Cartont, M. 1997. Optimization of the recognition of defects in flat steel products with the cost matrices theory. *NDT and E International*, 30: 3–10.

Ghanta, S., Karp, T. and Lee, S. 2011. Wavelet domain detection of rust in steel bridge images. In 2011*IEEE International Conference on Acoustics, Speech and Signal Processing (ICASSP)*, Prague, 1033–1036.

Ghorai, S., Mukherjee, A., Gangadaran, M. and Dutta, P.K. 2012. Automatic defect detection on hot-rolled flat steel products. *IEEE Transactions on Instrumentation and Measurement*, 62: 612–621.

Holba, M., Bilik, P. and Kelnar, M. 2015. Image processing in defectoscopy. *IFAC-Papers on Line*, 48: 65–70.

Shi, C., Wang, Y. and Zhang, H. 2011. Faults diagnosis based on support vector machines and particle swarm optimization. *International Journal of Advancements in Computing Technology*, 3(5): 70–79.

Song, K. and Yan, Y. 2013. A noise robust method based on completed local binary patterns for hot-rolled steel strip surface defects. *Applied Surface Science*, 285: 858–864.

Vapnik, V. 1995. Support-vector networks. *Machine Learning*, 20: 273–297.

Woźniak, M., Graña, M. and Corchado, E. 2014. A survey of multiple classifier systems as hybrid systems. *Information Fusion*, 16: 3–17.

Yamana, M., Murata, H., Onoda, T. and Ohashi, T. 2005. Development of system for crossarm reuse judgment on the basis of classification of rust images using support vector machine. In *17th IEEE International Conference on Tools with Artificial Intelligence (ICTAI'05*). Hong Kong, 5–406.

Zaidan, B.B., Zaidan, A.A., Alanazi, H.O. and Alnaqeib, R. 2010. Towards corrosion detection system. *International Journal of Computer Science Issues IJCSI*, 7: 46.

14 Effect of Y_2O_3 Doping on Mechanical Properties and Microstructure of Al_2O_3/10wt.% SiC Ceramic Composites Synthesized by Microwave Sintering

M. Madhan and G. Prabhakaran

CONTENTS

14.1 INTRODUCTION

Ceramics, such as alumina (Al_2O_3), are widely used in high-temperature engineering applications because of its exquisite mechanical properties and excellent heat, wear and chemical resistance. These enticing properties are due to the strong covalent and ionic bonds existing between its atoms (Ono et al., 2007). But the same strong covalent and ionic bonds are the cause of its intrinsic brittleness (Huang et al., 2015). Because of its brittleness, the machining performance cost is higher and low fracture toughness is lower when compared to metals. In order to minimize the brittleness for engineering applications, one way is to reinforce the alumina matrix with ceramic inclusions in the form of whiskers or particles (Chou et al., 1998; Wei et al., 1985). The mechanical properties depend mostly on the grain size and grain-size distribution; therefore, the control of grain size during the synthesis of ceramic composites plays a major role. In order to control the grain size, sintering additives are used, which influence the grain size by forming a second phase with the matrix material and creates grain boundary pinning. The selection of sintering additive material and its proportion plays a major role, which decides the final properties of the ceramic composites (Domanicka et al., 2014; Shi et al., 2018).

Structural ceramics are synthesized by sintering the compacted green samples using pressureless conventional electrical sintering, isostatic or hot isostatic pressing sintering, spark plasma sintering and microwave sintering. The properties of structural ceramics vary based on the sintering method. When compared to pressureless sintering, isostatic and hot isostatic sintering shows enhanced properties, but the manufacturing cost is higher and is not suitable for mass production. Structural ceramics synthesized by spark plasma sintering show enhanced mechanical properties, but suitable for simple symmetrical shapes and required expensive pulsed direct current (DC) generator (Suarez et al., 2013). In the case of microwave sintering, high dense structural ceramics are synthesized because of volumetric heating with minimum holding time (Madhan and Prabhakaran, 2019). In this chapter, the mechanism and methods of microwave sintering and mechanical properties and microstructure of microwave sintered Al_2O_3/10 wt.% SiC structural ceramics composite doped with various proportions of Y_2O_3 was investigated. The Y_2O_3 additive used in the Al_2O_3/10 wt.% SiC ceramic composite can have the primary function to form a second phase with Al_2O_3, which can be responsible for densification and microstructure control.

14.2 MICROWAVE SINTERING

14.2.1 Basics of Microwave Heating

Microwaves are electromagnetic waves that lie above radio waves and below the infrared frequency regions in the electromagnetic spectrum. Although most of the microwave's frequencies are used for the purpose of communications and radar, the frequencies selected for industrial, medical and scientific uses are 915 MHz, 2.45 GHz, 5.8 GHz and 20.2–21.1 GHz, respectively. Home appliance microwave ovens run at a frequency of 2.45 GHz because the water molecules existing in food absorbs microwaves at this frequency. The respective availability of 915 MHz and 2.45 GHz microwave ovens give rise to their application for ceramic processing (Clark and Sutton, 1996).

14.2.2 Interaction of Microwaves with Materials

Microwaves have components of electrical and magnetic field, phase angle, amplitude and the capability to propagate, i.e. energy is transferred from one end to the other. These properties regulate the synergy of microwaves with materials, and, in some materials, it produces heat. Based upon the electrical and magnetic properties of the materials, their synergy with microwaves are categorized into transparent, opaque and absorbent.

14.2.2.1 Transparent

Microwaves penetrate and are transmitted through the material entirely with no energy transfer occurring during the penetration, which is illustrated in Figure 14.1a. These kinds of materials are called low-loss insulators.

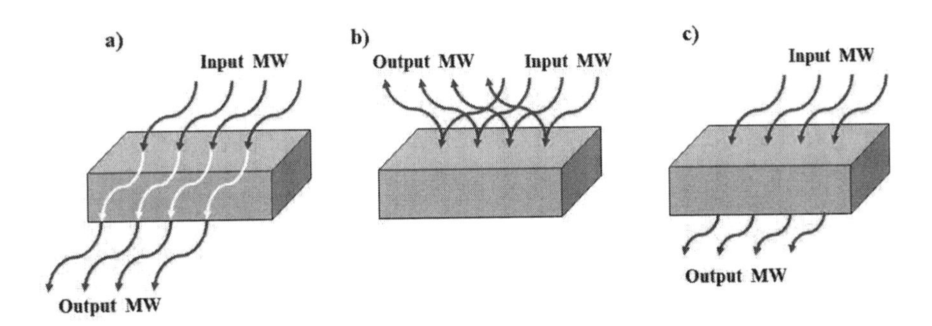

FIGURE 14.1 Representation of material/microwave interaction according to their behaviour: (a) transparent, (b) opaque and (c) absorbent.

14.2.2.2 Opaque

Microwaves are reflected back completely with no penetration into the material and no energy transfer, which is shown in Figure 14.1b. These are called conductors. Normally, metals are opaque to microwave energy.

14.2.2.3 Absorbent

Microwaves are absorbed by the material, and a transfer of electromagnetic energy happens, as shown in Figure 14.1c. Dielectric properties of materials decide the amount of microwave absorption.

A fourth type of interaction was proposed, in which multi- or mixed-phase materials with different degrees of microwave absorption are used. This type is called mixed absorption. At ambient temperature, most of the ceramics, such as alumina, silica, glasses and MgO, remain transparent to microwaves, but, when the materials are heated over a certain critical temperature, they begin to couple and absorb more effectively with microwave radiation. Other ceramics, for example SiC, can absorb microwave energy more efficiently at room temperature. Hence, by introducing a second phase component having microwave-absorbing ability to ceramics that behave as transparent at room temperature can significantly increase the interaction with microwaves permitting a hybrid heating of the material.

14.2.3 Heating Pattern

Figure 14.2 illustrates the difference between a conventional and a microwave process.

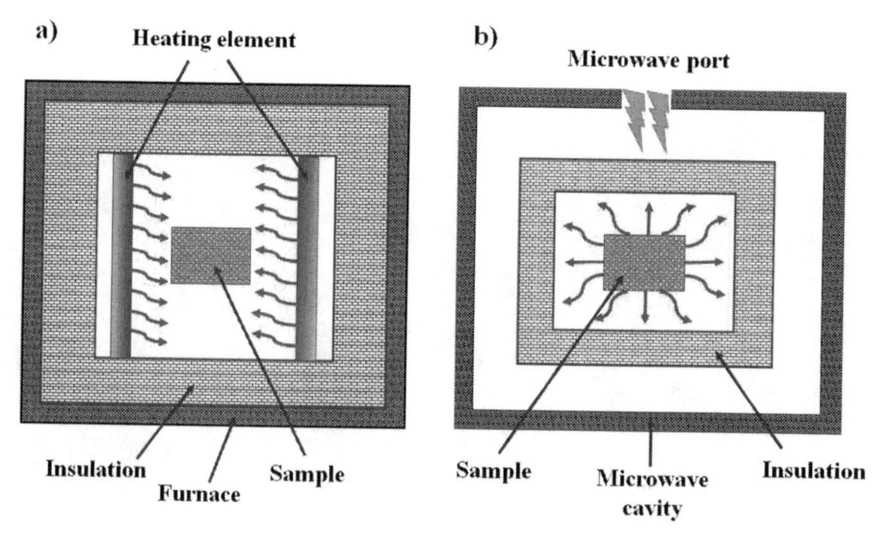

FIGURE 14.2 Heating pattern in (a) conventional furnace (b) microwave furnace.

In conventional furnaces, the heat is produced by the heating elements and supplied to the sample; when compared with the interior of the sample, most of the heat is concentrated on the surface of the sample. In a microwave furnace, microwave energy is absorbed by the material, and it converts the microwave energy into heat. The heating pattern during microwave processing is more internal in nature.

14.2.4 Microwave Heating Mechanisms

Numerous physical mechanisms were proposed to elucidate the interaction of microwave radiation with absorbing materials and the energy transfer that happens during this interaction. These mechanisms consist of resistive heating, bipolar rotation, dielectric heating and electromagnetic heating. The response to the incoming microwave radiation can be associated with any one mechanism or a combination of two or more mechanisms, depending on the material.

14.2.4.1 Resistive Heating

Resistive heating takes place in conductors or semiconductors with comparatively high electrical resistivity. These materials have free electrons or a high ionic content in which the ions obtain enough freedom; thus current can be generated.

14.2.4.2 Bipolar Rotation

When electrically unbiased polar molecules with positive and negative charges are divided, bipolar rotation occurs. Along the direction of increasing amplitude, these dipoles rotate inside a microwave field. Because of this rotation, friction amongst the molecules occurs, which generates heat evenly all over the material.

14.2.4.3 Electromagnetic Heating

Electromagnetic heating occurs in materials with magnetic properties that are extremely vulnerable to external electromagnetic fields, such as those induced by microwave radiation. This kind of heating can be explained as, in oscillating electrical fields, the rotation of magnetic pole of the material corresponding to the rotation of polar molecules.

14.2.4.4 Dielectric Heating

Dielectric heating is a combination of resistive heating and bipolar rotations. This is the prevailing mechanism during the microwave sintering of ceramics. The fundamentals of dielectric heating for microwave-absorbent materials are described below.

When the microwaves pervade the material, the motion in the bound charges (ions), free electrons and in the dipoles are induced by the electromagnetic field. A resistance is created for this motion because it causes a change in the natural equilibrium of the material system, and this resistance, because of elastic, frictional and inertial forces, causes the dissipation of energy. Accordingly, heating of the material occurs because of attenuation of the electric field related with microwave radiation.

14.2.5 Microwave Heating Methods

14.2.5.1 Microwave-Assisted Heating

Materials with very low effective dielectric values, such as ceramics, were developed by using microwave-assisted heating methods. These methods depend on a susceptor that produces heat by absorbing microwaves and supplies it to the sample. The main purpose of the susceptors is to raise the temperature of a sample high enough that the material starts to absorb microwaves and heat by itself, independent of the susceptor material. In the early stages of this method, the sample would experience a conventional heating, and at the later stage, samples are subjected to microwave heating. Some of the most normally used susceptors are SiC-based Al_2O_3 composite, ZrO_2 fibre boards, Pure-SiC rods and SiC grit layer. One of the downsides to using these susceptors is that there is a limited control over the amount of conventional heat supplied to the sample. To overcome this, a new design for microwave-assisted heating was developed by modifying the microwave cavity to include heating elements with an independent power supply. This arrangement substantially controls the amount of conventional heat needed for raising the temperature of the sample (Wroe & Rowley, 1996).

14.2.5.2 Direct Microwave Heating

Most of the research on microwave sintering has used microwave-assisted heating; only a small number of investigations have focused on the effects of direct microwave heating on sintering (Goldstein et al. 1999; Charmond et al. 2010). Direct microwave heating is only due to the interaction between the sample and microwave energy. The absence of a susceptor during microwave sintering minimizes the perturbation of the electromagnetic field distribution in the cavity, eradicates the presence of additional heating sources external to the sample and may potentially improve the overall sintering behaviour. The rise in temperature for a given material is directly proportional to the frequency, effective dielectric loss and magnitude of the internal electric field. For a given frequency, temperature increases by increasing the applied electric field of the direct microwave heating. As the temperature of the sample increases, further heating takes place by the rise in effective dielectric loss.

14.2.6 Advantages of Microwave Sintering

When compared to conventional sintering methods, microwave heating has been shown to accelerate the sintering process of the ceramics. This method has the following advantages:

(i) Reducing the thermal expansion mismatch stresses among numerous materials while producing a multicomponent system.
(ii) Suppressing the problem related with grain growth at high temperatures.
(iii) Reducing the consumption of energy to lower the overall manufacturing costs.

14.3 SYNTHESIS OF AL_2O_3/10 WT.% SIC WITH Y_2O_3 CERAMIC COMPOSITE SAMPLES USING MICROWAVE SINTERING

This section gives the details on synthesis of Y_2O_3 doped Al_2O_3/10 wt.% SiC using the microwave sintering method. Figure 14.3 illustrates an overview of the experimental procedure designed during this study. From the view point of synthesis of Y_2O_3 doped Al_2O_3/10 wt.% SiC, the main features for processing a sample are properties of raw materials, powder consolidation, sintering and characterization.

14.3.1 MATERIALS

Structural ceramic materials used for this study were chosen because of their common use, cost effectiveness, strong ionic interatomic bonding, high-temperature applications and chemical resistance. These structural ceramic materials do have some weakness, including low fracture toughness, brittle nature, sensitive to sudden load due to the presence of microcracks formed during manufacturing processes.

Considering the above-mentioned points, alumina (Al_2O_3) strengthened with silicon carbide (SiC) particles and Y_2O_3 as sintering additive for microstructure and control were selected as the raw materials. High-quality, commercially available

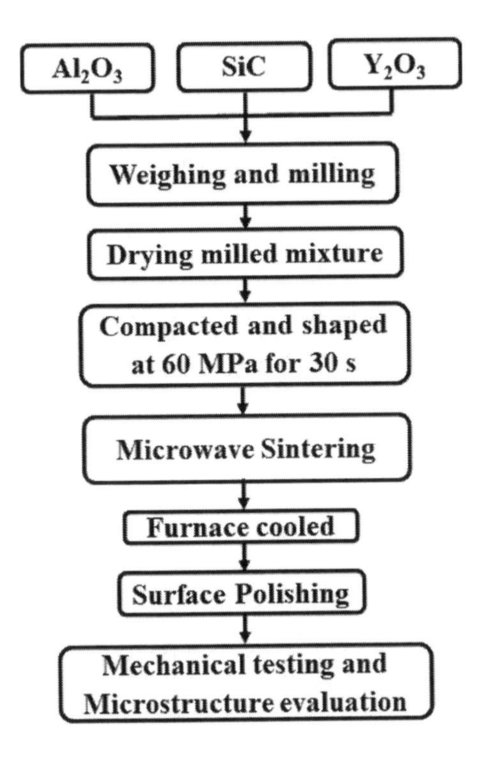

FIGURE 14.3 Process flow chart for sample preparation and characterization.

α-Al_2O_3, β-SiC and Y_2O_3 supplied by Sigma Aldrich, India were used in which α-Al_2O_3 has 99.5% purity, 101.96 g/mol and average grain size of 3 μm; SiC has 99% purity, 40.10 g/mol and average grain size of 1 μm; and Y_2O_3 has 99.9% purity. Composition of Al_2O_3/10 wt.% SiC with Y_2O_3 ceramic composites are given in Table 14.1.

14.3.2 Weighing and Milling

Al_2O_3, SiC and Y_2O_3 powders are weighed based on the composition of the samples separately by using an electronic physical balance (Model – AB204-S, Mettler Toledo). For each proportion of Al_2O_3/10 wt.% SiC with Y_2O_3 ceramic composites, the powders are mixed and homogenized with isopropyl alcohol in a planetary ball mill (VB Ceramics, Chennai), using a tungsten carbide lined vial with tungsten carbide balls at 300 rpm for 180 minutes. Figure 14.4 and 14.5 shows the planetary mill and tungsten carbide vial with tungsten carbide balls respectively. The mixture was dried in a vacuum evaporator at 80°C and crushed in mortar to eliminate soft agglomerates and sieved through a 100 μm mesh screen.

14.3.3 Powder Compaction and Green Pellet Fabrication

The milled, dried and sieved powders are compacted into pellets of cylindrical shape 15 mm in diameter and 5 mm thick. The compaction was done in uniaxial pressing technique. The following are the three steps involved in uniaxial pressing technique.

(i) filling of the mould with powder,
(ii) pressing and
(iii) pellet ejection.

Figure 14.6 shows the circular die used for pressing Al_2O_3/10 wt.% SiC with Y_2O_3 powders. The consequence of this process is that the powder particles (inside the die) will undergo simultaneous rearrangement and compaction, resulting in a compaction of the powders into green pellets. The pressure applied for fabricating the pellet is 60 MPa for 30 seconds. Four green pellets for each composition were compacted and fabricated.

TABLE 14.1
Composition of Al_2O_3/10wt.% SiC with Y_2O_3 (wt. %)

Sample Name	Al_2O_3	SiC	Y_2O_3
AS	90	10	–
AS0.5Y	89.5	10	0.5
AS1Y	89	10	1
AS3Y	87	10	3

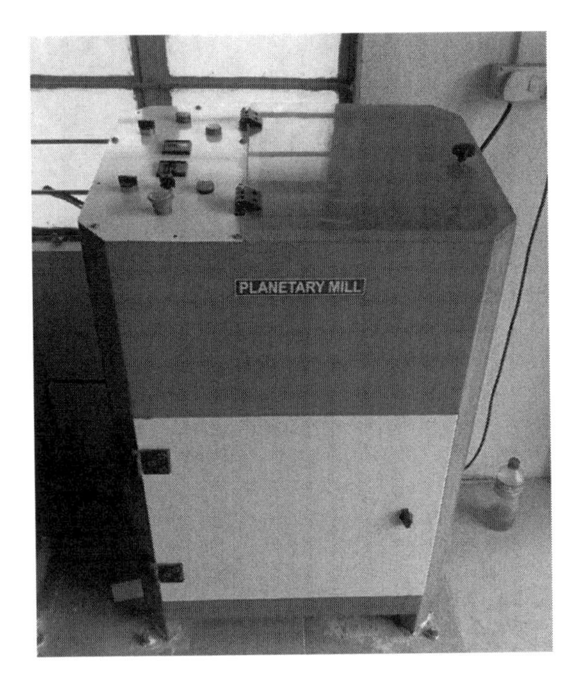

FIGURE 14.4 Planetary mill.

FIGURE 14.5 Tungsten carbide lined jar with tungsten carbide balls.

FIGURE 14.6 Circular die.

14.3.4 Sintering

Sintering of ceramic materials is the process of consolidating of ceramic powder particles by heating the green, compact pellets to a high temperature that is lower than the melting point. During this process, the individual particles diffuse to the neighbouring powder particles and form a solid part. In this study, all sets of compacted green pellets were sintered at 1500°C in a microwave furnace at 2.45 GHz (magnetron heating element) using susceptor materials as auxiliary heating elements. Input power ranging from 0.9 to 2.4 kW and a holding time of 15 minutes were used. The temperature in the microwave furnace was measured by a noncontact type of infrared sensor and controlled by Eurotherm (Model 2416) microprocessor-based proportional-integral-derivative (PID) controller with a digital indicator. Figure 14.7 represents the schematic sketch of the microwave furnace with sample arrangement and Figure 14.8 shows the actual microwave furnace. After sintering, the pellets were cooled in the furnace itself and the surface of the pellets was polished in lapping machine using diamond paste.

14.4 MECHANICAL TESTING AND CHARACTERIZATION

The following are the experimental test conducted to determine the various properties and characterization of Al_2O_3/SiC with Y_2O_3 ceramic composites synthesized by microwave sintering.

(i) Relative density
(ii) Average matrix grain size
(iii) Vickers hardness
(iv) Fracture toughness
(v) X-ray diffraction (XRD)
(vi) Microscopy using scanning electron microscope (SEM)

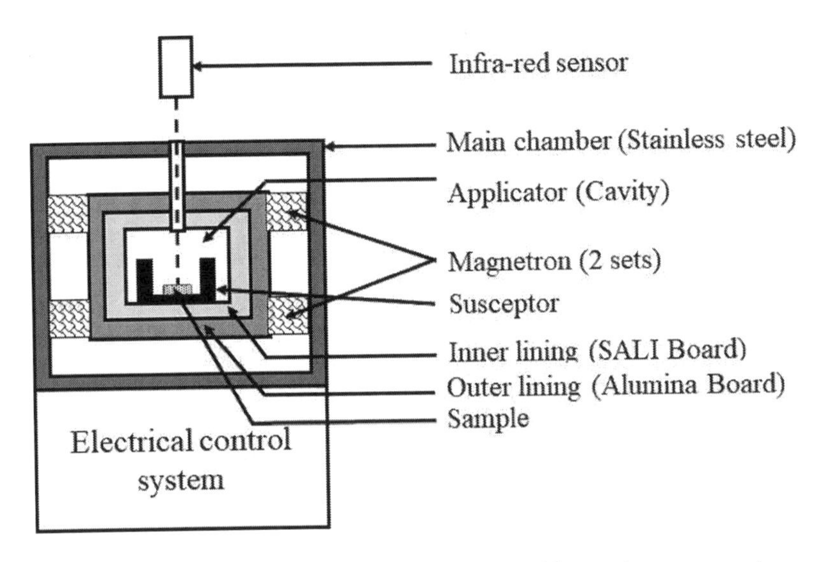

FIGURE 14.7 Schematic sketch of microwave furnace with sample arrangement.

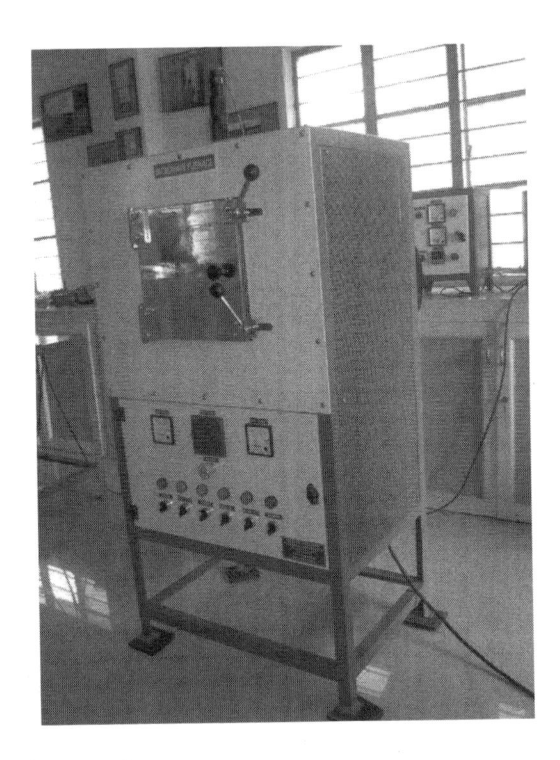

FIGURE 14.8 Microwave furnace.

14.4.1 Relative Density

Relative density was calculated using theoretical density and actual density using Eq. 14.1. Theoretical density was calculated using the rule of mixture and actual density and using Archimedes' principle by immersing in distilled water based on ASTM B311 standard. Relative density was calculated from the theoretical and actual density using the Eq. 14.1.

$$\text{Relative density} = \frac{\text{Actual density}}{\text{Theoretical density}} \times 100 \tag{14.1}$$

14.4.2 Average Matrix Grain Size

The average matrix grain size is calculated by using the linear intercept method, a technique used to calculate the grain size of the sample by marking arbitrarily positioned lines on the micrograph. The number of times the grain boundary intersects each line segment was computed, and the ratio of line length to intercepts was calculated using Eq. 14.2. Similarly, five lines are made and by taking the average for all five lines, the average matrix grain size was measured. Similarly, the average matrix grain size for all the four compositions was calculated.

$$\text{Grain size} = \frac{\text{Length of the line}}{\text{Number of intercepts}} \tag{14.2}$$

14.4.3 Vickers Hardness

In this study, Vickers hardness (H_V) of the samples was measured by using Vicksys Computerized Vickers Hardness Tester (Model: VM-50 PC) based on the ASTM C1327 standard with indentation load of 5 kg for 30 seconds. The hardness was calculated using Eq. 14.3.

$$H_V = \frac{P}{A} = \frac{2P\sin\left(\frac{\theta}{2}\right)}{d^2} = \frac{2P\sin 68^\circ}{d^2} = 1.854 \times \frac{P}{d^2} \tag{14.3}$$

where, H_v = Vickers hardness, P =Applied load and d = diagonal length of the indentation.

14.4.4 Fracture Toughness

Fracture toughness was determined using the indentation technique, which was generally used by the researchers because of its advantages over conventional methods. In the indentation technique, the experimental procedure is forthright, including nominal specimen preparation and the need for a minor amount of material. In this study, fracture toughness was calculated using the Vickers indentation method, given by Eq. 14.4

$$K_{IC} = 0.203 H_V a^{1/2} \left(\frac{c}{a}\right)^{-3/2} \quad (14.4)$$

where, H_V=Vickers hardness, $2a$=diagonal length of the indentation and $2c$=total crack length.

14.4.5 X-Ray Diffraction (XRD)

XRD analysis for the Al_2O_3/10 wt.% SiC with Y_2O_3 ceramic composites was performed using an XRD-Smart Lab (9 kW), Japan diffractometer using Cu K beta radiation with 45 kV and 30 mA as working parameters. The scan range, stepping angle and scan speed were 20°–80°, 0.02° and 4° per minute respectively. The goal of the XRD experiment is to identify phase changes in Al_2O_3/10 wt.% SiC with Y_2O_3 ceramic composites.

14.4.6 Microscopy Using Scanning Electron Microscope (SEM)

The high spatial resolution attainable with the electron microscope makes it useful to study the microstructure of Al_2O_3/10 wt.% SiC with Y_2O_3 ceramic composites samples. Before taking the SEM images, the sample was dried, and a sputter carbon coating was made on the surface of the samples. In this study, the image attained from field emission scanning electron microscope (SUPRA 55 – Carl Zeiss, Germany) was used to study the microstructure and grain properties.

14.5 RESULTS AND DISCUSSION

Table 14.2 shows the theoretical density, actual density and relative densities for various wt.% of Y_2O_3 doped Al_2O_3/10wt.% SiC structural ceramic composites.

TABLE 14.2
Density Results for Various Wt.% of Y_2O_3 Doped Al_2O_3/10wt.% SiC Structural Ceramic Composites

Sample Name	Theoretical Density (g/cm³)	Actual Density (g/cm³)	Relative Density (%)
AS	3.89	3.86	99.2
AS0.5Y	3.90	3.88	99.4
AS1Y	3.90	3.88	99.5
AS3Y	3.92	3.90	99.5

Theoretical density of the composite was calculated by the densities 3.99, 3.21 and 5.01 g/cm^3 of Al_2O_3, SiC and Y_2O_3 respectively. Actual density was determined using the Archimedes' principle by immersing it in distilled water based on ASTM B311 standard.

As can been predicted from Table 14.2, AS3Y specimen with the maximum sintering additives composition has the highest density with a value of 3.90 g/cm^3 with 99.5% of its relative density. Meanwhile, AS and AS0.5Y with no and the lowest sintering additive content yields a density value of 3.86 g/cm^3 with 99.2% relative density and 3.88 g/cm^3 with 99.4% relative density, respectively. Three compositions of Al_2O_3/10 wt.% SiC with Y_2O_3 sintering additive shows higher relative density than the relative density of Al_2O_3/10 wt.% SiC without sintering additives, suggesting that all samples with Y_2O_3 dopant have achieved high densification development. The density increases as the wt.% of Y_2O_3 increases because Y_2O_3 reacted with Al_2O_3 and forms $Al_5Y_3O_{12}$ (YAG) and is present in the matrix as a solid precipitate, which hinders the grain growth of the Al_2O_3 matrix (de Souza Lima et al., 2013; Abal, 2018). Density increases because of the volumetric heating in microwave sintering (Wang et al., 2006; Madhan and Prabhakaran, 2019).

Table 14.3 shows the variation of the average matrix grain size with the inclusion of Y_2O_3 in Al_2O_3/10%SiC structural ceramic composites. Addition of Y_2O_3 to the Al_2O_3/10wt.%SiC composite led to a decrease of the average matrix grain size from 3.2±0.4 to 2.9±0.6 μm, because of the formation of $Al_5Y_3O_{12}$ (YAG) phase formation. This behaviour can be accompanied to the YAG phase, which remains on grain boundaries and triple-points junctions. YAG phase formation impacts the Al_2O_3 particles' surface energy, modifying the sintering mechanism (Bucevac et al., 2020). The consequence is reduction of alumina grain size due to the contact with a lower surface energy phase. Several strengthening mechanisms may result from decreased grain size, depending on the state yttria phase appears in the ceramic body.

The variations of Vickers hardness and fracture toughness with inclusion of Y_2O_3 in Al_2O_3/10 wt.% SiC are given in Table 14.4. Vickers hardness increases initially for 0.5 and 1% inclusion of Y_2O_3 and future increase in Y_2O_3 decreases the hardness. The highest hardness was obtained for AS1Y structural ceramic composite with the

TABLE 14.3
Average Matrix Grain Size for Various Wt.% of Y_2O_3 Doped Al_2O_3/10wt.% SiC Structural Ceramic Composites

Sample Name	Average Matrix Grain Size (μm)
AS	3.2±0.4
AS0.5Y	3.1±0.5
AS1Y	2.9±0.5
AS3Y	2.9±0.6

TABLE 14.4
Vickers Hardness and Fracture Toughness for Various Wt.% of Y_2O_3 Doped Al_2O_3/10wt.% SiC Structural Ceramic Composites

Sample Name	Vickers Hardness (GPa)	Fracture Toughness (MPa $m^{1/2}$)
AS	24.10 ± 0.55	5.73 ± 0.17
AS0.5Y	24.58 ± 0.39	5.82 ± 0.19
AS1Y	24.83 ± 0.61	5.95 ± 0.28
AS3Y	23.90 ± 0.49	5.69 ± 0.15

value of 24.83 ± 0.61 GPa and the lowest hardness for AS3Y ceramic composite with the value of 23.90 ± 0.49 GPa, and it is lower than the AS ceramic composite without Y_2O_3 dopant. Vickers hardness increased initially because of the formation of eutectic YAG phase and presence of SiC particles. Further increase in Y_2O_3 percentage increases the glassy precipitate YAG phase, which diminishes the hardness, preceding reports indicates the same phenomenon, that increase in Y_2O_3 percentage increases the glassy precipitate YAG phase (de Sousa Lima et al., 2015).

Fracture toughness increases initially for 0.5 and 1% inclusion of Y_2O_3 in Al_2O_3/10wt.%SiC but is reduced with 3% addition of Y_2O_3 in Al_2O_3/10wt.%SiC ceramic composites. The highest fracture toughness was obtained for AS1Y ceramic composite and the lowest for AS3Y ceramic composite. The presence of SiC particles and the formation of $Al_5Y_3O_{12}$ (YAG) phase hinders the grain growth of the Al_2O_3 matrix and reduces the grain size, which increases the fracture toughness. But in AS3Y ceramic composite, the formation of more glassy precipitate of $Al_5Y_3O_{12}$ (YAG) phase reduces the fracture toughness. Reduced holding time and volumetric heating by microwave sintering reduces the grain growth and internal stress due to thermal mismatch during the formation of $Al_5Y_3O_{12}$ (YAG) phase along with inert SiC phase.

XRD pattern for all the four ceramic composites are shown in Figure 14.9. XRD of AS ceramic composite indicates only Al_2O_3 and SiC phases because of the inert nature of SiC with Al_2O_3. But XRD of the other three ceramic composites AS0.5Y, AS1Y and AS3Y with Y_2O_3 indicates YAG phase formed by the reaction of Y_2O_3 with Al_2O_3 matrix.

Figure 14.10 shows the SEM images of all four composites. The Al_2O_3 (black phase), SiC (smaller white phase) and $Al_5Y_3O_{12}$ (YAG) (larger white phase) precipitate-like structures presented in the micrographs. It was very difficult to distinguish SiC phase and $Al_5Y_3O_{12}$ (YAG) phase because both are brighter in colour, but $Al_5Y_3O_{12}$ (YAG) phase differentiated by the formation of large white grains. It was clearly observed from SEM micrographs that density increased as the Y_2O_3 percentage increased because of the formation of $Al_5Y_3O_{12}$ (YAG) phase. The average matrix grain size also reduced because of the inert SiC phase and formation of $Al_5Y_3O_{12}$ (YAG) phase because of grain boundary pinning.

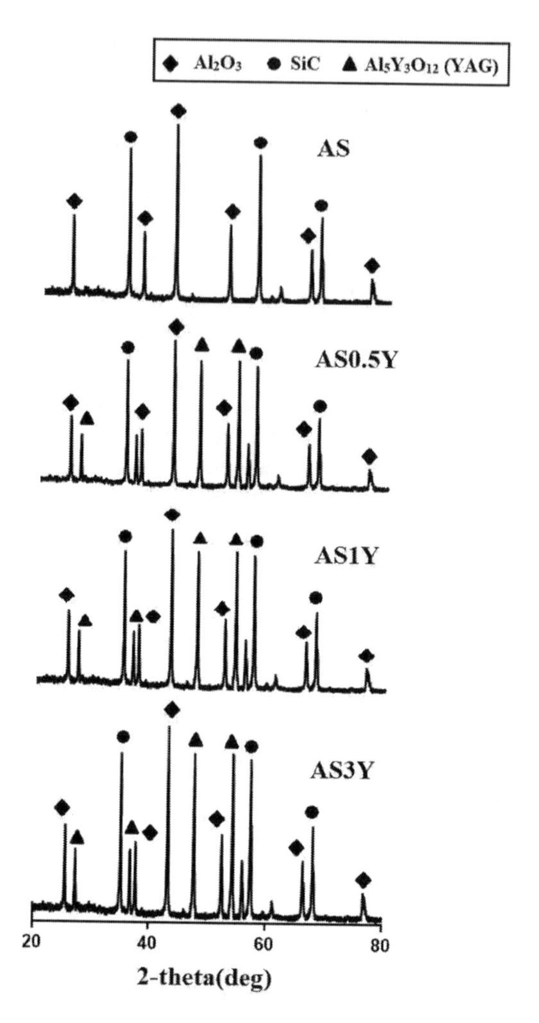

FIGURE 14.9 XRD pattern.

14.6 CONCLUSION

Al_2O_3/10 wt.% SiC doped with Y_2O_3 (0.5, 1 and 3 wt.%) were synthesized by microwave sintering at 1500°C with a minimum holding time of 15 minutes and their mechanical properties were studied. When compared to AS ceramics, density increased as Y_2O_3 sintering additives increased because of the formation of $Al_5Y_3O_{12}$ (YAG) phase and the presence of SiC, which hinders the grain growth and reduces the average Al_2O_3 matrix grain size. Vickers hardness and fracture toughness increased initially for AS0.5Y and AS1Y ceramic composites; with a further increase in Y_2O_3, both hardness and fracture toughness decrease because of the formation of more $Al_5Y_3O_{12}$ (YAG) bright glassy phase. The higher hardness and fracture toughness of 24.83 ± 0.61 GPa and 5.95 ± 0.28 MPa $m^{1/2}$ was obtained for

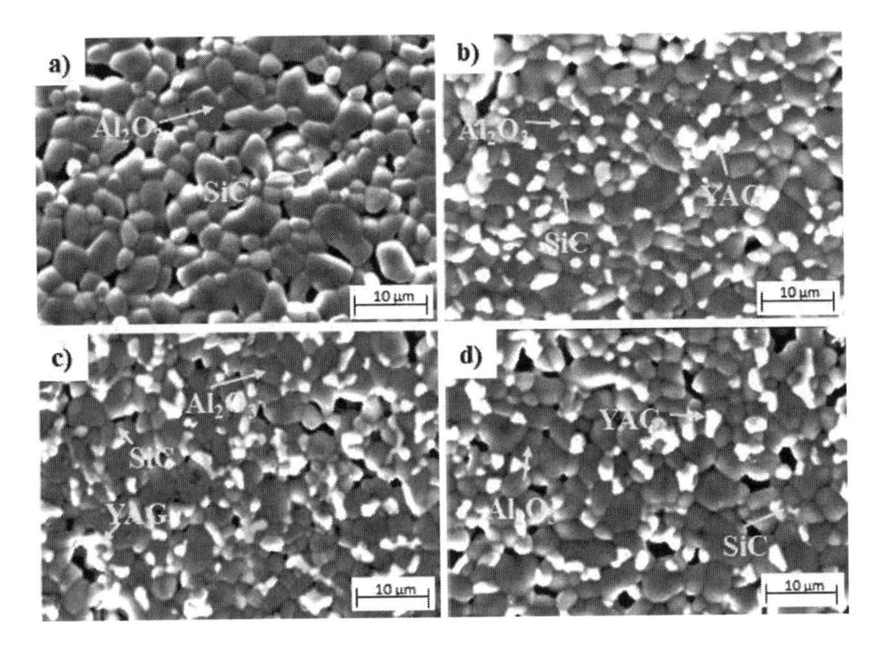

FIGURE 14.10 SEM images (a) AS, (b) AS0.5Y, (c) AS1Y and (d) AS3Y.

AS1Y ceramic composites. XRD pattern showed the formation of $Al_5Y_3O_{12}$ (YAG) phase along with Al_2O_3 and SiC phases. SEM images clearly indicated the presence of $Al_5Y_3O_{12}$ (YAG) phase and a reduction in matrix grain size with an increase in wt.% of Y_2O_3.

REFERENCES

Abal, S. and Karacam, C.U. 2018. The effect of the addition of Y_2O_3 on the microstructure of polycrystalline alumina ceramics. *Proceedings MDPI Journal*, 2: 1407.

Chou, A., Chan, H.M. and Harmer, M.P. 1998. Effect of Annealing Environment on the Crack Healing and Mechanical Behavior of Silicon Carbide-Reinforced Alumina Nanocomposite, *Journal of the American Ceramic Society*, 81: 1203–1208.

Bucevac, D., Omerasevic, M., Egelja, A., Radovanovic, Z., Kljajevic, L., Nenadovic, S. and Krstic, V. 2020. Effect of YAG content on creep resistance and mechanical properties of Al_2O_3-YAG composite. *Ceramics International*, 46: 15998–16007.

Charmond, S., Carry, C.P. and Bouvard, D. 2010. Densification and microstructure evolution of Y-tetragonal zirconia polycrystal powder during direct and hybrid microwave sintering in a single-mode cavity. *Journal of the European Ceramic Society*, 30: 1211–1221.

Clark, D.E. and Sutton, W.H. 1996. Microwave processing of materials. *Annual Review of Materials Science*, 26: 299–331.

de Sousa Lima, E., Itaboray, L.M., Santos, A.P.O., Santos, C. and Cabral, R.F. 2015. Mechanical properties evaluation of Al_2O_3-YAG ceramic composites. *Materials Science Forum*, 820: 239–243.

de Souza Lima, E., Leme Louro, L.H., de Freitas Cabral, R., de Campos, J.B., de Avillez, R.R. and da Costa, C.A. 2013. Processing and characterization of Al_2O_3-yttrium aluminum garnet powders. *Journal of Materials Research and Technology*, 2: 18–23.

Domanická, A., Klement, R., Prnová, A., Bodišová, K. and Galusek, D. 2014. Luminescent rare-earth ions doped Al_2O_3–Y_2O_3–SiO_2 glass microspheres prepared by flame synthesis. *Ceramics International*, 40: 6005–6012.

Goldstein, A., Travitzky, N., Singurindy, A. and Kravchik, M. 1999. Direct microwave sintering of yttria-stabilized zirconia at 2·45 GHz. *Journal of the European Ceramic Society*, 19: 2067–2072.

Huang, M., Li, Z., Wu, J., Khor, K.A., Huo, F., Duan, F., Lim, S.C., Yip, M.S. and Yang, J. 2015. Multifunctional alumina composites with toughening and crack-healing features via incorporation of NiAl particles, *Journal of the American Ceramic Society*, 98: 1618–1625.

Madhan, M. and Prabhakaran, G. 2019. Microwave versus conventional sintering: Microstructure and mechanical properties of Al_2O_3–SiC ceramic composites. *Boletín de la Sociedad Española de Cerámica y Vidrio*, 58: 14–22.

Ono, M., Nakao, W., Takahashi, K., Nakatani, M. and Ando, K. 2007. A new methodology to guarantee the structural integrity of Al_2O_3/SiC composite using crack healing and a proof test. *Fatigue and Fracture of Engineering Materials and Structures*, 30: 599–607.

Shi, S., Cho, S., Goto, T., Kusunose, T. and Sekino, T. 2018. Combinative effects of Y_2O_3 and Ti on Al_2O_3 ceramics for optimizing mechanical and electrical properties. *Ceramics International*, 44: 18382–18388.

Suarez, M., Fernandez, A., Menendez, J.L., Torrecillas, R., Kessel, H.U., Hennicke, J., Kirchner R. and Kessel T. 2013. Challenges and opportunities for spark plasma sintering: A key technology for a new generation of materials. *Sintering Applications*, 13: 319–342. doi: 10.5772/53706.

Wang, J., Binner, J. and Vaidhyanathan B. 2006. Evidence for the microwave effect during hybrid sintering. *Journal of the American Ceramic Society*, 89: 1977–1984.

Wroe, R. and Rowley, A.T. 1996. Evidence for a non-thermal microwave effect in the sintering of partially stabilized zirconia. *Journal of Materials Science*, 31: 2019–2026.

Wei, C. and Becher, P.F. 1985. Development of SiC Whisker Reinforced Ceramics, *American Ceramic Society Bulletin*, 64: 289–304.

15 Multiobjective Optimization of BSL 165 Aluminium Composite For Aeronautical Applications

R. Suresh Kumar, T. Ramakrishnan and S. Balasubramani

CONTENTS

15.1 INTRODUCTION

One can find the applications of aluminium alloys in various forms in several industries such as aerospace, medical and automotive because of its unique and superior properties over other materials. The wide application of aluminium alloys thus makes it important to have extensive knowledge and understanding of its behaviour while it undergoes machining processes. From a machining perspective, end milling is a process that is being adopted in various manufacturing sectors to accomplish the needs required for manufacturing. In simple terms, end milling is a process of machining in which gradual removal of material from the workpiece is executed by providing a minimum feed rate at a higher speed. The quality of the machined surfaces is evaluated based upon the surface roughness achieved. On the other hand, production time is evaluated based on the material removal rate and is maximized by providing a higher feed rate. This condition stands contradictory to the former condition stated. This contradiction makes it important to have an extensive study of the behaviour of controllable and uncontrollable factors involved during machining.

Factors such as feed rate, spindle speed and depth of cut are considered to be controllable factors, as they are directly governed by the user. Whereas, noncontrolled factors such as nonhomogeneity of the workpiece, tool wear, machine-motion errors, formation of chips, vibrations and chatter are few examples that cannot be controlled directly but can be minimized by providing an apt level of controllable factors. A detailed study on the machining characteristics of surface roughness while machining AL2014-T6 was performed by Ming-Yung et al. in slot end milling(Ming-Ying and Chang 2004). It was concluded that feed rate influenced the surface roughness to a greater extent, whereas prediction accuracy is influenced by vibrations.Palanisamy et al. (2007)applied genetic algorithm and experimental validation for optimizing the end-milling operations. They concluded that surface roughness was greatly influenced by feed rate followed by the depth of cut. Martellotti (1941) and Quintana et al. (2010) proposed a theoretical approach toward the prediction of surface roughness. In the above research work, the analysis has been restricted to a single objective function in which the possible interaction effects on the other responses are not considered and studied. Mansour et al. (2002) and Alauddin et al. (1997) studied the influence of spindle speed, feed rate, axial depth of cut by developing analytical models for the prediction of surface roughness and tool life. Chang et al. (2007) and Coker et al. (1996) performed the optimization of surface roughness by using experimental investigations. Gologlu et al. (2008) and Dhokia (2008) performed the optimization by using the design of experiments for optimizing the machining parameters to predict the surface roughness. The survey also provides light on the techniques adopted to arrive at the optimum level of machining in which one can determine optimization techniques, such as neural network, genetic algorithm, fuzzy logic and neural fuzzy (Lou and Chen 1999; Tsai et al. 1999; Benardo and Vosniakos 2002; Brecher et al. 2011; Chen and Lou 2000; Ali and Zhang 1999) to arrive at an optimized set of parameters for machining. Zain et al. (2008) made a literature survey on the various studies on the prediction of an optimized set of cutting parameters in terms of surface roughness using a genetic algorithm in end-milling process and was found to be very limited. Brezocnik et al. (2003) studied the effect of cutting parameters and vibration between the cutting tool and workpiece for optimizing the surface roughness by using genetic algorithm. Oktem et al. (2006) investigated the effect of cutting parameters for the optimization of surface roughness by applying genetic algorithm. The roughness predicted by genetic algorithm was found to be lower than the one achieved by experimental results. In the above-noted studies, one can find the application of higher optimization tools, but the work considered was constrained to one objective function, whereas a machining process is focused on providing higher quality in reduced time of machining. Therefore, it becomes mandatory to study the effect of accompanying responses to arrive at the best possible machining parameters.

From the literature sources, it is evident that the machining of aluminium metal matrix composite is an area of interest for which very few studies have been conducted in optimizing the machining parameters in the end-milling process. In the present study, an investigation is carried out on the effect of process parameters, such as spindle speed, feed rate and depth of cut on noise factors, such as surface roughness and material removal rate (MRR) in end milling using response surface methodology (RSM) approach. The experiments were carried out using a 3-axis vertical computer numerical control (CNC) machine and 12 mm diameter high-speed steel (HSS) end-mill cutters. The validity of the mathematical model developed is ensured by conducting confirmatory runs and was found to be within the desired level of expectation.

15.2 OPTIMIZATION BY USING GENETIC ALGORITHM

Genetic algorithm (GA) works on the principles of evolutionary biology, such as genetic inheritance, natural selection, crossover and mutations to figure out the best solutions for the given problems. For a specified problem, individuals are represented using abstract representations called chromosomes. The iterative process involves a working set of individuals, commonly known as a population, that works toward an objective function. The evolutionary process starts from a randomly generated population, called generations. New population is created by randomly selecting multiple individuals from each generation. GA solves the problem assigned in the form of coded series, a string of binary numbers (chromosomes) containing the information pertaining to the set of possible parameters. Thus, the objective function assigned is then taken to evaluate the fitness of chromosomes. Sorted individuals are then reproduced in pairs by the application of genetic operators. The reproduced individuals are perturbed by mutation to form a new generation. This process is repeated until the satisfying result is attained.

15.3 SURFACE ROUGHNESS

In industries, average roughness (R_a), root-mean-square (RMS) and maximum peak-to-valley roughness (R_{max}) are used to evaluate the standard measure in terms of surface roughness. The modelling and analysis of variables influencing a response is extensively performed by RSM. By performing multivariate analysis on the experimental data one can obtain a relevant model that establishes the relationship between the variables as well as the responses considered. The theoretical equation derived from previous studies is given by Eq.15.1.

$$\text{Ra} = \varepsilon A^{k1} B^{k2} C^{k3} D^{k4} \tag{15.1}$$

Where R_a = predicted surface roughness (microns);ε = response error; A = spindle speed (rpm);B = depth of cut (mm); C = feed rate (mm/rev); D =coolant flow rate (l/min); k_1, k_2, k_3, k_4 = modal parameters estimated from experimental data.

The above equation can also be written as shown in Eq. 15.2:

$$\ln \text{Ra} = \ln \varepsilon + k1 \ln A + k2 \ln B + k3 \ln C + k4 \ln \tag{15.2}$$

15.4 MATERIAL REMOVAL RATE (MRR)

The initial weight of the workpiece before machining was weighed using an electronic balance (0.0001gm accuracy). After every experimental cut, the workpiece is removed and weighed on a digital weighing machine. During every experimental run, the machining time is recorded using a stop watch. MRR is computed as:

$$Mrr = wrw / (\rho * t) \tag{15.3}$$

Where wrw = workpiece removal weight; P= density of the workpiece; t= machining time.

15.5 EXPERIMENTAL OUTLINE

The entire study is carried out in the following sequences:

1. The experiments are conducted on a 3-axis CNC vertical machining centre with a high-speed steel end-mill cutter of 12 mm diameter having four flutes with 22 mm overhung length under wet condition.
2. The process parameters considered are spindle speed (rpm), feed rate (mm/rev), depth of cut (mm) and coolant flow rate (l/min).
3. The surface roughness is measured by using a surface roughness tester.
4. The experiment is conducted at three levels, four-factor central composite rotatable designs having 25 sequences of experimental runs.
5. The second-order quadratic model is developed for the prediction of surface roughness and MRR and is checked for its adequacy using analysis of variance (ANOVA).
6. GA is utilized to determine the optimized set of machining parameters that leads to the minimum value of surface roughness (R_a) and the maximum value of MRR.
7. Validation of the result is confirmed by performing experimental runs.

15.6 SELECTION OF CONTROLLABLE PARAMETERS

Various parameters affect the surface roughness, namely spindle speed, feed rate, depth of cut, tool rake angle and coolant flow. In this study, the controllable factors considered are spindle speed (rpm), depth of cut (mm), feed rate (mm/rev) and coolant flow (l/min). The ranges of the selected parameters were set based on the manufacturer's specifications and research carried out so far.

15.6.1 Selection of Tool

An HSS end-mill cutter of 12 mm dia, four flutes is used for machining end-milling operation. To reduce the adverse effect of chatter caused by the tool overhung length, it is maintained to a maximum length of 22 mm.

15.6.2 Limits of Process Parameters

The upper limit of a given factor was coded as ($X_{\max}$) and the lower limit was coded as ($X_{\min}$). The coded values for intermediate values were calculated using Eq. 15.4.

$$Xi = 2\{2X - (X_{\max} + X_{\min})\}X_{\max} - X_{\min} \tag{15.4}$$

Where, X_i = required coded value of variable X and the value of X varies from X_{min} to X_{max}. The levels assigned for the process parameters considered are given in Table 15.1.

15.7 DESIGN MATRIX

The experimental runs are performed based on Box-Behnken (BB) design. Initially, four factors with three levels are assigned as input variables with levels represented by +1 for upper limits and −1 for lower limits. Two responses are assigned, namely roughness and MRR. The experimental sequences generated in response are reflected in Table 15.2.

15.8 EXPERIMENTATION

The workpiece taken for experimentation is a rectangular block of 75 mm × 30 mm × 12 mm prepared from aluminium alloy BSL 165. The experiment was conducted on 3-axis CNC vertical machining centre with an HSS end-mill cutter of

TABLE 15.1
Parameters and Levels

Parameters	Level 1	Level 2	Level 3
Spindle Speed (rpm)	6000	9000	12000
Depth of Cut (mm)	1	2	3
Feed Rate (mm/rev)	0.10	0.12	0.14
Coolant Flow Rate (l/min)	4	6	8

TABLE 15.2
BB Experimental Sequences with Responses

Spindle Speed (rpm)	Depth of Cut (mm)	Feed Rate (mm/rev)	Coolant Flow Rate (l/min)	Roughness (microns)	MRR (mm^3/min)
9000	1	0.12	8	0.225	233.165
12000	3	0.12	6	0.637	254.814
6000	2	0.1	6	0.851	248.114
9000	2	0.12	6	0.575	240.775
6000	3	0.12	6	0.986	268.981
9000	3	0.12	4	0.788	257.456
6000	2	0.12	8	0.569	250.112
9000	2	0.14	8	0.397	242.815
9000	2	0.1	4	0.708	234.112
9000	2	0.1	8	0.708	233.156
9000	1	0.14	6	0.047	235.551
12000	2	0.12	4	0.629	252.421
6000	2	0.12	4	0.569	249.112
9000	3	0.1	6	0.921	256.118
9000	3	0.12	8	0.788	258.114
9000	3	0.14	6	0.610	262.336
9000	1	0.1	6	0.358	222.651
9000	1	0.12	4	0.225	232.221
12000	2	0.14	6	0.599	238.775
9000	2	0.14	4	0.397	241.336
12000	2	0.1	6	0.614	235.551
6000	1	0.12	6	0.014	230.123
12000	1	0.12	6	0.484	212.365
6000	2	0.14	6	0.243	254.814
12000	2	0.12	8	0.629	236.118

12 mm diameter under flooded condition. The surface roughness was measured by using a Surftest SJ-201 surface roughness tester.

15.9 RESPONSE SURFACE MODEL FOR THE PREDICTION OF SURFACE ROUGHNESS

Design-Expert software is used to perform rigorous analysis on the experimental data, and a second-order quadratic model is developed for the prediction of surface roughness. ANOVA is performed to check the adequacy of the model and is illustrated in Table 15.3.

From Table 15.3, it is evident that 'model F value' of 696871 with a 'model P value' less than 0.0001 implies that the selected model is significant. If the values are greater than 0.10, then the model is said to be insignificant. The P value <0.0001 represents that there is only a 0.01% chance that such a model could occur because

TABLE 15.3
ANOVA for Roughness

ANOVA for Response Surface Quadratic Model						
Source	**Sum of Squares**	**df**	**Mean Square**	**F Value**	**Prob > F**	
Model	1.544732177	14	0.110338013	696871.6586	< 0.0001	significant
A-Spindle Speed	0.010860083	1	0.010860083	68590	< 0.0001	
B-Depth of Cut	0.949781333	1	0.949781333	5998618.947	< 0.0001	
C-Feed Rate	0.290474083	1	0.290474083	1834573.158	< 0.0001	
D-Coolant Flow Rate	0	1	0	0	1.0000	
AB	0.167281	1	0.167281	1056511.579	< 0.0001	
AC	0.08791225	1	0.08791225	555235.2632	< 0.0001	
AD	0	1	0	0	1.0000	
BC	0	1	0	0	1.0000	
BD	0	1	0	0	1.0000	
CD	0	1	0	0	1.0000	
A^2	0.001603843	1	0.001603843	10129.5356	< 0.0001	
B^2	0.01336177	1	0.01336177	84390.12384	< 0.0001	
C^2	0.001418843	1	0.001418843	8961.114551	< 0.0001	
D^2	4.90196E-09	1	4.90196E-09	0.030959753	0.8638	
Residual	1.58333E-06	10	1.58333E-07			
Cor Total	1.54473376	24				
Std. Dev.	0.000397911		R-Squared	0.999998975		
Mean	0.54264		Adj R-Squared	0.99999754		
C.V. %	0.073328765		Pred R-Squared	N/A		
PRESS	N/A		Adeq Precision	3152.502951		

of noise. Surface plots are plotted to provide a clear view of the relationship between the response and the process parameters. The measures of R^2, Adj R^2 and Pred R^2 indicate the goodness of fit for the models and are close to 1.

15.9.1 Surface Roughness

$$\begin{aligned} \text{Ra} = 1.42904 - 1.997E - 004.A + 1.170.B - 16.566.C + 1.25E \\ - 004.D - 6.816E - 005.A.B + 2.470E - 003.A.C + 2.648E \\ - 009.A^2 - 0.068.A^2 - 56.041.C^2 - 1.041E - 005.D^2 \end{aligned} \quad (15.5)$$

15.10 RESULTS AND DISCUSSION ON SURFACE ROUGHNESS

15.10.1 Parameter Interaction Effects

Figure 15.1 shows the surface interaction plot of spindle speed and depth of cut over surface roughness. Minimum roughness is achieved when the depth of cut is maintained at the lower limit and spindle speed is varied between 6000 rpm to 10000 rpm.

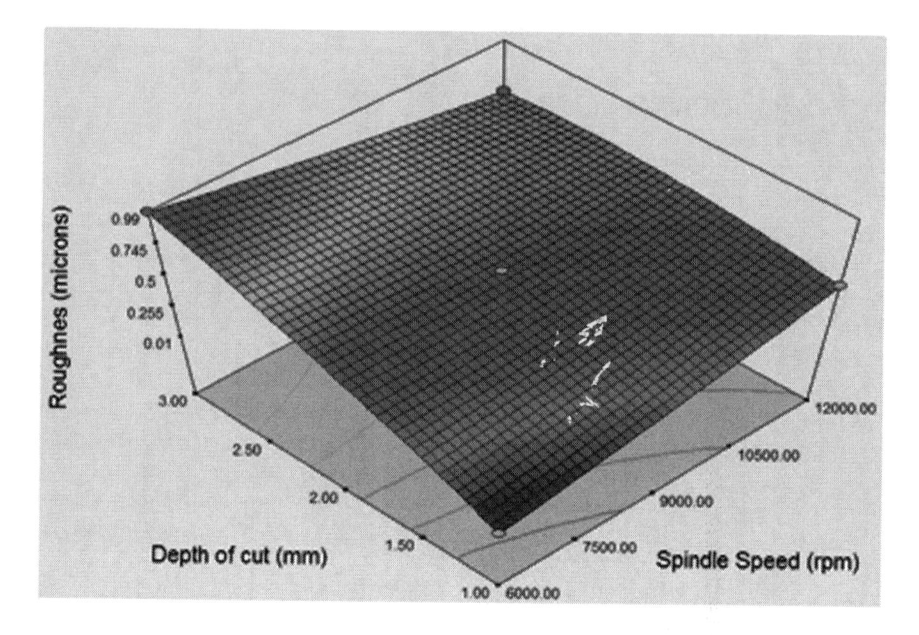

FIGURE 15.1 Surface interaction plot of spindle speed and depth of cut parameters on surface roughness.

Similarly, the better surface roughness is possible, when the depth of cut is maintained between 1 mm to 1.5 mm. Therefore, for the workpiece considered, it can be concluded that spindle speed and depth of cut are to be at lower level to achieve the desired output.

Figure 15.2 shows the surface interaction plot of spindle speed and feed rate over surface roughness. From the graph, one can identify that spindle speed and depth of cut are inversely proportional in achieving the best surface finish. This plays a prominent role in determining the optimized set of machining parameters, as one needs to identify the optimum level of setting between feed rate and spindle speed.

Figure 15.3 shows the surface interaction plot of spindle speed and coolant flow rate. From the graph, it is evident that coolant follows a steady pattern with spindle speed. When compared with other parameters, the effect of coolant flow rate is found to be minimal. At the same time, the relationship interpreted by the graph shows that better results can be attained at the lower limit of coolant flow rate and medium level of spindle speed.

Figure 15.4 shows the surface interaction plot of feed rate and depth of cut. From the graph, it is evident that when feed rate is maintained between 0.13 to 0.14 mm/rev and depth of cut between 1 mm to 1.5 mm, the desired roughness can be achieved. One can conclude that for the considered workpiece, higher feed rate and lower depth of cut is to be maintained to obtain the desired response.

Figure 15.5 shows the surface interaction plot of depth of cut and coolant flow rate. From the graph, one can identify that the flow of coolant follows a steady pattern with depth of cut. When compared with other parameters, the effect of coolant flow rate is found to be the most insignificant. At the same time, the relationship

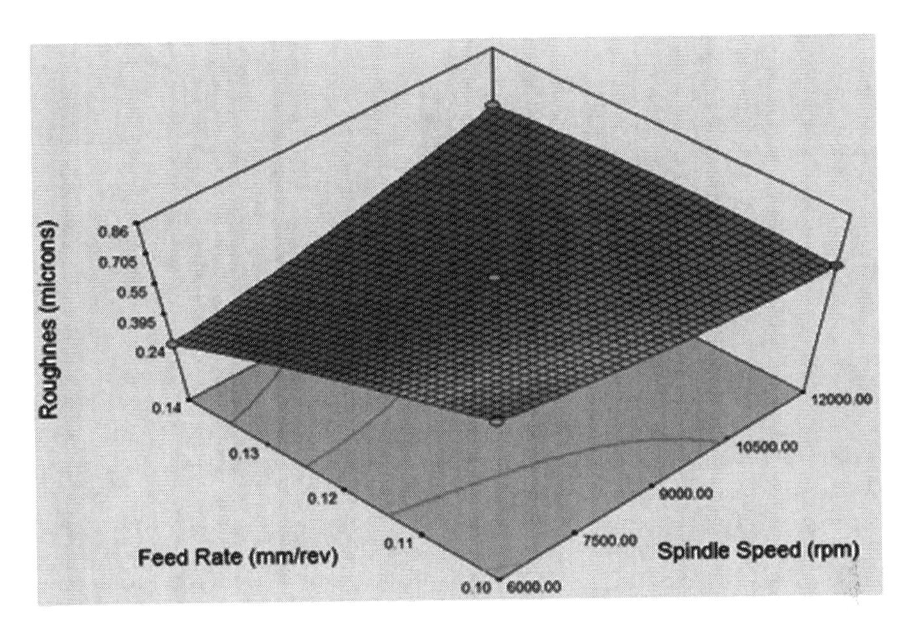

FIGURE 15.2 Surface interaction plot of spindle speed and feed rate parameters on surface roughness.

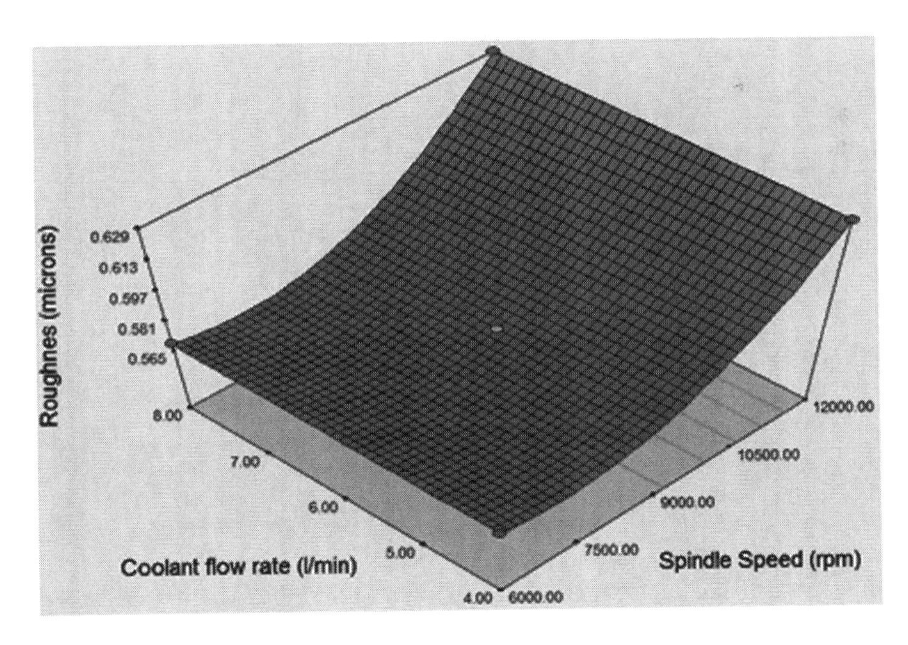

FIGURE 15.3 Surface interaction plot of spindle speed and coolant flow rate parameters on surface roughness.

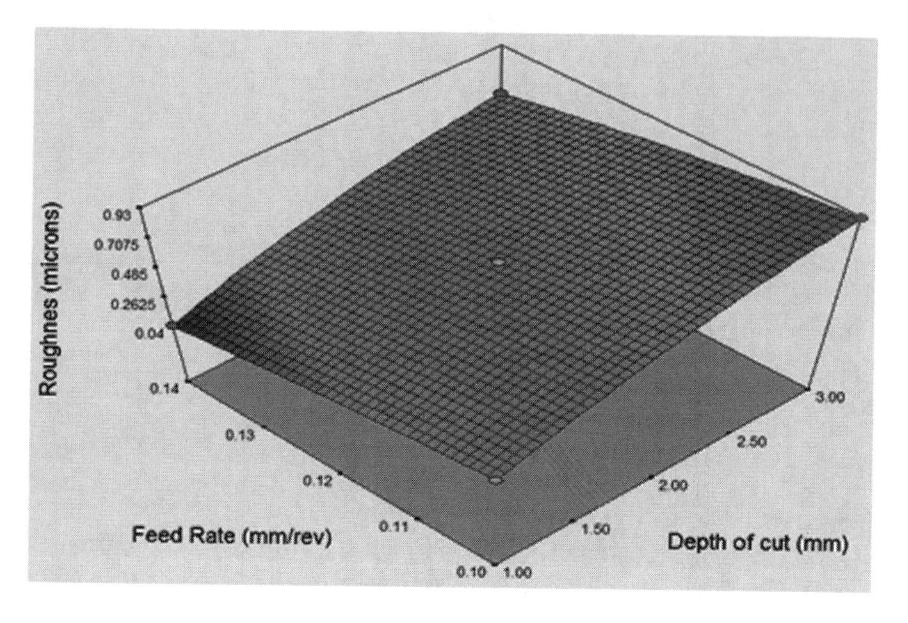

FIGURE 15.4 Surface interaction plot of feed rate and depth of cut parameters on surface roughness.

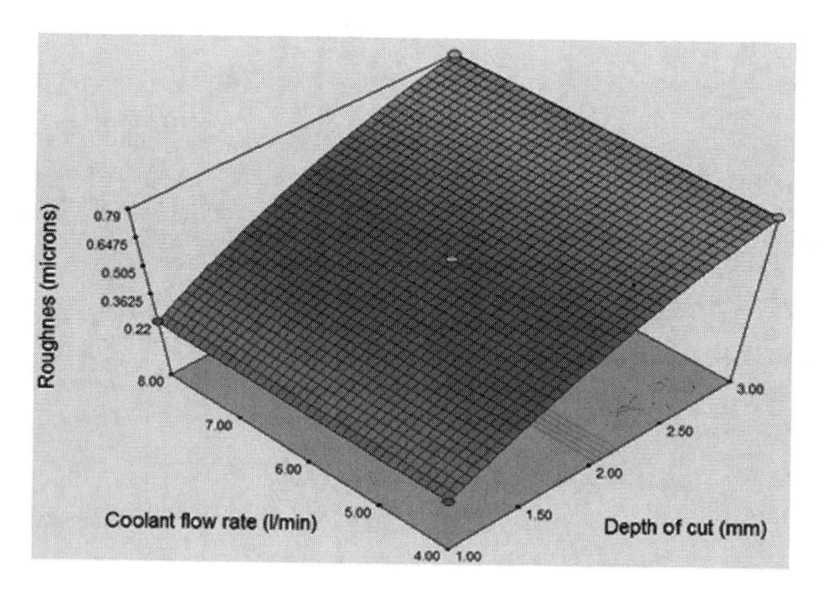

FIGURE 15.5 Surface interaction plot of depth of cut and coolant flow rate parameters on surface roughness.

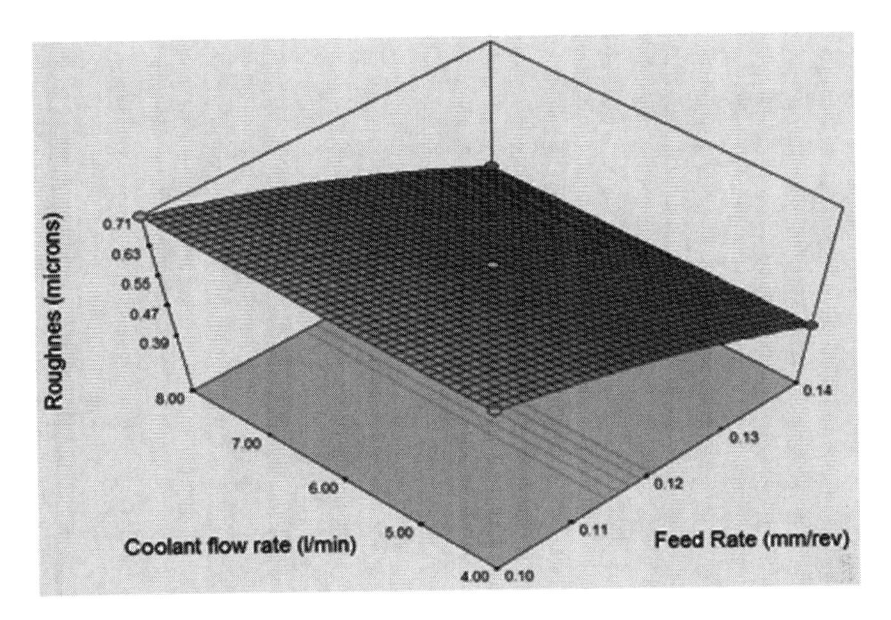

FIGURE 15.6 Surface interaction plot of coolant flow rate and feed rate parameters on surface roughness.

interpreted by the graph shows that better results can be attained at the lower limits of coolant flow rate and depth of cut.

Figure 15.6 shows the surface interaction plot of feed rate and coolant flow rate. From the graph, one can identify that the flow of coolant follows a steady pattern with feed rate. When compared with other parameters, the effect of coolant flow rate is found to be almost the same with few deviations.

15.11 RESPONSE SURFACE MODEL FOR THE PREDICTION OF MRR

ANOVA is performed to check the adequacy of the model created as shown in Table 15.4. From Table 15.4, it is evident that 'model F value' of 6.68091 with a 'model P value' less than 0.0024 implies that the selected model is significant. If the values are greater than 0.10, then the model is said to be trivial. The P value <0.0001 represents that there is only a 0.01% chance that such a model could occur because of noise. Surface plots are plotted to provide a clear view of the relationship between the response and the process parameters. The measures of R^2, Adj R^2 and Pred R^2 indicate the goodness of fit for the models and are closer to 1.

15.11.1 Material Removal Rate (MRR)

$$\begin{aligned}\text{MRR} = {} & 170.43558 - 1.86861E - 003.A + 16.51383.B + 738.54167.C \\ & + 0.33692.D + 2.99250E - 004.A.B - 0.014483.A.C - 7.20958E \\ & - 004.A.D - 83.52500.B.C - 0.035750.B.D + 15.21875.C.D \\ & + 2.97542E - 007.A^2 + 1.75225.B^2 - 1421.25000.C^2 + 0.3066.D^2\end{aligned} \tag{15.6}$$

TABLE 15.4
ANOVA for MRR

ANOVA for Response Surface Quadratic Model						
Source	**Sum of Squares**	**df**	**Mean Square**	**F Value**	**Prob > F**	
Model	3821.467932	14	272.9619952	6.680911902	0.0024	Significant
A-Spindle Speed	422.5957453	1	422.5957453	10.34328952	0.0092	
B-Depth of Cut	3063.781504	1	3063.781504	74.98792755	< 0.0001	
C-Feed Rate	175.7588021	1	175.7588021	4.301804257	0.0648	
D-Coolant Flow Rate	14.47164033	1	14.47164033	0.354202255	0.5650	
AB	3.22382025	1	3.22382025	0.078904974	0.7845	
AC	3.020644	1	3.020644	0.073932111	0.7912	
AD	74.84845225	1	74.84845225	1.831961681	0.2057	
BC	11.162281	1	11.162281	0.27320366	0.6126	
BD	0.020449	1	0.020449	0.000500502	0.9826	
CD	1.48230625	1	1.48230625	0.036280353	0.8527	
A^2	20.2475704	1	20.2475704	0.495571678	0.4975	
B^2	8.669308412	1	8.669308412	0.21218663	0.6549	
C^2	0.912542824	1	0.912542824	0.022335044	0.8842	
D^2	4.645064162	1	4.645064162	0.113690788	0.7430	
Residual	408.5699664	10	40.85699664			
Cor Total	4230.037899	24				
Std. Dev.	6.391947797		R-Squared	0.903412221		
Mean	243.24424		Adj R-Squared	0.76818933		
C.V. %	2.627789993		Pred R-Squared	N/A		
PRESS	N/A		Adeq Precision	9.2377362		

15.12 RESULTS AND DISCUSSION ON MRR

15.12.1 Parameter Interaction Effects

Figure 15.7 shows the surface interaction plot of spindle speed and depth of cut. From the graph, it is evident that the MRR is maximum when spindle speed is at a lower limit and depth of cut at a maximum level. Moreover, it is noteworthy that when the spindle speed is at a higher level, MRR decreases considerably even when the depth of cut is maintained at the higher limit.

Figure 15.8 shows the surface interaction plot of the spindle speed and feed rate. From the graph, it is evident that the MRR decreases as the spindle speed is increased. Also, it was observed that an increase in the feed rate does not affect much considerably; therefore, to increase the MRR, spindle speed is to be maintained between 6000 rpm to 7500 rpm with a feed rate of higher limit.

Figure 15.9 shows the surface interaction plot of the coolant flow and spindle speed. From the graph, one can identify that the flow of coolant follows a steady pattern with feed rate. When compared with other parameters, the effect of the coolant flow rate is found to be almost the same with few deviations.

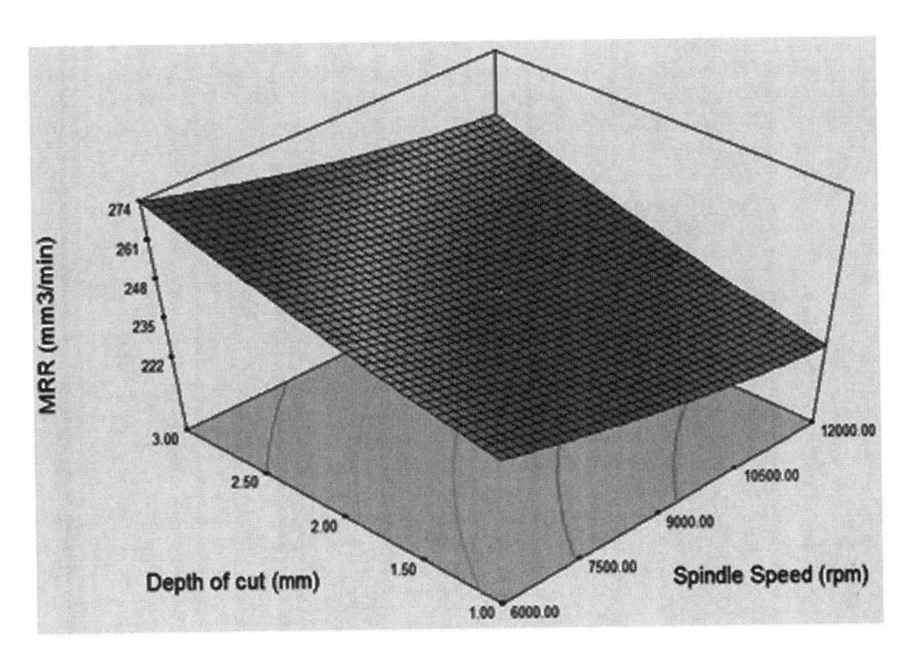

FIGURE 15.7 Surface interaction plot of spindle speed and depth of cut parameters on the MRR.

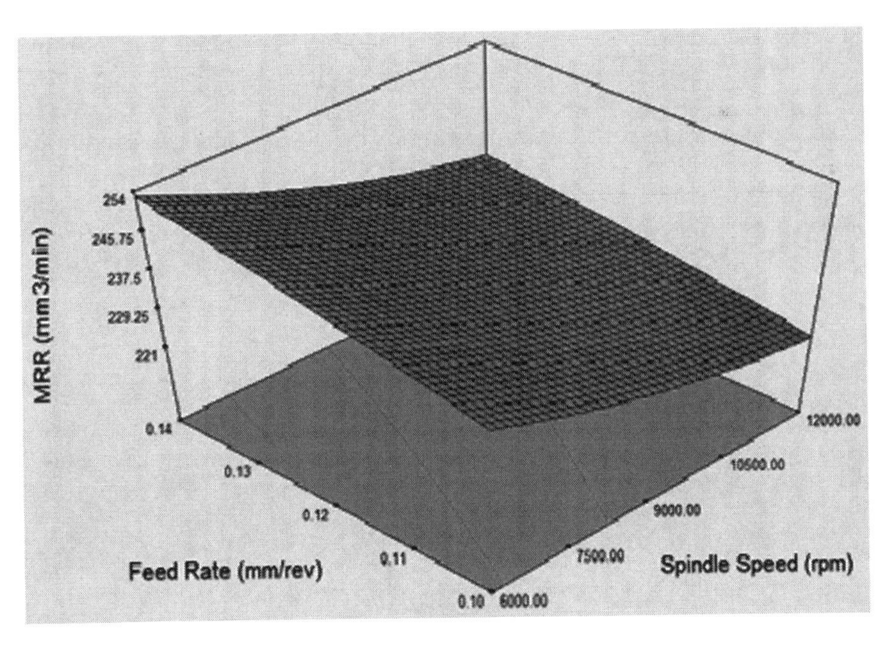

FIGURE 15.8 Surface interaction plot of spindle speed and feed rate parameters on the MRR.

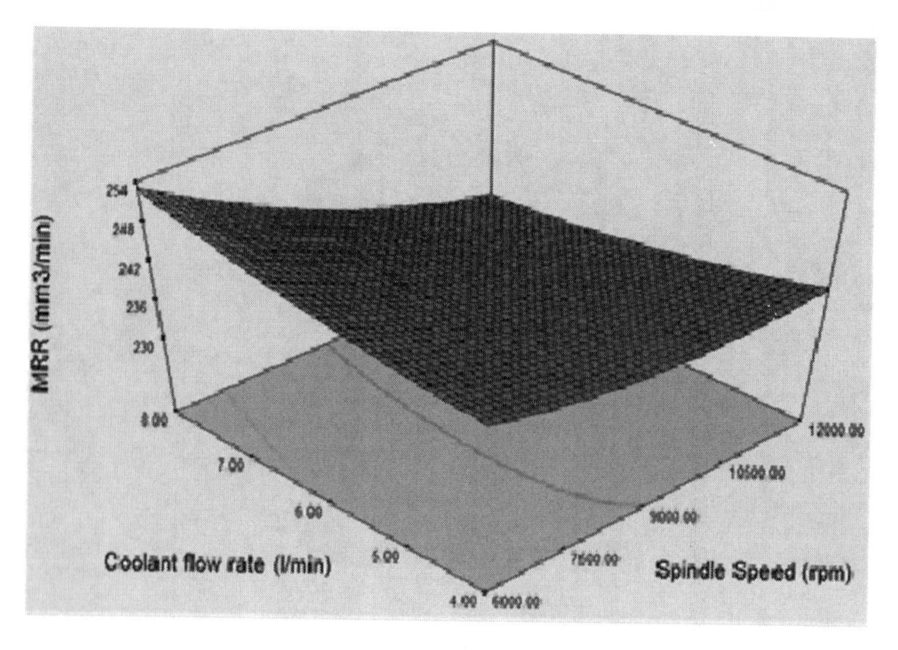

FIGURE 15.9 Surface interaction plot of the coolant flow rate and spindle speed parameters on the MRR.

Figure 15.10 shows the surface interaction plot of the feed rate and depth of cut. From the graph, feed rate and depth of cut play a prominent role in increasing the MRR. Depth of cut and feed rate are directly proportional to the increase in the MRR; therefore, it becomes necessary to determine the best combination of the above based on the spindle speed at which the machining operation is to be executed.

Figure 15.11 shows the surface interaction plot of the coolant flow and depth of cut. From the graph, one can identify that the flow of coolant follows a steady pattern with depth of cut. The increased level of both parameters enhances the MRR.

15.12.2 An Optimized Set of Parameters

The optimized set of parameters attained by the above analysis is shown in Table 15.5.

Figure 15.12 shows the effect of the parameters on the responses. The combined measure of the oughness and MRR is found to be 0.847775, whereas it is found to be 1 when roughness alone is considered. This clearly proves that the effect of MRR considerably reduces the effectiveness.

15.13 GENETIC ALGORITHM OPTIMIZATION

GA has been implemented to further optimize the results attained by the BB design. This has been made to ensure the availability of any further set of optimized

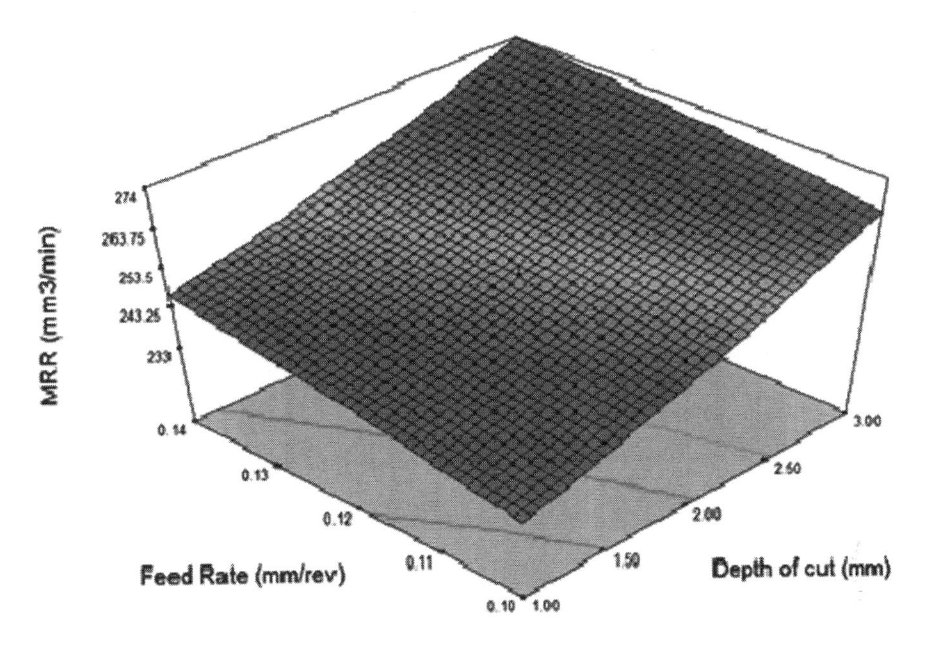

FIGURE 15.10 Surface interaction plot of feed rate and depth of cut parameters on the MRR.

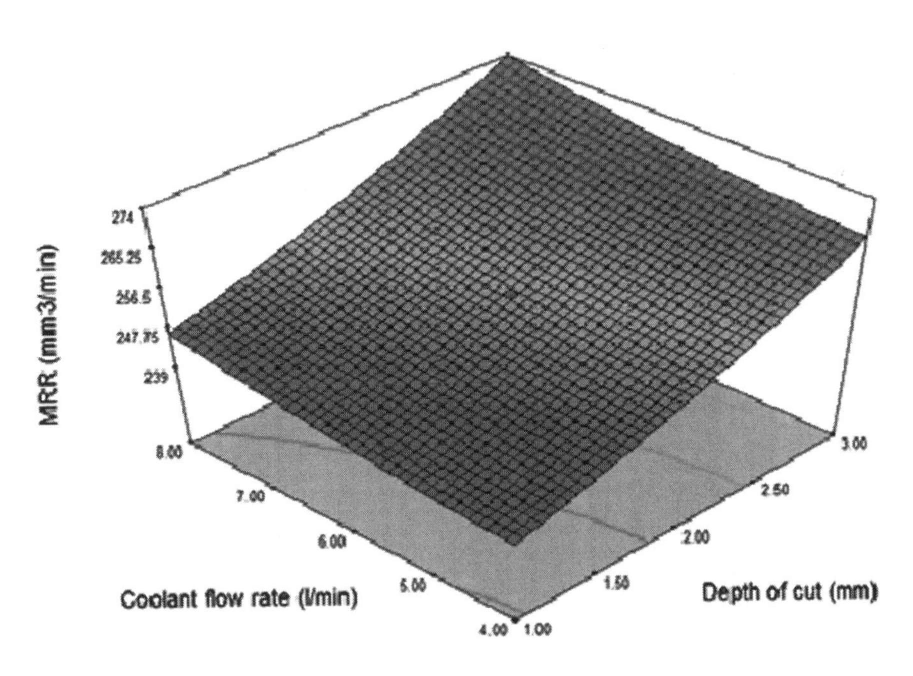

FIGURE 15.11 Surface interaction plot of the coolant flow rate and depth of cut parameters on the MRR.

TABLE 15.5
Optimized Set by BB Design

Spindle Speed (rpm)	Depth of Cut (mm)	Feed Rate (mm/rev)	Coolant Flow Rate (l/min)	Roughness (R_a) (microns)	MRR (mm^3/min)
6000.02	1.56	0.14	7.99	0.01399	253.056

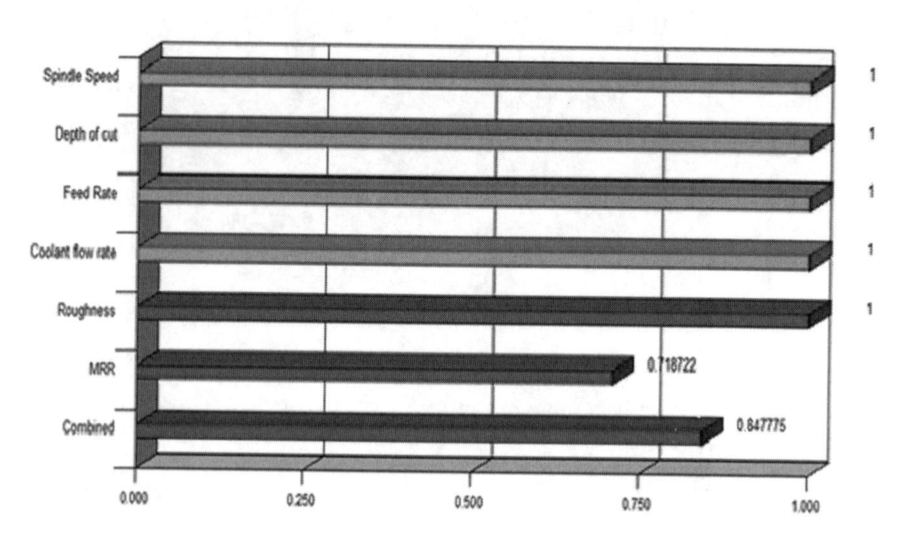

FIGURE 15.12 Bar chart – Effect of parameters.

parameters. The following are the objective functions assigned to perform the task stated as above:

Minimize: R_a (A, B, C, D)
Maximize: MRR (A, B, C, D); within the below specified ranges
$6000 \text{ rpm} \leq A \leq 12000 \text{ rpm}$; $1 \text{ mm} \leq B \leq 3 \text{ mm}$;
$0.10 \text{ mm/rev} \leq C \leq 0.14 \text{ mm/rev}$; $4 \text{ l/min} \leq D \leq 8 \text{ l/min}$

For the initial settings, the population size is taken as 100, stochastic function is opted with mutations performed based on Gaussian mutation with a mutation ratio fixed at 0.1. The scattered crossed function with crossover rate fixed at 1.0 is chosen for 1000 generations. The optimized level predicted by GA is shown in Table 15.6 and is found to be lower than the prediction value of BB design and experimental values.

15.13.1 Validation of the Model

The surface roughness and the MRR as per BB design is shown in Table 15.2. The regression equations for roughness and MRR are shown in Eq. 15.5 and Eq. 15.6

TABLE 15.6
Predicted Level by GA

Spindle Speed (rpm)	Depth of Cut (mm)	Feed Rate (mm/rev)	Coolant Flow Rate (l/min)	Roughness (R_a) (microns)	MRR (mm^3/min)
5915	1.15	0.13	7.44	0.0126	252.013

respectively. Table 15.7 shows the response comparison between predicted and measured values. From the Table 15.7, it is evident that the GA-predicted optimum conditions are found to be within ±1.5% ariations, which proves the fit of the model; therefore, it is concluded that the experimental results of responses as predicted by GA is efficient to be considered for machining.

15.14 CONCLUSION

BB design is followed in conducting the experiments to develop second-order mathematical models for predicting the responses (namely roughness and MRR) in end milling of BSL 165. An HSS end-mill cutter of a 12 mm diameter, four flutes having 22 mm overhung length is considered. The experimental sequences are generated by varying the process parameters, such as spindle speed, depth of cut, feed rate and coolant flow rate. The two stage approach of generating the mathematical model and optimizing the model by genetic algorithm has resulted in attaining the optimized set of machining parameters.

The second-order mathematical model developed is able to provide accurately the predicted values of responses close to actual values found in the experiments. The equations are found to be adequate with a confidence level of 95%.

The optimized parameters by BB design is achieved when the machining parameters are set at 6000.02 rpm, 1.56 mm, 0.14 mm/rev with coolant flow rate at 7.99 l/min, resulting in 0.01399 microns of surface roughness and 253.056 mm³/min of MRR.

On the other hand, GA recommends 0.0126 μm as the best minimum predicted surface roughness value and 251 mm³/min as the maximum material when the machining parameters are set at 5915 rpm, 1.15 mm, 0.13 mm/rev and with a coolant flow rate of 7.44 l/min. The confirmatory test showing the predicted values was found to be in agreement with the observed values.

In comparison, one could figure out that the objective set is well achieved in the latter set of optimized parameters. It is also interesting that the optimized level is achieved at a spindle speed lower than the value recommended by the BB design. To arrive at the optimized level in multiobjective functions, it is necessary to have thorough knowledge and understanding to interpret accurately. Though the optimization of two contradictory responses is complicated, it is very much needed to maintain the quality of the machined components in today's competitive production world.

TABLE 15.7
Predicted Measured Values of the Responses

Spindle Speed (rpm)	Depth of Cut (mm)	Feed Rate (mm/rev)	Coolant Flow Rate (l/min)	R_a- GA Prediction Value (microns)	R_a- Measured Value (microns)	MRR – GA Prediction (mm^3/min)	MRR –Measured Value (mm^3/min)
5915	1.15	0.13	7.44	0.0126	0.0128	252.013	252.110
5915	1.15	0.13	7.44	0.0124	0.0126	251.121	251.112
5915	1.15	0.13	7.44	0.0125	0.0126	252.015	251.912
5915	1.15	0.13	7.44	0.0126	0.0125	251.876	251.845
5915	1.15	0.13	7.44	0.0126	0.0125	251.872	252.110

REFERENCES

Alauddin, M., El Baradie, M. A. and Hashmi, M. S. J. 1997. Prediction of tool life in end milling by response surface methodology. *J Mater Process Technol*,71: 456–465.

Ali, Y. and Zhang, L. 1999. Surface roughness prediction of ground components using a fuzzy logic approach. *J Mater Process Technol*,89: 561–568.

Benardos, P. G. and Vosniakos, G. C. 2002. Prediction of surface roughness in CNC face milling using neural networks and Taguchi's design of experiments. *Robot Comput Integr Manuf*,18: 343–354.

Brecher, C., Quintana, G., Rudolf, T. and Ciurana, J. 2011. Use of NC kernel data for surface roughness monitoring in milling operations. *Int J Adv Manuf Technol*,53: 953–962.

Brezocnik, M. and Kovacic, M. 2003. Integrated genetic programming and genetic algorithm approach to predict surface roughness. *Mater Manuf Process*,18: 475–491.

Chang, H., Kim. J., Kim, I. H., Jang, D. Y. and Han, D. C. 2007. In-process surface roughness prediction using displacement signals from spindle motion. *Int J Mach Tool Manuf*,47: 1021–1026.

Chen, J. C. and Lou, M. S. 2000. Fuzzy-nets based approach using an accelerometer for in-process surface roughness prediction system in milling operations. *Comput Integr Manuf Syst*,13: 358–368.

Coker, S. A. and Shin, Y. C. 1996. In-process control of surface roughness due to tool wear using a new ultrasonic system. *Int J Mach Tool Manuf*,36: 411–422.

Dhokia, V. G., Kumar, S., Vichare, P. and Newman, S. T. 2008. An intelligent approach for the prediction of surface roughness in ball-end machining of polypropylene. *Robot Comput Integr Manuf*,24: 835–842.

Gologlu, C. and Sakarya, N. 2008. The effects of cutter path strategies on surface roughness of pocket milling of 1.2738 steel based on Taguchi method. *J Mater Process Technol*,206: 7–15.

Lou, S. J. and Chen, J. C. 1999. In-process surface roughness recognition (ISRR) system in end-milling operations. *Int J Adv Manuf Technol*,15: 200–209.

Mansour, A. and Abdalla, H. 2002. Surface roughness model for end milling: a semi-free cutting carbon casehardening steel (EN 32) in dry condition. *J Mater Process Technol*,124: 183–191.

Martellotti, M. E. 1941. An analysis of the milling process. *Transl ASME*,63: 667.

Ming-Yung, W. and Chang, H. Y. 2004. Experimental study of surface roughness by the slot end milling AL2014-T6. *Int J Mach Tool Manuf*,44: 51–57.

Oktem, H., Erzurumlu, T. and Erzincanli, F. 2006. Prediction of minimum surface roughness in end milling mold parts using neutral network and genetic algorithm. *Mater Des*,27: 735–744.

Palanisamy, P., Rajendran, I. and Shanmugasundaram, S. 2007. Optimization of machining parameters using genetic algorithm and experimental validation for end-milling operations. *Int J Adv Manuf Technol*,32: 644–655.

Quintana, G., Ciurana, J. and Ribatallada, J. 2010. Surface roughness generation and material removal rate in ball end milling operations. *Mater Manuf Process*,25: 386–398.

Tsai, Y., Chen, J. C. and Lou, S. 1999. An in-process surface recognition system based on neural networks in end milling cutting operations. *Int J Mach Tool Manuf*,39: 583–605.

Zain, A. M., Haron, H. and Sharif, S. 2008. An overview of GA technique for surface roughness optimization in milling process. In *IEEE proc Inter Symp on Infor Technol, IT Sim*, 4: 1–6.

16 Abrasive Wear Behaviour of Titanium and Ceramic-Coated Titanium

R. Sathiyamoorthy, K. Shanmugam and K. Murugan

CONTENTS

16.1 INTRODUCTION

Wear occurs in many different industrial situations and results in high costs due to equipment failure, replacement of worn parts and downtime during repairs. In addition, wear may influence the quality of the products involved. Wear is defined as the progressive removal of material from a surface due to mechanical movement with or without chemical processes. Among the various wear mechanisms, abrasive wear is the most important one because of its destructive character and its high occurrence frequency (50% of total wear failures). In abrasive wear, detachment of material from surfaces in relative motion is caused by hard particles between the opposing surfaces or fixed in one of them. Its control and minimization depends essentially on not only the appropriate selection of materials, but also understanding the mechanisms that are responsible for the abrasive wear of these materials. Its control and minimization depends essentially on not only the appropriate selection of materials, but also understanding the mechanisms that are responsible for the abrasive wear of these materials.

The dry sand rubber wheel (DSRW) test has been employed to examine the abrasion behaviour of a wide range of materials. In many programmes, the test is used simply to provide a quantitative ranking of the abrasion resistance of different materials. In DSRW abrasion, the operative mechanisms of wear depend largely upon the material properties (e.g. hardness, ductility, toughness) along with the manner in which the particles move through the contact between the wheel and specimen. The particles may embed into the moving rubber wheel and slide across the sample material through the contact region (generally termed two-body abrasion) or pass through the contact region by rolling between the rubber wheel and the sample (generally termed three-body abrasion). It has been observed that the manner in which the particles move through the contact affects the mode of wear and, ultimately, the rate of material removal. The motion of particles through the contact zone has itself been shown to depend upon a number of parameters associated with the system, amongst them particle shape, applied load and the hardness of the test surface and counter body (Trezona et al. 1999; Shipway 2004).

16.2 ABRASIVE WEAR BEHAVIOUR OF TITANIUM

The standard samples of titanium were made and the samples were exposed to the ASTM G65 standard test. The wear rate of titanium substrate, shown in Figure 16.1, shows a higher wear rate compared to coatings. The mass loss for these samples was calculated by weighing them before and after the test. Figure 16.2 shows the wear scars left on the titanium substrate.

The abraded surface of titanium is shown in Figure 16.2 and ploughing and grooving are observed as the major wear mechanism. When the particles entrained in between the rubber wheel and counter surface, the particle may slide, roll or embed with the rubber wheel, causing two-body abrasion. From the scanning electron microscope (SEM) image of the abraded titanium surface, we could see three possibilities, some part of the surface show the evidence of particle rolling with significant indentation of the surface and little directionality. Some part of worn surfaces shows the evidence of particle sliding (grooving) and embedded particles with rubber wheel plough the surface and also causes grinding on the surface. The

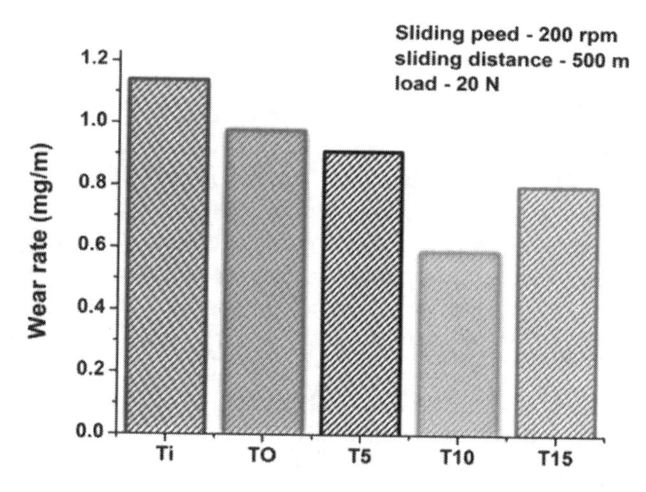

FIGURE 16.1 Wear rate of titanium and coatings.

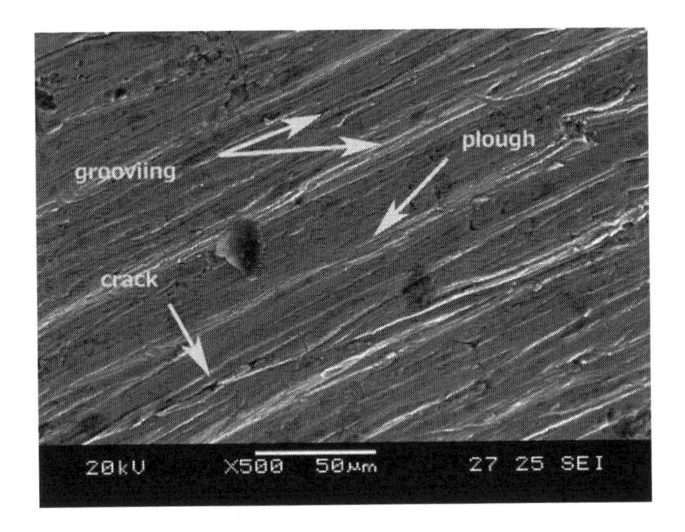

FIGURE 16.2 SEM micrograph of abraded titanium surface.

abrasive grain may rub, plough or cut and that the wear rate of a material is either controlled by the fraction of grains which cut it.

For abrasive cutting to be possible, the abrasive must be hard enough to indent the counterface. The abrasive cutting occurs when the hardness of the abrasive (H_a) is greater than 1.2 Hs, where H_s is hardness of surface. In this study, titanium has hardness of 220 $HV_{0.3}$,which is significantly lower than quartz sand of hardness 1743 HV. The motion of particles through the contact zone depends upon both material type and the applied load. It is proposed that the particle will move through the contact by sliding (grooving) if the clockwise movement is less than the counterclockwise movement i.e. (Fang et al. 1993).

$$F_p h < P_p e$$

Where F_p is the lateral force on the particle, P_p is the transverse force on the particle and the dimensions e and h are distances of forces. It has been shown for a given applied load on a particle that the high sample hardness favours particle sliding (grooving), while low sample hardness favours particle rolling. When analyzing the abraded surface of titanium, this is not justifiable because the major area was ploughed, indicating that other factors, such as particle shape, size, material properties and material removal mechanism, were affecting the wear mechanism.

The abrasive shape clearly has an effect on the distribution of attack angles presented to the counterface as well as the wear rate and wear mechanism. The quartz particles used in this study are angular so that the attack angles will tend to rise on counterface, indicating that higher penetration of the abrasive particle into the surface leads to increased ploughing.

When the particle entrained in between the rubber wheel and titanium substrate, three main zones may be distinguished: the entrance zone, where the particles are crushed (breaking occurs); the central zone, where the pressure is maximal (distance between wheel and specimen is minimal); and the exit zone. In this study, with applied constant load, the changes of particle size and shape may happen at the particle rubber interface due to particle crushing. This phenomenon causes the lateral forces on the indenting particles and the friction coefficient. Because it has been argued that a low coefficient of friction favours particle sliding (grooving), particle sliding is thus shown to be favoured by constant applied load. It has been reported that when there is adhesion between rubber and a rigid body with tangential motion between them, there will be an asymmetrical distribution of contact forces across the contact area (Bui and Ponthot 2002). When the rubber wheel loaded, the contact particle increased in size, and the distance between the counter surface and rubber wheel decreased. This increases the frictional and lateral forces, which changes particle motion during sliding. The temperature of the interface between rubber and counterface increases during testing, which decreases the rubber wheel hardness, and thus causes a decrease in elastic moduli and high strain associated with the normal and lateral forces on the particles (Nahvi et al. 2009).

The shape of the particles has an important role in determining the wear mechanism and wear rate. Figure 16.3 shows SEM images of particles that indicate angular shape. Such particle protrusions contribute to the severity of wear. Sharp protrusions promote rapid material removal (Stachowiak and Stachowiak 2001). It has been confirmed in laboratory tests by many researchers that an increase in particle angularity results in a significant increase in abrasive wear rate, although it is difficult to define a quantity that describes particle shape well in this context.

FIGURE 16.3 SEM morphology of AFS 50/70 quartz sand.

16.3 ABRASIVE WEAR BEHAVIOUR OF TITANIA (TIO_2)

The abrasive wear resistance of titania (T0) coating is shown in Figure 16.1. The wear rate is lower than titanium substrate. The SEM morphology of abraded surface(Figure 16.4) shows brittle fracture and chipping is the major wear mechanism. During wear of ceramic material, such as titania, both plastic deformation and brittle fracture of the ceramic might occur. This behaviour depends on the nature of the ceramic material that is being tested. In ceramic material, critical depth of cut is an important parameter. When critical depth of cut is reached, the behaviour of ceramic will change from plastic deformation to brittle fracture and chipping (Lima and Marple 2008).

An equation for critical depth of cut in a ceramic material is (Malkin and Hwang 1996):

$$\text{Critical depth of cut } d_c = \beta \frac{E/H}{\left(\dfrac{H}{Kc}\right)^2}$$

Where d_c is the critical depth of cut, E is the elastic modulus, H is the hardness, Kc is fracture toughness and β is constant. From this equation, it is obvious that the critical depth of cut is directly proportional to the square of the toughness to hardness ratio for each ceramic. From the Figure 16.4 SEM image of abraded titania coating, we can understand that the major material removal mechanism is chipping and brittle fracture. In other words, it can be said that depth cut is significantly lower.

Titania has higher volume percentage of porosity compared to other coating. The porosity in thermal sprayed coating is divided into two categories: coarse and fine pores. The coarse porosity happens because of failure in completely filling the gaps

FIGURE 16.4 SEM morphology of abraded surface of T0 coating.

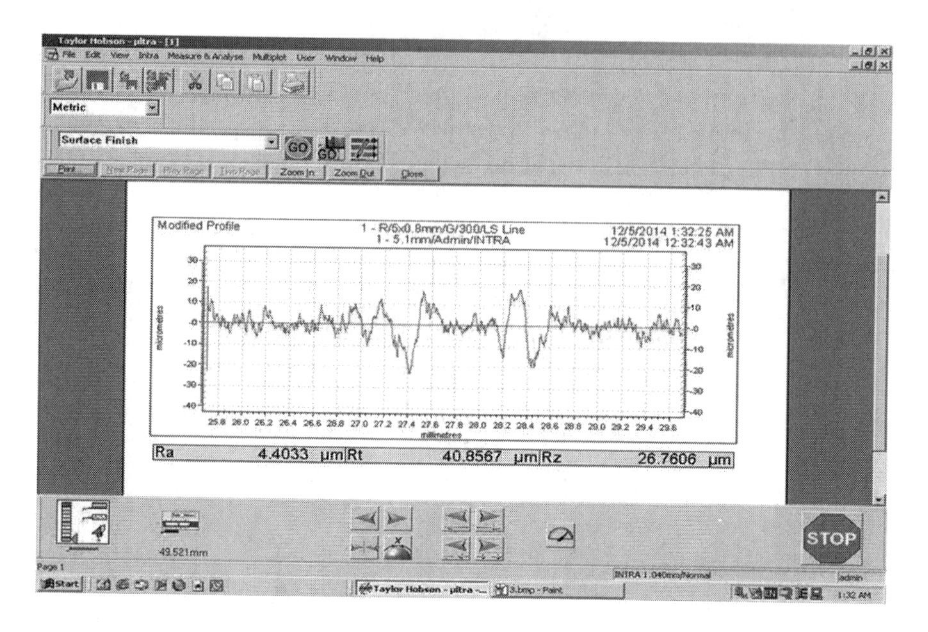

FIGURE 16.5 Surface roughness profile of T0 as sprayed coating.

from splats (splat stacking), unmelted particles and gas entrapment (Bolelli et al. 2006). In fine porosity, the same horizontal lines in the cross section of the coatings are the boundaries between different layers of splats, which form the coating. These lines are called the fine pores, which always present in the thermal sprayed coatings because of the nature of the process (Lima and Marple 2005). This porosity weakens the mechanical interlocking between under laying splats, which allows the intersplat crack growth during abrasion, leading to splat ejection or pullout.

The surface roughness of the sprayed titania coating (shown in Figure 16.5) is measured using a surface roughness tester (Talysurf, Taylor Hobson, UK) is R_a 4.403 µm. The worn surface depth and roughness parameters are higher than other coatings. The surface roughness R_a is 3.588 µm, which is comparatively higher than the other coatings. Using confocal microscopy, images were taken from the scar; these images are focused on every point of the surface, which provides qualitative information on the smoothness of the surface. By looking at the images shown in Figure 16.4, it is obvious that the scar left on the titania surface shows signs of breakage and chipping (Figure 16.6).

16.4 ABRASIVE WEAR BEHAVIOUR OF TiO_2+5 SIC COATING

Figure 16.1 shows the wear rate of T5 coating, which is comparatively lower than T0 coating and higher than T10 and T15 coatings. The SEM morphology of abraded surface (Figure 16.7) shows the microcracking and brittle fracture is more predominant where hard phase regions were substantially removed. This, in turn, easily attacks the region and removes the material. Figure 16.8 shows the surface roughness profile of as sprayed T5 coating, the average surface roughness R_a is 7.93 µm.

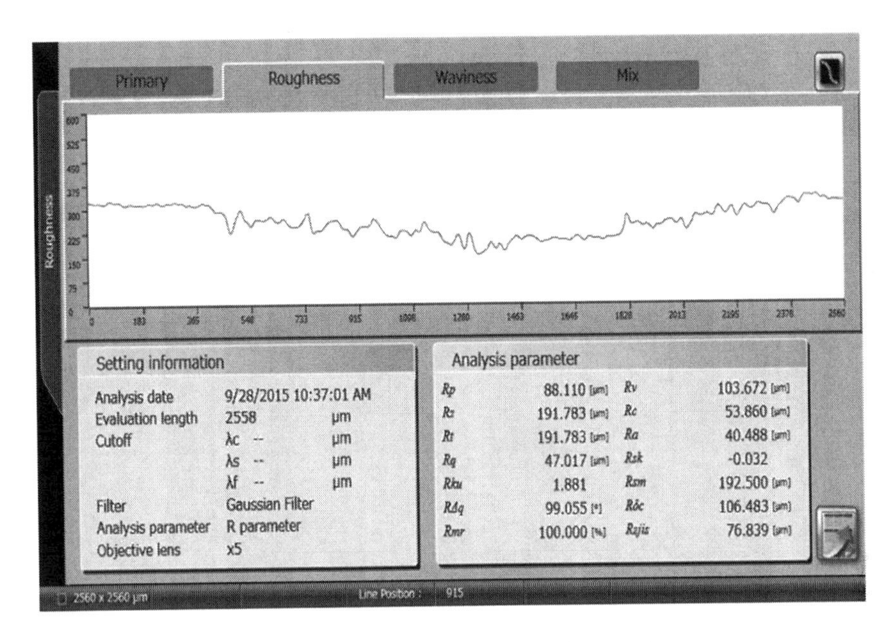

FIGURE 16.6 Surface roughness profile of T0 worn surface coating.

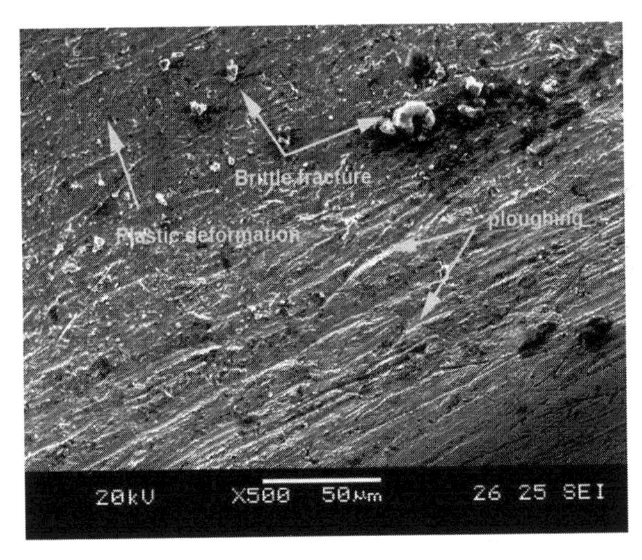

FIGURE 16.7 SEM morphology of worn surface of T5 coating.

The confocal topography and surface roughness profile of abraded surface is shown in Figure 16.8. Here, we see the average surface roughness R_a is 5.562 µm. The surface roughness is higher than the other SiC reinforced coatings, and it is thought that the soft phase TiO_2 removed from the surface and hard phase SiC fractured and still adhered (embed with the coating) with surface by being keyed into the microcavities.

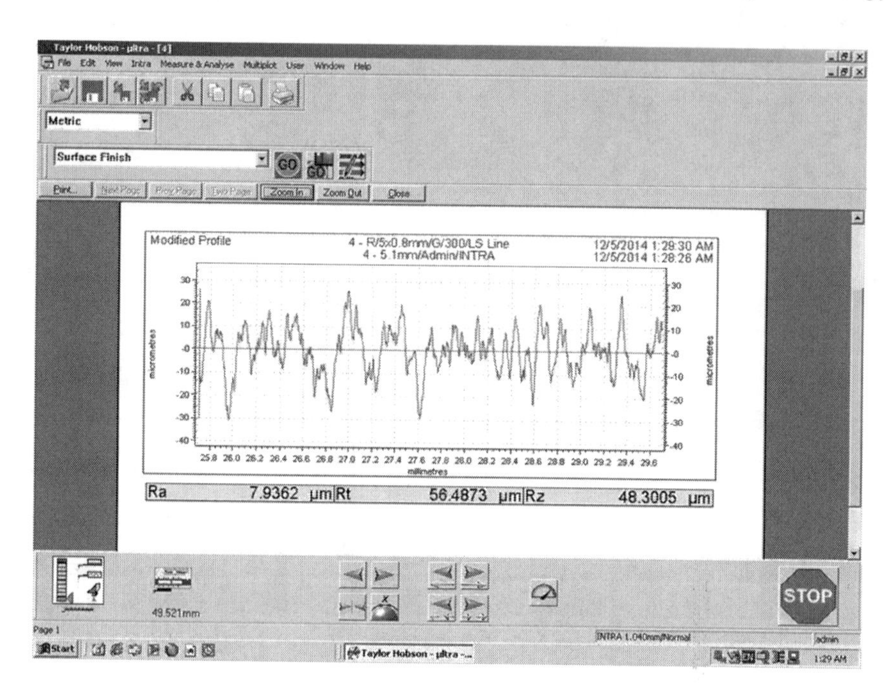

FIGURE 16.8 Surface roughness profile of T5 as sprayed coating.

Wear rate is sensitive to the ratio of abrasive hardness H_a to the surface hardness H_s. Abrasion under conditions in which $H_a/H_s > 1.2$ is sometimes termed 'Hard abrasion', in contrast to soft abrasion, in which $H_a/H_s < 1.2$. It is evident that for all coatings the abrasive wear is 'hard' abrasion in this study. Figure 16.7 shows the evidence of particle sliding (grooving) and also evidence of small particle rolling across the surface of the sample. The carbide cracking and pullout can be seen in worn surface. The coating matrix seems to wear at a higher rate, leaving unprotected carbide particles and also some splat pullouts due to subsurface cracking.

Cyclic indenting contact of abrasive particles during three-body abrasion process causes compressive stress in the surface of coating. The titania is initially compressed out of the surface by these stresses ahead of and to the sides of indenter. The next stage is probably damage to the secondary hard phases, such as SiC and TiC grains, which are heavily loaded locations because the titania has flowed plastically. The SiC grains break into small fragments and are gradually pulled out from the surface. This secondary hard phase, present in the coating, is at the edge of a defect (a crack or an area of surface damage); these grains will experience greater load as the abrasive particles indent than will the grains away from the defect.

These grains will be the first to be damaged, resulting in growth of the defect. Microcracks from around the pits propagate through the coating preferentially in the SiC rich phase or along the splat boundaries. Elastic-plastic indentation of the abradent into the coating can also cause subsurface cracks to form close to the surface of the coating, resulting in detachment of fragments of coating (Stewart et al. 1999) (Figure 16.9).

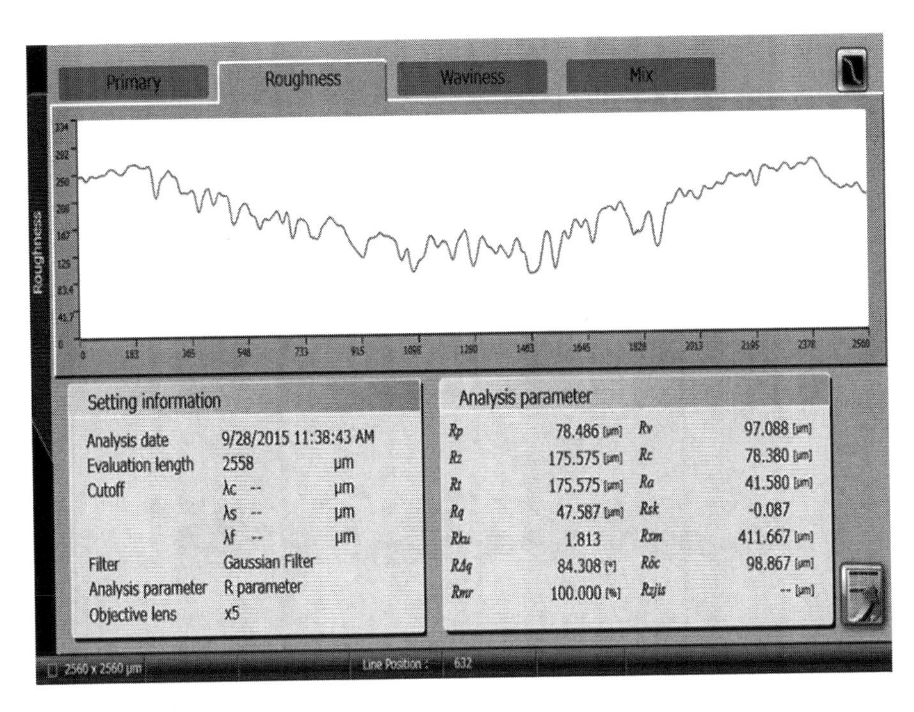

FIGURE 16.9 Surface roughness profile of T5 worn surface coating.

16.5 ABRASIVE WEAR OF TIO_2 + 10 SIC COATING

The abrasive wear rate of TiO_2 + 10 SiC (denoted as T10) is superior to all the coatings shown in Figure 16.1. The abraded surface analyzed using SEM (Figure 16.10) shows plastic deformation and ploughing as the major wear mechanism of material removal. Figure 16.11 shows that the surface roughness of the sprayed coating has an average surface roughness R_a value of 8.0765 μm. The surface roughness value of abraded surface, analyzed using confocal microscopy, is shown in Figure 16.12, which shows average roughness value of 3.2381 μm. The depth of the wear scar also observed in confocal microscopy shows the depth 38.228 μm is comparatively lower than other coatings. This smooth wear scar observed in SEM is conformed in surface roughness profile. This is due to effective melting of SiC particles and agglomeration of partially melted SiC with TiO_2, which leads to low wear rate of the coating. During wear testing, crack arresting effect is caused by the presence of secondary phases and unmelted SiC particles throughout the coating. Low porosity of the coating hinders the crack propagation, and higher bond strength of the coating gives superior wear resistance.

The material removal in this coating is plastic deformation caused by hard quartz sand; this plastic deformation occurs mostly by plastic ploughing and cutting and followed by some local associated fracture in more brittle composite (Hutchings and Shipway 1992). During abrasion, the hard and sharp quartz sand causes plastic deformation of the surface, resulting in the formation of grooves with material pileup at the groove edges in the first stage. Fatigue of the surface layers most probably

FIGURE 16.10 SEM morphology of abraded surface of T10 coating.

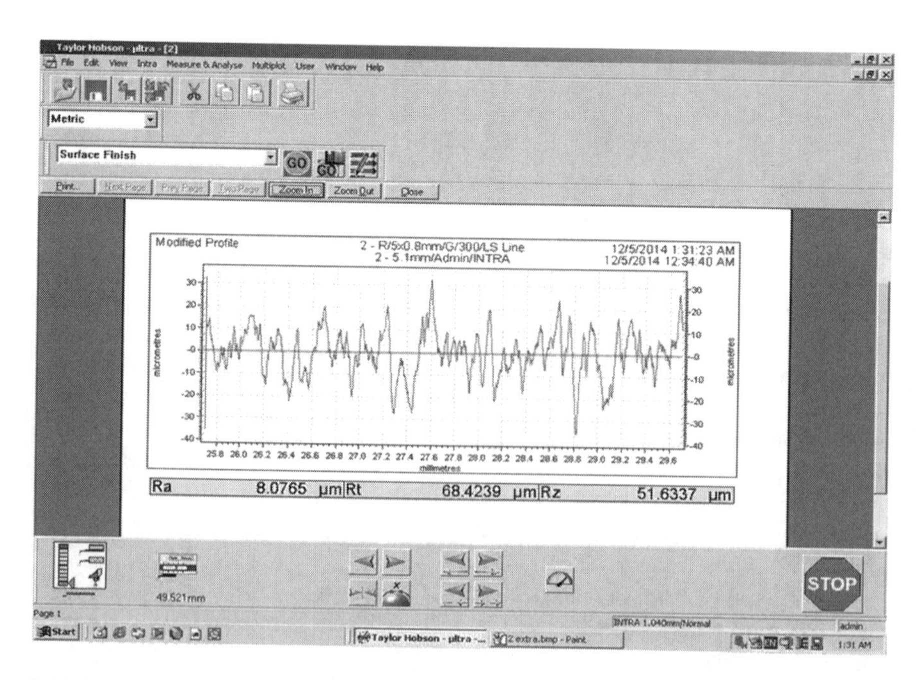

FIGURE 16.11 Surface roughness profile of T10 as sprayed coating.

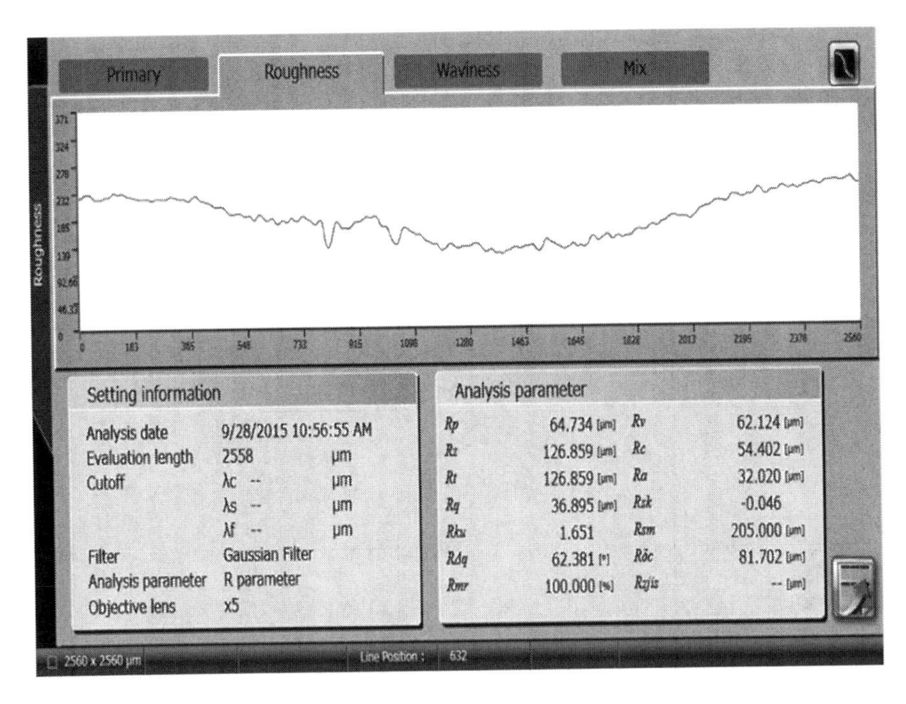

FIGURE 16.12 Surface roughness profile of T10 worn surface coating.

occurs through mechanical deformation of those layers and results in a spalling type failure, while subsurface material deformation leads to cracking, which propagates into coating as is seen for some cases at the second stage.

In wear scars of all coatings abraded with quarts sand, two-body abrasion can be clearly observed. The SEM images of wear scars show grooves on the surfaces, the grooves are macro- and microgrooves. The dimensions of deformation caused by individual abrasive particles is substantially greater than the size of the carbide grains, whereas in the second type (microgrooves), the carbide grains are comparable in size with scale of the abrasion damage or larger. In the first type of groove, the behaviour of coatings against deformation is very much like a homogeneous solid, while in the second, coatings respond heterogeneously. In the macroscale mode, when coating behaves homogeneously (like solid material), plastic deformation occurs in TiO_2 coating matrix, with macrogrooves being formed parallel to the sliding direction. The displacement and removal of material in this mode depends on the depth of grooves and consequently bulk hardness of the coatings. To remove material by abrasion, penetration of abrasives into material and high enough shear force (parallel to the surface) acting on the penetrating particle are necessary.

In the microscale mode, in which coatings respond heterogeneously during abrasion, grooves were formed parallel to sliding direction in microscale mode for all coating. Although wear is generally by cutting mode, the obvious clean cutting across the TiO_2 coating matrix with some cracks, delamination and large grooves on abraded surfaces. SEM micrographs of abraded surface reveal a number of fragmented carbides and voids resulting from pullout of splats.

Porosity has an important role in formation of grooves and consequent removal of material by the microscale mode. In the case of hard abrasive particles moving in between the soft rubber wheel and hard coating surfaces, entrapment of the corner of an abrasive particle (which is sharp) into the surface can occur (Ghabchi et al. 2010), or coating collapse may take place near pores (Hartfield-Wünsch and Tung 1994). In these cases, a corner of abrasive particle can start to scratch from deep indentation; therefore open surface pores will serve as origins for wear scratches made by individual abrasive particles. However, penetration of abrasive particle into the dense region of coating will be more difficult. Thus, a bigger portion of pores leads to more grooves and results in a higher wear rate. Formation of grooves in T10 coating with much smaller width than actual abrasive particle size provides evidence of the suggested mechanism, which is based on entrapment of a corner abrasive particle into the surface open porosities.

In the second stage of wear, fracture is the predominant source of material degradation. At the first stage of wear, when a sharp particle embeds into the surface and slides, a plastic groove forms. The penetration of the surface abrasive particles is observed to be different for each coating. Lateral cracks grow upward to the surface from base of the surface indented region, driven by the residual stresses associated with deformation. In thermal sprayed coatings, two types of cracks are formed: lateral cracks parallel to the surfaces and median cracks perpendicular the surface. The formation of a vertical crack is the initial stage of the material loss procedure and is caused by the indentation of an abradent into the coating. These cracks run down through the coating and end when they reach splat boundary. Immediately after, they propagate parallel to the coating surface until they find a path that leads back to the surface. Generally, this process results in a high material removal rate in the coating. Although plastic deformation and fracture are the primary material removal mechanisms with abrasion, the wear mechanisms for all coatings are not same. Because the hardness of quartz sand is higher than that of all coatings, the plastic deformation is dominant wear mechanism in the first stage of abrasion.

During two-body abrasion (when abrasive particles are temporarily embedded in the rubber wheel), ploughing and grooving marks appear. The depth of grooves depends on the bulk hardness of coating at this stage, and the penetration degree of the abrasive into the coating is important. In this study, T10 coating has higher hardness, low porosity and higher bond strength. There is no evidence that cracking or fracture occurred for this coating. Moreover, the higher hardness of coating causes a decrease in the rate of wear under abrasion. The low porosity also protects the coating surface from grooving. The surface profile of the coating shown in Figure 16.11 reveals lower depth compared to other coatings.

16.6 ABRASIVE WEAR OF TIO_2+15 SIC COATING

The wear rate of TiO_2+15 SiC coating (denoted as T15) is shown in Figure 16.1, is high compared to T10 coating and lower than T0 coating. SEM micrographs of abraded coating are shown in Figure 16.13. The fracturing and fragmentation of large ceramic grains with cracks and deep grooves are clearly pronounced on the surface of coating. The surface roughness profile of the coating, shown in Figure 16.14, reveals R_a is 8.777 µm. The surface roughness of abraded surface is observed form confocal microscopy is 5.382 µm and is shown in Figure 16.15. The wear scar depth is also higher than T10 coating. The main wear mechanism operating in this coating is brittle fracture of TiO_2 splats and the loosening of SiC grains through low cycle fatigue during multipassing of abrasive particles and direct breakaway of splats triggered by crack intersection. We could understand that the higher wear rate of this coating is due to high porosity, low hardness and low bond strength of the coating. From the SEM and wear profile, it can be seen that worn surface of T15 coating is characterized by deep grooves, which are formed as the abrasive particles plough across the surface and eventually remove the material.

The higher porosity initiates intergranular and transgranular cracks during indentation of abrasives and propagates through unmelted particles, causing an increase in wear rate. We could also understand that the lower bond strength due to poor adhesion of splat to splat and splat to substrate leads to low abrasion resistance.

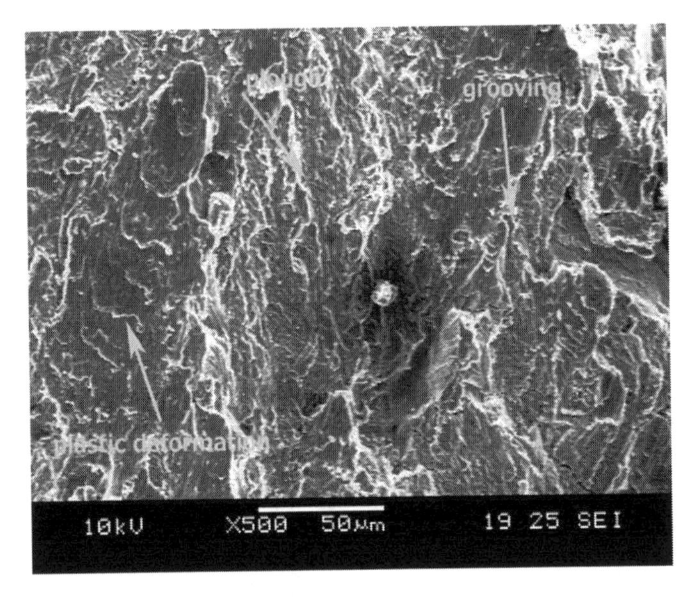

FIGURE 16.13 SEM morphology of abraded surface of T15 coating.

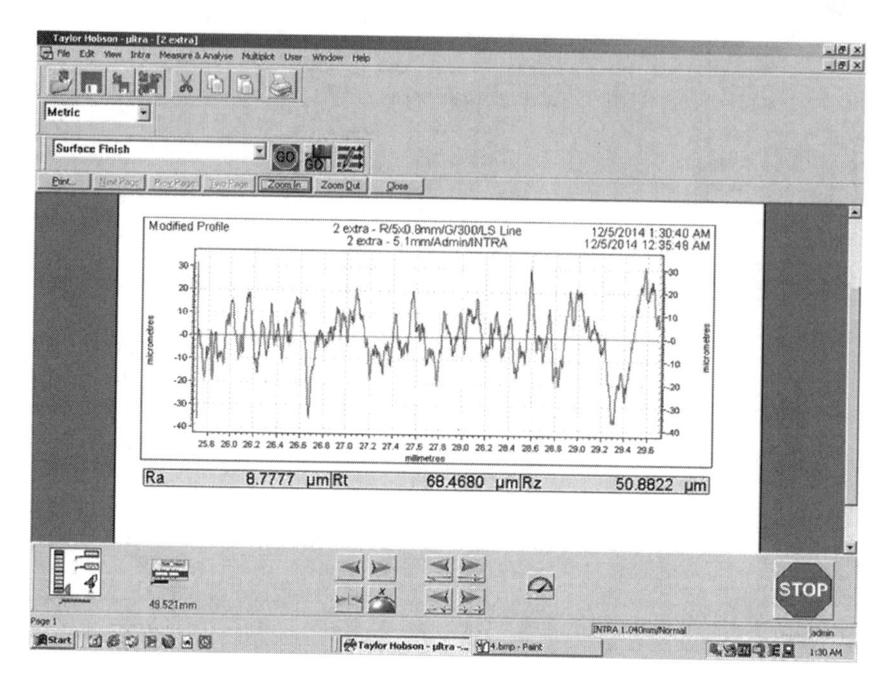

FIGURE 16.14 Surface roughness profile of T15 as sprayed coating.

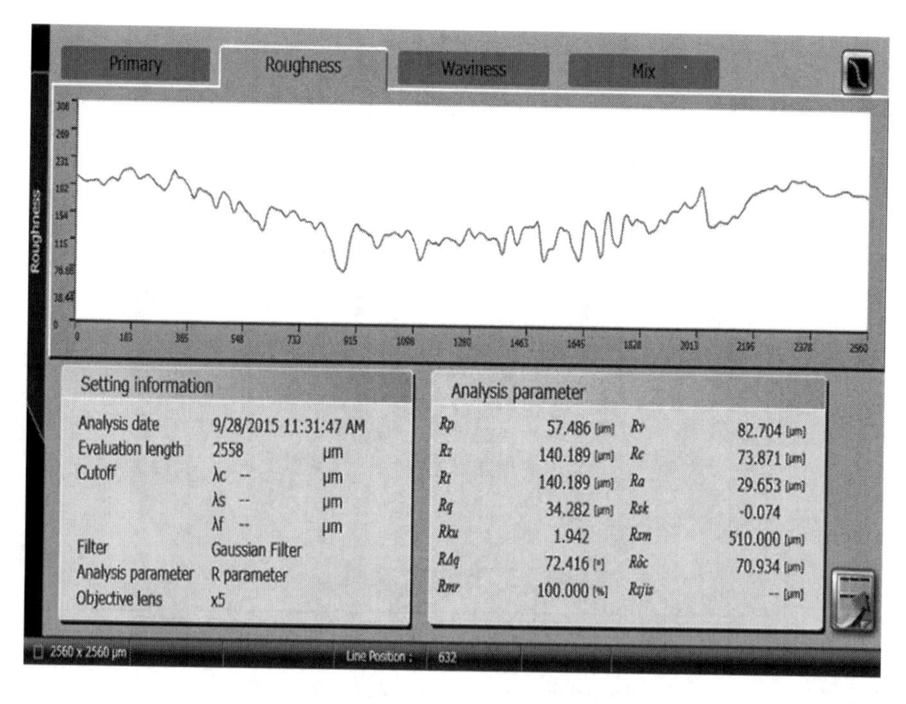

FIGURE 16.15 Surface roughness profile of T15 worn surface coating.

16.7 SUMMARY

In this part of the work, an abrasive wear test was carried out on titanium base metal as well as on sprayed coatings as per ASTM G65 standard. The abrasion of titanium in dry sand rubber wheel shows higher material loss. The abrasive wear rate of titania coating is higher than SiC reinforced titania coatings. Titania shows the higher values of average roughness, and the major wear mechanism was fragmentation and brittle fracture. Among these coatings, T10 coating shows very good wear resistance compared to other coatings and base metal. This is justified that the mechanical and microstructural characteristics of T10 coating is superior to other coatings, such as high microhardness, low porosity level and higher bond strength.

REFERENCES

Bolelli, G., Cannillo, V., Lusvarghi, L. and Manfredini, T. 2006. Wear behaviour of thermally sprayed ceramic oxide coatings. *Wear*, 261(11):1298–1315.

Bui, Q.V. and Ponthot, J.P. 2002. Estimation of rubber sliding friction from asperity interaction modeling. *Wear*, 252(1):150–160.

Fang, L., Kong., X.L., Su, J.Y. and Zhou, Q.D. 1993. Movement patterns of abrasive particles in three-body abrasion. *Wear*, 162:782–789.

Ghabchi, A., Varis., T., Turunen., E., Suhonen., T., Liu., X. and Hannula, S.P. 2010. Behavior of HVOF WC-10Co4Cr coatings with different carbide size in fine and coarse particle abrasion. *Journal of Thermal Spray Technology*, 19(1–2):368–377.

Hartfield-Wünsch, S.E. and Tung, S.C. 1994. *The effect of microstructure on the wear behavior of Thermal Spray Coatings*. Materials Park, OH (United States): ASM International.

Hutchings, I. M., & Shipway, P. 1992. Tribology. Friction and wear of engineering materials. Institution of Mechanical Engineers. *Part C: Journal of Mechanical Engineering Science*, 207.

Lima, R.S. and Marple, B.R. 2005. Enhanced ductility in thermally sprayed titania coating synthesized using a nanostructured feedstock. *Materials Science and Engineering: A*, 395(1):269–280.

Lima, R.S. and Marple, B.R. 2008. Process–property–performance relationships for titanium dioxide coatings engineered from nanostructured and conventional powders. *Materials and Design*, 29(9):1845–1855.

Malkin, S. and Hwang, T.W. 1996. Grinding mechanisms for ceramics. *CIRP Annals – Manufacturing Technology*, 45(2):569–580.

Nahvi, S.M., Shipway, P.H. and McCartney, D.G. 2009. Particle motion and modes of wear in the dry sand–rubber wheel abrasion test. *Wear*, 267(11):2083–2091.

Shipway, P.H. 2004. A mechanical model for particle motion in the micro-scale abrasion wear test. *Wear*, 257(9):984–991.

Stachowiak, G.B. and Stachowiak, G.W. 2001. The effects of particle characteristics on three-body abrasive wear. *Wear*, 249(3):201–207.

Stewart, D.A., Shipway, P.H. and McCartney, D.G. 1999. Abrasive wear behaviour of conventional and nanocomposite HVOF-sprayed WC–Co coatings. *Wear*, 225:789–798.

Trezona, R.I., Allsopp, D.N. and Hutchings, I.M. 1999. Transitions between two-body and three-body abrasive wear: influence of test conditions in the microscale abrasive wear test. *Wear*, 225:205–214.

17 Impact of Friction Stir Welding Process on Mechanical and Microstructural Behaviour of Aluminium Alloy Joint – A Review

K.P. Yuvaraj, P. Ashoka Varthanan and K.P. Boopathiraja

CONTENTS

17.1 INTRODUCTION

Aluminium is a soft, durable, lightweight, nonmagnetic, nonsparking, malleable metal with an appearance starting from silvery to dull grey, depending on the surface roughness. It is also insoluble in alcohol, though it is often soluble in water in specific forms. The metallic element has a fraction of the density and stiffness of steel. It is ductile so it can be easily machined, cast, drawn and extruded to different shapes. Corrosion resistance is often outstanding because of a thin surface layer of corundum that forms once the metal is exposed to air, effectively preventing further oxidization. Strong aluminium alloys exhibit less corrosion resistance because of galvanic reactions with the alloyed copper. The vital factors in choosing metal and its alloys are their high strength-to-weight ratio, corrosion resistance to several chemicals, high thermal and electrical conductivity, nontoxicity, reflectivity, appearance, ease of formability, machinability and their nonmagnetic nature. The melting temperature of those alloys ranges from 482–660°C. The thermal conductivity is six times that of steel, and the thermal expansion is double that of steel. The aluminium alloys can be divided into two groups based on production method: wrought alloys and cast alloys. Furthermore, alloys can be classified into two types: heat treatable and nonheat treatable. The common wrought alloys and their designation system is mentioned in Table 17.1. It consists of a four-digit designation for different grades of alloys. The first digit indicates the major alloying element. Normally, the second digit will be zero. The last two digits of the designation indicate the presence of a particular alloy within the family.

The low melting point of cast aluminium alloys makes it suitable for the casting process. The strength is enhanced because of the presence of small alloying elements and suitable casting characteristics. Moreover, the formation of age hardening precipitates is permitted by the addition of alloying elements such as copper, silicon, magnesium and zinc. The designation system for the cast aluminium alloys is indicated in the Table 17.2. The first digit of the designation represents the alloy group. The second and third digit specifies the particular alloy or aluminium purity. The final digit indicates the form of the product (casting or ingot), which is indicated separately as a decimal point.

The most extensively used techniques for welding aluminium alloys are gas metal arc welding (GMAW), variable polarity plasma arc (VPPA), gas tungsten arc welding (GTAW) and electron beam welding (EBW). These processes enable users to obtain optimum mechanical properties with minimal distortion because of the high heat concentration provided by these sources. The first challenge within the fusion welding of aluminium alloys is cracking of the weld due to solidification. It mainly occurs in aluminium alloys because the weld metal composition differs from the filler metal composition and the dilution quantity. Therefore, one should carefully select the filler metal composition and the welding process parameters. The suggestions

TABLE 17.1
Wrought Aluminium Alloys Classification

Sl.no	Designation of Material	Primary Alloying Element
1	1xxx	Aluminum, 99% and greater
2	2xxx	Copper
3	3xxx	Manganese
4	4xxx	Silicon
5	5xxx	Magnesium
6	6xxx	Magnesium and Silicon
7	7xxx	Zinc
8	8xxx	Other element

TABLE 17.2
Cast Aluminium Alloys Classification

Sl.no	Designation of Material	Primary Alloying Element
1	1xxx	Aluminum, 99% and greater
2	2xxx	Copper
3	3xxx	Silicon with Cu and Mg
4	4xxx	Silicon
5	5xxx	Magnesium
6	7xxx	Zinc
7	8xxx	Tin
8	9xxx	Other element
9	6xxx	Un used series

for the selection of filler metals exist for various categories and grades of aluminium alloys. For some dissimilar aluminium alloy mixtures, no filler metals exist that will manufacture crack-free welds. Even though there is an affordable filler metal choice, satisfactory joint efficiencies are unattainable. For the above-mentioned reasons, industries typically avoid fusion welding of dissimilar aluminium alloys. With the development of solid-state welding processes, such as friction stir welding (FSW), aluminium alloys have been successfully welded. The FSW process can be used as an alternative method in order to improve weldability without great loss of strength and corrosion properties of the joint.

FSW is an eco-friendly, innovative, solid-state welding process that was established by The Welding Institute (TWI) of the United Kingdom. The FSW process is free from the dangerous fumes and it avoids the formation of solidification porosity and cracking. Since the time of its invention, this method has been frequently improved and its scope of application has been expanded. The FSW process is preferred mostly for automotive and aerospace industries, which demand high structure performance. Joints fabricated using the FSW process are free from severe distortions and residual stresses when compared to the conventional fusion welding process. Dendrite structure formed during the fusion welding significantly declines the strength of the weld joint. The FSW process consists of a cylindrical tool with a shoulder and pin arrangement. This FSW tool is rotated at a constant speed and traversed through the line at the butt joint as shown in Figure 17.1. Frictional heat is generated between the rotating tool shoulder and workpiece. The heat generated from the friction softens the material in the weld zone without reaching its melting point. The tool pin acts as a stirring element that moves the plasticized material from retreating side to advancing side and vice-versa. The tool axial force is applied to maintain the adequate frictional contact between the tool shoulder and workpiece. Moreover, it is employed to forge the plasticized material moving around the tool pin in the weld zone.

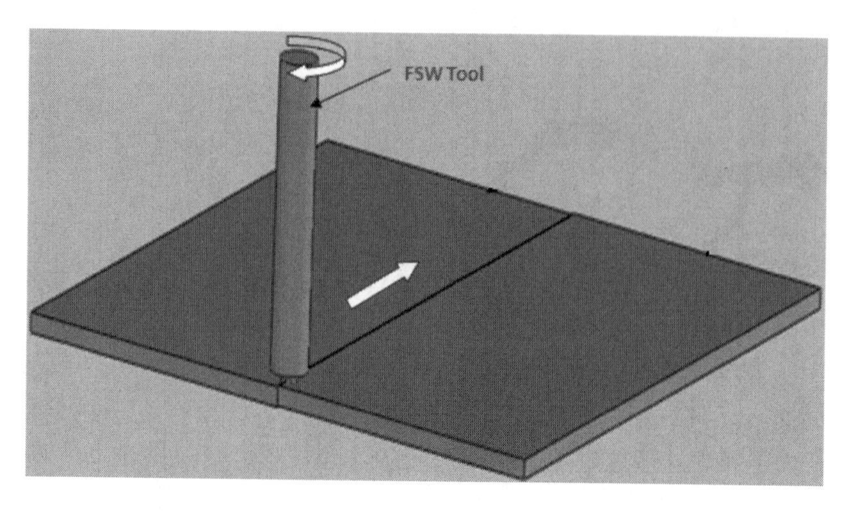

FIGURE 17.1 Friction stir welding process.

Friction stir welding parameters are classified into two categories: machine parameters and tool parameters. They have a significant effect on the flow of material pattern, temperature distribution and microstructural changes in the material.

The FSW process parameters include:

- Tool rotational speed and welding speed,
- Tool tilt and plunge depth,
- Tool design,
- Tool shoulder diameter,
- Tool pin diameter,
- Pin length,
- Tool inclined angle and
- Ratio between shoulder and pin diameters.

The weld joint obtained using this method is also very strong and finds application in the high load and safety applications, such as in space vehicles and critical defence systems. The ultimate tensile strength and shear properties of the weld joint will improve if the metals are joined using this procedure. FSW process does not require additional material to make the joint, which is also a weight reduction factor that is deemed vital in weight-sensitive industries, such as aerospace. The advantages of the FSW process have been consistently studied and improved. This welding technology has been applied to different alloys of aluminium over the course of time and optimal parameters for joining different materials under different conditions has been specified by various experimental researchers. The literature review presented in this chapter consists of studies on the effect of FSW process, postweld heat treatment (PWHT) in FSW joints, friction stir welding of dissimilar joints and aluminium metal matrix and the effect of underwater welding process.

17.2 LITERATURE SURVEY

17.2.1 Effect of Friction Stir Welding Parameters

Jamalian et al. (2016) worked to identify the optimum tool parameters for the welding of Al 5086 H34 alloys to produce the maximum joint efficiency. They conducted full factorial experiments with five rotational speeds and three traverse speeds. They found that the optimum FSW parameters for maximizing the tensile strength are 1250 rpm tool rotational speed and 80 mm/min traverse speed.

Muthu and Jayabalan (2016) investigated the effect of different tool pin profiles on the efficiency of the weld formed between the AA 1100 H14 aluminium and commercially available copper. In this study, the tool speed was kept constant at 1075 rpm and the traverse speed was set at 70 mm/min. It was observed that the whorl pin profile produced a better mixture between the two dissimilar materials. They identified that the maximum weld strength of 116 MPa was obtained when the plain tapered tool profile is used.

Costa et al. (2015) investigated the effect of FSW process parameters and the tool profile on friction stir lap welding using AA 5754 H22 and work-hardened aluminium alloy joint. They identified that the slight deviations in the mechanical properties do not have any significance on the tensile shear strength of the weld joint; however, the effective plate thickness has a significant effect on the weld strength. The FSW tool with a conical pin profile produced a weld with the maximum strength.

The optimum process parameters for welding the dissimilar AZ13B Mg and AA6061-T6 joint was identified by Fu et al. (2015). They found that FSW joints with better mechanical properties were obtained when the tool rotation rate is around 600–800 rpm, the traverse speed is 30–60 mm/min with a tool offset of 0.3 mm and the magnesium alloy is placed on the advancing side. They suggest that the extent of the homogeneity produced in the weld is affected significantly by the heat input and the intermixing between the component metals in the weld.

Kumar et al. (2014) employed the FSW process as a technique to weld Al4.5%Cu/TiC metal matrix composites. The maximum hardness occurs in the stir zone and gradually decreases while moving away from the stir zone. The FSW process parameters that exhibited better mechanical properties included 500 rpm tool rotational speed and 20 mm/min traverse speed using a tool with a flat shoulder and inclined pin.

Mastaniah et al. (2016) identified the optimum welding parameters for obtaining defect-free welds in FSW of dissimilar alloys AA2219 and AA5083. The intermixing of metals in the nugget zone influenced the strength of the obtained weld. The FSW parameters that exhibited maximum tensile strength, which is 97% of the strength of base material, were 800 rpm tool rotational speed and 210 mm/min traverse speed.

Mehta and Badheka (2017) investigated the effect of tool pin profiles on the strength and the hardness of FSW electrolytic tough pitch copper and AA6061-T651 dissimilar joint. They have identified that the deterioration in the strength of the weld is primarily due to the formation of intermetallics, which are brittle in nature. The FSW parameters determined in this study, 8 mm diameter cylindrical pin profile,

26.64 mm shoulder diameter, 50 mm/min traverse speed and 1500 rpm tool rotational speed, exhibited the maximum tensile strength of 89 MPa.

The optimum tool pin geometries for the FSW of AA5456 aluminium alloy with different tempers and different thickness was identified by Salari et al. (2014). They introduced a new pin profile that produced a weld with better efficiency when compared to other tool geometries under the same condition of rotational and transverse speed. The maximum strength was obtained when the stepped cone tool was used at rotational and transverse speed of 600 rpm and 30 mm/min, respectively.

Saravanan et al. (2016) studied the effect of various tool parameters on the quality of the weld produced between dissimilar aluminium alloys AA6061-T6 alloy and AA7075-T6 alloy. The FSW parameters, which were tool rotational speed of 1100 rpm, welding speed of 26 mm/min, axial load of 7kN and the D/d ratio of 3, revealed the maximum tensile strength of the joint. They stated that the fine grain structure in the weld stir zone (WSZ) was the major cause for the improvement in the tensile strength of the weld.

The effect of tool parameters on the quality of the dissimilar lap weld produced between the differently tempered alloys AA5456 H321 and AA5456 aluminium alloys was investigated by Shirazi et al. (2015). They suggested that the optimum parameters for improving the mechanical properties were 600 rpm and 100 mm/min tool rotational and transverse speeds, respectively. They found that the hooking and kissing bond defect has the maximum effect on the weld strength of the material.

Singarapu et al. (2015) attempted to identify the optimum welding parameters for the FSW of AZ13B alloy. They modelled regression equations to signify the effect of various parameters on the weld quality. The stipulated range of welding parameters for producing a weld with no macrostructural defects were 900–1400 rpm at a welding speed of 25–75 mm/min. The tool rotational speed was found to have the maximum influence on the ultimate tensile strength and the microhardness of the weld produced.

The optimum welding parameters for the friction stir linear lap welding of the AA6061 aluminium alloy to NZ30K magnesium alloy was identified by Tan et al. (2017a). In this research work, the effect of temperature on the microstructure of the weld region was extensively studied, and the welding parameters were correlated based on the observed heat inputs. It was found that the shear failure load was maximum when the tool rotation speed of 900 rpm and the tool traverse speed of 120 mm/min were used.

Verma et al. (2018) carried out an investigation on the dissimilar welding of aluminium AA6061 and AZ13B magnesium alloy in a butt joint configuration. The final composition of the weld was obtained using energy-dispersive spectrometer. Scanning electron microscopy (SEM) analysis was carried out to identify the microstructural details and tensile fracture failure mode. Higher rotational speed and the low traverse speed of the tool exhibited better joint properties and simultaneously improved the corrosion resistance.

From the review, it is found that there exists limited research work focusing on the FSW parameters such as tool tilt angle, tool pin diameter and tool offset. Moreover, very few attempts have been made to study the influence of these FSW parameters on mechanical and microstructural behaviour of the weld joint.

17.2.2 Postweld Heat Treatment of FSW Joints

Sivaraj et al. (2014) reported the effect of PWHT on the mechanical and microstructural behaviour of FWS AA7075-T651 joint. The FSW joints fabricated with the combination of solution treatment and artificial ageing exhibited the maximum tensile strength of 445 MPa and improved the joint efficiency by 9% when compared to the as-welded (AW) joint without PWHT.

Kumar et al. (2015) investigated the impact of PWHT on mechanical, microstructure and corrosion behaviour of FSW AA7075 joint. In this work, retrogression and reaging (RRA) and peak ageing (T6) was studied. Peak aged (T6) condition exhibited the maximum tensile strength and hardness of 540 MPa and 158Hv, respectively. RRA condition resisted the pitting corrosion and reduced the loss of weld strength due to the presence of discontinuous grain boundary precipitates.

Güven et al. (2015) studied the influence of PWHT on microstructure and mechanical behaviour of FSW AA6061 plate with two different temper conditions: O and T6 respectively. Abnormal grain growth (AGG) was observed in the weld zone of FSW AA6061 plate with the O temper condition, but PWHT improved the mechanical behaviour of the O temper condition when compared to base metal and AW joints. Results revealed that the O temper condition required high tool rotational speed and welding speed along with PWHT to yield better mechanical properties than T6-temper condition.

The impact of different PWHT on the tensile behaviour of FSW AA6061 aluminium alloy was studied by Elangovan and Balasubramanian (2008). Age hardening, solution treatment and a combination of both were applied to the welded joint. When compared to other welded joints, age-hardened specimens exhibited better tensile properties. The joint efficiency was increased to 77% in age-hardened specimens, and it looked better when compared to the 66% joint efficiency of AW joint.

Boonchouytan et al. (2014) attempted to study the impact of heat treatment on FSW AA6061 at three different tool rotational speeds of 710, 1000 and 1400 rpm. Artificially aged weld joint (WA) along with tool rotational speed of 1400 rpm, welding speed 160 mm/min and tool tilt 30 exhibited the maximum tensile strength of 179.8 N/mm^2 and the hardness value of 92.7Hv. The stir zone of weld joint with maximum tensile strength displayed fine grain structure with Mg_2Si precipitates uniformly distributed in the aluminium matrix.

The influence of PWHT on microstructure and mechanical behaviour of FSW AA7039 joint was reported by Sharma et al. (2013). The experiments were conducted with FSW parameters including tool rotational speed of 635 rpm, welding speed of 75 mm/min and tool tilt angle of 2.50. The weld joint with natural ageing process exhibited the maximum tensile strength of 392.8 N/mm^2 and joint efficiency of 174.2%. Naturally aged welded specimens with maximum tensile properties showed larger and deeper dimples with little flat regions.

Chen et al. (2015) made an attempt to investigate the effect of PWHT and three different welding heat inputs on hardness and microstructural behaviour of FSW AA2024-T3 alloy. After PWHT, stir zone hardness value decreased because of overaging effect. The joint fabricated with the tool rotational speed of 500 rpm and 100 mm/min exposed high resistance to overaging effect and exhibited the maximum hardness of 131.78Hv. The strengthening precipitates morphology dominated the hardness distribution in the stir zone.

The impact of PWHT, along with three different tool rotation speeds (850, 1070 and 1350 rpm) and welding speeds (90,140 and 224 mm/min) on mechanical properties and microstructure behaviour of AA6082 joint, was studied by El-Danaf and El-Rayes (2013). The reprecipitation of second phased particles during PWHT retains the tensile strength and hardness of the weld joint. The joint fabricated with the tool rotational speed of 850 rpm, welding speed of 224 mm/min and PWHT of 175°C for 12 hours exhibited the maximum tensile strength of 227 MPa and hardness of 162Hv.

Singh et al. (2011) studied the influence of PWHT on mechanical and microstructural behaviour of FSW AA7039 joint. The presence of very fine equiaxed grain structure in the weld nugget zone of AW joint improved the hardness to 143Hv, which was better than the base metal. The joint fabricated with higher welding speed of 12 mm/min and AW condition exhibited the maximum tensile strength of 312 MPa along with the maximum joint efficiency of 92.11%.

The mechanical properties and microstructural behaviour of FSW AA7075 alloy with different temper conditions, such as prior and after PWHT was investigated by Güven et al. (2014). The AA7075-O and AA7075-T6 FSW joints were fabricated by using two different sets of parameters, including 1000 rpm/150 mm/min and 1000 rpm/400 mm/min, respectively. The FSW joints fabricated with AA7075-O plate exhibited maximum tensile strength and hardness in both AW and PWHT conditions.

Bayazid et al. (2016) observed the influence of cyclic solution treatment (CST) on mechanical properties and microstructural behaviour of FSW AA7075 alloy. The CST consists of repeatedly heating the specimen to between 4000–4800°C for 0.25 hour holding time. There was no considerable change observed in the grain size after the CST process, but the change of $MgZn_2$ precipitates present in the welding area to $MgAlCu/Al_7Cu_2Fe$ improved the tensile strength and hardness distribution of the weld joint.

From the literature review, it was observed that very few research works were found reporting the effect of the PWHT process by varying the artificial aging temperature and holding time.

17.2.3 Friction Stir Welding of Dissimilar Joint

Chang et al. (2011) made an attempt to improve the tensile strength of the FSW AA6061-T6 and AZ31 Mg dissimilar joint by using Ni foil placed between the faying edges. From their observations, they suggested the hybrid FSW process with Ni between the faying edge for the best tensile strength of the weld joint. The optimum process parameters, reported as 800 rpm rotational speed, 35 mm/min welding speed and 2 kW laser power, exhibited the maximum tensile strength of 169 MPa.

The impact of FSW process parameters on tensile strength of the dissimilar AA6061-T6 and AZ31B Mg joint was studied by Dorbane et al. (2016). Additionally, the effect of ambient temperature and material position on the weld joint efficiency was also studied. The FSW parameters of 1400 rpm rotational speed, 500 mm/min traverse speed and aluminium material in advancing side of the weld exhibited the maximum joint efficiency of 79% at an ambient temperature of 200°C.

The FSW process parameters for improving the tensile strength of FSW dissimilar AA6061-T6 alloy and AZ31B Mg joint was optimized by Fu et al. (2015). The two major factors observed to impact the quality of the weld were found to be the heat input and good intermixing of the two materials. The FSW parameters of tool rotational and traverse speed of 600–800 rpm and 30–60 mm/min with a tool offset of 0.3 mm exhibited the maximum tensile strength. The reduction in the tensile strength of the material was observed to be due to the presence of intermetallics.

Jafari et al. (2017) made an attempt to weld the dissimilar 304L stainless steel to copper joint using FSW process. The grain size in the WSZ was less than the grain size in the heat-affected zone (HAZ) and thermomechanically affected zone (TMAZ) region. Moreover, the tensile fracture surface revealed that the specimen always breaks at the HAZ of the copper side. The improvement in the tensile strength was found to be due to grain refinement, which has been attributed to the dynamic recrystallization and the work hardening that takes place during FSW process.

The influence of FSW parameters on mechanical behaviour of the FSW copper and aluminium dissimilar joint was studied by Mehta and Badheka (2016). The FSW parameters, such as tool design, tool offset, material position, welding speeds and other factors, were employed to improve the quality of the weld joint. Finally, based on their study they found that sufficient analysis has not been made into determining the tool design and welding parameters for different plate thicknesses.

Murugan et al. (2018) made an attempt to fabricate the dissimilar joint of commercially available pure aluminium and 304 stainless steel by FSW process. The tool rotational speed of 400 rpm, tool traverse speed of 60 mm/min with a tool offset of 1.25 mm toward the aluminium exhibited the maximum tensile strength of approximately 90% of the strength of aluminium. The highly refined aluminium grains led to the improvement in properties.

The mechanisms behind the FSW of dissimilar AA6061-T6 and copper joint were investigated by Ouyang et al. (2006). The temperature distribution during the welding process was correlated with the microstructures of the different zones in the weld joint. Moreover, it was observed that better dissolution of the aluminium inside the copper takes place when the aluminium plate was kept on the advancing side of the weld. Heat input in the weld has been found to be a major factor in affecting the quality of the weld produced.

Rafiei et al. (2017) made an attempt to fabricate the dissimilar aluminium and A316L austenitic steel joint by FSW process. They correlated the available thermal data to the microstructural behaviour of the joint. Moreover, it was found that the rotational speed of 250 rpm and the transverse speed of 16 mm/min reveal the maximum tensile strength. The low transverse speed helped in maintaining the temperature for a longer period of time, thereby improving the intermixing of the material.

The effect of intermetallic formation on the mechanical and microstructural behaviour of FSW AA6061-T6 and Mg dissimilar joint was investigated by Shi et al. (2017). The weld strength was minimum when the tool rotational speed of 800 rpm was employed. They also identified that the increase in the heat input led to a thinner lamellae of Al being transported into the Mg alloy. They have suggested that the distribution of the intermetallics in the band is curved, short and loosely distributed instead of agglomerating in a same position.

Shen et al. (2015) made an attempt to optimize the FSW parameters to fabricate the dissimilar lap joint of AA5754 and DP600. The maximum tensile strength of the joint was obtained at the tool rotational speed of 1800 rpm, traverse speed of 45 mm/min and penetration depth of 0.389 mm. Moreover, it was found that the shear failure occurs in the material when the penetration depth is less than 0.17 mm, which has been ascribed to the presence of an intermetallic compound Fe_4Al_{13}.

The dissimilar joint of AA6013-T4 and X5CrNi18-10 stainless steel using FSW process was fabricated by Uzun et al. (2005). They performed various microstructural analyses and considered mechanical properties such as the hardness and the fatigue of the material as a measure to classify good welds. The distribution of hardness in the weld was erratic because of the variation in the distribution of the steel particles. The fatigue strength of the weld was found to be 30% less than the base material.

Taban et al. (2010) made an attempt to fabricate the dissimilar joint between AA6061-T6 and AISI 1018 using the FSW process. It was found that the FSW joint fabricated under higher pressure exhibited the maximum tensile strength of 250 MPa. The presence of the rare FeAl intermetallic was identified, which was attributed to the combination of the high temperature and the heavy loads applied, which evidently activated the intermetallic formation condition.

The dissimilar butt joint between the SS400 mild steel plate and AA5083 alloy of 2 mm thickness was produced by Watanabe et al. (2006). The maximum tensile strength of the weld was obtained with the tool rotational speed of 250 rpm, transverse speed of 25 mm/min and tool offset of 0.2 mm toward the aluminium side. They have identified that the path of fracture was along the distribution of intermetallics present in the weld.

Yan et al. (2010) made an attempt to fabricate the dissimilar joint of AA5052 and AZ31 magnesium alloy by optimizing the FSW process parameters to maximize the tensile strength of the joint. They observed that the microhardness of the weld was almost twice that of the base material, and the fracture occurs at the advancing side of the weld. The hardness of the material peaked approximately at the centre of the stir zone, and the fracture occurred at the aluminium side of the weld where the variation in the hardness was maximum.

From the review, it was observed that FSW parameters such as tool rotational speed, traverse speed and axial force play a dynamic role in deciding the mechanical properties of the weld joint. Also, it was found that few studies on material position and tool offset exist.

17.2.4 Friction Stir Welding of Metal Matrix Composite

Ceschini et al. (2007) studied the various effects of FSW parameters on the mechanical properties of AA7005 aluminium matrix imbibed with 10% (by volume) $Al_2 0_3$ particles (W7A10A). The refinement of grain structure and hardness was the same as the base material except in the TMAZ, where the hardness was slightly higher than that of the base material. The fatigue test was carried out for the same specimens and identified that the low-cycle fatigue life of the material was lower than the high-cycle fatigue strength of the material.

The influence of process parameters on FSW of AA6063 aluminium matrix reinforced with various proportions of B_4C was investigated by Chen et al. (2009). The tool wear was considered to be a significant factor in the optimization of the FSW parameters. Moreover, it has been observed that, after T6 treatment, there was a rapid growth of the grain structure in the matrix as well as in the dispersed phase. The presence of equiaxed grains along the direction of the weld exhibited the better joint efficiency.

The optimum parameters for the FSW of A15052/SiC metal matrix composite were identified by Dolatkhah et al. (2012). The FSW parameters, such as tool rotational speed of 1120 rpm and tool traverse speed of 20 mm/min, exhibited better mechanical properties. They also observed that increasing the number of passes and changing the direction of tool rotation between the passes also impacted the hardness of the materials to a higher extent. Furthermore, the reduction of grain size from 243 μm to 0.9 μm improved the hardness of material.

Fallahi et al. (2017) investigated the influence of process parameters on FSW AA5038 and austenitic steel (A316) dissimilar joint reinforced with SiC nanoparticles. The number of passes of the tool was found to be an important parameter in improving the hardness of the Stir Zone (SZ). The improvement in the properties was ascribed to the effect of increasing heat input due to low traverse speeds, and the increasing number of passes favoured the homogeneous distribution of nanoparticles without agglomerations.

The influence of process parameters on friction stir vibration welding of AA5052 alloy reinforced with SiO_2 particle was studied by Fouladi and Abbasi (2017). The optimal tool parameters, such as 1200 rpm tool rotation, 20 mm/min traverse speed and vibration frequency of 33 Hz, exhibited better mechanical properties. The improvement in the mechanical properties was identified to be due to the increased density of the dislocations and enhanced dynamic recrystallization of fine grain structure.

Gopalakrishnan and Murugan (2011) utilized the FSW process to fabricate the aluminium metal matrix joint reinforced with different weight proportions of TiC (3–7%). The FSW parameters, such as tool rotational speed, welding speed, tool pin profile and % of TiC, were considered for producing the joint. The tool pin profile and welding speed had the most significant effect in determining the tensile strength of the joint.

The effect of process parameters on mechanical behaviour of FSW hot-rolled Al4.5%/Cu/TiC was investigated by Kumar et al. (2014). They identified that the tool geometry played a substantial role in determining the mechanical properties of the joint. The optimum welding parameters, such as 500 rpm tool rotation speed and 7° tool concavity angle, exhibited the maximum tensile strength with 89% joint efficiency. The microhardness was observed to be maximum at the WSZ and minimum at the HAZ because of the variation of grain size.

Karzakizis et al. (2018) made a comparative study on the effect of the SiC and TiC nanopowder on mechanical behaviour of the weld joint fabricated by using FSW process. The presence of intermetallic in the WSZ was independent of the reinforcing material. They suggested that SiC reinforcement can be utilized when a higher elongation of the material is required, and the TiC reinforcement can be used when the material is required to have a higher wear resistance.

The optimum process parameters for improving the mechanical behaviour of FSW AA6061 reinforced with Al_2O_3/2Op was identified by Marzoli et al. (2006). The process parameters, such as the welding speed and tool rotational speed, were considered for fabricating the joint. Defect-free welds without any agglomerations of the reinforcing particles and other conventional defects were considered to be conforming the quality of weld. All the specimens were found to fail near the HAZ and the maximum weld efficiency was found to be 80%.

Selvakumar et al. (2017) made an attempt to fabricate the molybdenum material reinforced with AA6082 by using FSW process. The dispersed phase played a vital role in improving the tensile strength of the joint to 37.38% without much loss in ductility. The critical process parameters were identified to be the tool geometry, traverse speed and the tool rotational speed. They observed that the molybdenum particles were smoothly bonded with the matrix and caused grain refinement in the matrix to improve the mechanical properties of the joint.

The combined effect of P-modification and process parameters on FSW of Al-Mg_2Si-Si alloys was studied by Qin et al. (2017). The FSW parameters, such as tool rotational speed of 800 rpm, traverse speed of 80 mm/min and tool tilt angle of 2.5°, were employed to fabricate the joints. The structure of the Mg_2Si in the base alloy was found to be equiaxed at an approximate size of 44 μm before the addition of the P-modifier. They observed that the addition of P-modifier reduced the particle size to 9–11 μm and improved the mechanical behaviour of the weld joint.

Uzun et al. (2007) utilized the FSW process for joining the AA2124 aluminium alloy reinforced with SiC/25P. The tool rotational speed and the traverse speed were kept constant at 800 rpm and 120 mm/min, respectively. The electrical conductivity experiment revealed the presence of SiC along the SZ band. The average hardness of the parent material (250Hv) was higher than the hardness found at the weld nugget (240Hv). The hardness was found to be minimum at the HAZ because of the annealing effect.

The influence of tool profile on the mechanical properties of Al-10wt% TiB_2 aluminium metal matrix composite subjected to FSW was investigated by Vijay and

Murugan (2010). The FSW parameters, such as tool rotational speed, traverse speed and the axial load, were kept constant at 2000 rpm, 30 mm/min and 19.6kN, respectively. The tapered hexagonal pin and square pin tool revealed defect-free welds. The FSW tool with a tapered pin profile exhibited a smaller SZ when compared to the straight profile tools.

From the literature review, it was observed that tool geometry plays a vital role in deciding the strength of the weld joint. Also, it was found that a limited number of studies exist on modifying the tool shoulder concave angle along with tool pin profile.

17.2.5 Under Water Friction Stir Welding Process

Heirani et al. (2017) have attempted to optimize the process parameters for underwater friction welding of AA5083 alloy. The presence of fine grain structures in the SZ was the probable cause for the observed increase in the tensile strength of the joint. The depth-to-width ratio (DWR) of the HAZ and the microhardness of the HAZ were the entities that varied with the welding parameters, and they account for the microhardness and the strength of the weld joint.

The suitable FSW tool profile for improving the strength of AA2519T-87 joint using underwater FSW process was identified by Sabari et al. (2016). Considering the different tool profiles that were used for the underwater welding of AA2519-T87 alloy, the taper threaded cylindrical tool exhibited the maximum joint efficiency of 76%. Moreover, this taper threaded tool exhibited higher hardness WSZ.

The optimum FSW process parameters for improving the mechanical properties of FSW AA5052 aluminium alloy was identified by Shanavas et al. (2018). They found that the mechanical behaviour of the weld joint produced by the underwater FSW process was better than the joint produced by the air-cooled FSW process. The optimum welding parameters were obtained at 700 rpm tool rotational speed, 65 mm/min traverse speed and 7kN axial load.

Tan et al. (2017b) made an effort to study the influence of FSW parameters on mechanical properties and microstructural behaviour of FSW AA3003 joint. A conical tool with a median diameter of 3.5 mm, operating at 800 rpm rotational speed and a 200 mm/min transverse speed was used. The strength and overall properties were better in the hot bands than in the annealed hot bands. They also observed that the grain size decreases with decreasing ambient temperature, and the grain size in the annealed hot band was larger than the grain size in the hot bands.

The influence of tool rotation speed and weld speed on mechanical behaviour of underwater FSW AA7055-T6 joint was investigated by Wang et al. (2015). It was observed that the tool traverse speed had more influence over the thermal cycle when compared to the rotational speed. Defects can be eliminated by decreasing the weld speed and increasing the tool rotational speed, which simultaneously influenced the strength of the joint in the same mode.

Yazdipour and Heidarzadeh (2016) attempted to identify the optimum process parameters for improving the strength of FSW AA5083-H321 and 316L stainless steel dissimilar joint with constant tool rotational speed of 280 rpm. The optimal process parameters for maximizing the tensile strength were identified as 160 mm/min transverse speed and 0.4 mm tool offset. They also observed that increasing the tool traverse speed declines the tensile strength of the weld joint because of micro-cracking and the existence of coarse particles in the WSZ.

The effect of underwater FSW of the spray formed AA7055 alloy and the mechanism behind the better mechanical behaviour was studied by Wang et al. (2016). Most of the studies often neglect the effect of the dynamic nature of stirring water. In this work, a comparison study of mechanisms and modes of failure between the FSW joint and underwater FSW joints was made. The strength of the weld joint produced by underwater FSW process was approximately 30% more than the joint fabricated with conventional FSW process. The $MgZn_2$ intermetallic, a strengthening phase, was finely distributed, which accounted for the joint strength.

Zhang et al. (2014) identified a challenge in the dissimilar weld produced by using the FSW process and have tried to alleviate the problem using the underwater FSW process. They identified that the presence of Al-Cu intermetallic compound formed during underwater welding was significantly lower than those produced during the conventional welding methods. The width of the intermetallic layer was 18 μm in the classical FSW sample, whereas the underwater FSW process reduced this to 2 μm; this was attributed to the presence of water, which prevented oxidation of the weld. Its interaction with the thermal cycle of the weld simultaneously decreased the peak temperatures as well.

The mechanical behaviour of the underwater FSW AA7055 alloy was studied by Zhao et al. (2014). They observed that the joints welded by employing the underwater FSW process do not have an s-line defect. Moreover, the tensile strength of the welded joint is 75% of the base metal. The water environment reduced the residual stresses and even produced a little compressive stress in the weld zone. Furthermore, the joint efficiency was improved by the presence of an intermetallic compound, such as MgZn2, in the WSZ.

Yong Zhao et al. (2015) have utilized underwater FSW process to produce a dissimilar AZ31 magnesium alloy and Al 6013 alloy joint. SEM, energy-dispersive X-ray spectroscopy (EDAX) line scan and X-ray diffraction techniques were used to observe the microstructure of the welds produced. They found that the welds endure a brittle fracture during the tensile test of the specimen along the weld line where the hardness gradient is high. The optimum process parameters are a rotational speed of 1200 rpm and a translational speed of 80 mm/min. The hardness of the weld (142 HV) was only slightly higher than the base material. The tensile strength of the joint was approximately 152.3 MPa, equivalent to 63.3% the strength of AZ31 Mg alloy; however, because of the presence of undesirable intermetallic compounds, the ductility was comparatively less.

From the review, it was observed that the integration of conventional FSW process with other sources such as underwater (UW), ultrasonic vibration (UV) and laser enhanced the mechanical behaviour of the weld joint. Moreover, these investigations will open new opportunities of exploring potential operating conditions to replace conventional FSW process.

17.3 CONCLUSION

From the literature survey, the following things were observed:

- FSW is the best choice for manufacturers fabricating these joints with good mechanical properties and microstructural behaviour.
- RSM is the best optimization technique to optimize the FSW parameters to ensure the durability and structural integrity of the welded joint.
- The PWHT process was found to be a promising method for improving the mechanical and microstructural behaviour of dissimilar alloy joints after welding process.
- Material position, rotational speed and tool offset dominates the strength of the dissimilar joints.
- Tool geometry with different shoulder concave angles play a vital role in deciding the strength of the aluminium metal matrix weld joint.
- Integration of conventional the FSW process with the underwater welding concept enhanced the mechanical behaviour of the weld joint because of the presence of intermetallic compound in the WSZ.

REFERENCES

Bayazid, S.M., Farhangi, H., Asgharzadeh, H., Radan, L., Ghahramani, A. and Mirhaji, A. 2016. Effect of cyclic solution treatment on microstructure and mechanical properties of friction stir welded 7075 Al alloy. *Materials Science and Engineering. part A*, 649(1): 293–300.

Boonchouytan, W., Chatthong, J., Rawangwong, S. and Burapa, R. 2014. Effect of heat treatment T6 on the friction stir welded SSM 6061 aluminum alloys. *Energy Procedia*, 56(1): 172–180.

Ceschini, L., Boromei, I., Minak, G., Morri, A. and Tarterini, F. 2007. Effect of friction stir welding on microstructure, tensile and fatigue properties of the AA7005/10 vol.% Al_2O_3p composite. *Composites Science and Technology*, 67(3–4): 605–615.

Chang, W.S., Rajesh, S.R., Chun, C.K. and Kim, H.J. 2011. Microstructure and mechanical properties of hybrid laser-friction stir welding between AA6061-T6 Al alloy and AZ31 Mg Alloy. *Journal of Materials Science and Technology*, 27(3): 199–204.

Chen, X.G., da Silva, M., Gougeon, P. and St-Georges, L. 2009. Microstructure and mechanical properties of friction stir welded AA6063-B4C metal matrix composites. *Materials Science and Engineering. part A*, 518(1–2): 174–184.

Chen, Y., Ding, H., Li, J.Z., Zhao, J.W., Fu, M.J. and Li, X.H. 2015. Effect of welding heat input and post-welded heat treatment on hardness of stir zone for friction stir-welded 2024-T3 aluminum alloy. *Transactions of Nonferrous Metals Society of China (English Edition)*, 25(8): 2524–2532.

Costa, M.I., Verdera, D., Costa, J.D., Leitao, C. and Rodrigues, D.M. 2015. Influence of pin geometry and process parameters on friction stir lap welding of AA5754-H22 thin sheets. *Journal of Materials Processing Technology*, 225(1): 385–392.

Dolatkhah, A., Golbabaei, P., Besharati Givi, M.K. and Molaiekiya, F. 2012. Investigating effects of process parameters on microstructural and mechanical properties of Al5052/SiC metal matrix composite fabricated via friction stir processing. *Materials and Design*, 37(1): 458–464.

Dorbane, A., Mansoor, B., Ayoub, G., Shunmugasamy, V.C. and Imad, A. 2016. Mechanical, microstructural and fracture properties of dissimilar welds produced by friction stir welding of AZ31B and Al6061. *Materials Science and Engineering. part A*, 651(1): 720–733.

Elangovan, K. and Balasubramanian, V. 2008. Influences of post-weld heat treatment on tensile properties of friction stir-welded AA6061 aluminum alloy joints. *Materials Characterization*, 59(9): 1168–1177.

El-Danaf, E.A. and El-Rayes, M.M. 2013. Microstructure and mechanical properties of friction stir welded 6082 AA in as welded and post weld heat treated conditions. *Materials and Design*, 46(1): 561–572.

Fallahi, A.A., Shokuhfar, A., Ostovari Moghaddam, A. and Abdolahzadeh, A. 2017. Analysis of SiC nano-powder effects on friction stir welding of dissimilar Al-Mg alloy to A316L stainless steel. *Journal of Manufacturing Processes*, 30(1): 418–430.

Muthu, M.F.X. and Jayabalan, V. 2016. Effect of pin profile and process parameters on microstructure and mechanical properties of friction stir welded Al-Cu joints. *Transactions of Nonferrous Metals Society of China (English Edition)*, 26(4): 984–993.

Fouladi, S. and Abbasi, M. 2017. The effect of friction stir vibration welding process on characteristics of SiO_2 incorporated joint. *Journal of Materials Processing Technology*, 243(1): 23–30.

Fu, B., Qin, G., Li, F., Meng, X, Zhang, J. and Wu, C. 2015. Friction stir welding process of dissimilar metals of 6061-T6 aluminum alloy to AZ31B magnesium alloy. *Journal of Materials Processing Technology*, 218(1): 38–47.

Gopalakrishnan, S. and Murugan, N. 2011. Prediction of tensile strength of friction stir welded aluminium matrix TiC particulate reinforced composite. *Materials and Design*, 32(1): 462–467.

Güven, İ., Erim, S. and Çam, G. 2014. Effects of temper condition and post weld heat treatment on the microstructure and mechanical properties of friction stir butt-welded AA7075 Al alloy plates. 70(1–4): 201–213.

Heirani, F., Abbasi, A. and Ardestani, M. 2017. Effects of processing parameters on microstructure and mechanical behaviors of underwater friction stir welding of Al5083 alloy. *Journal of Manufacturing Processes*, 25(1): 77–84.

İpekoğlu, G., Erim, S. and Çam G. Investigation into the influence of post-weld heat treatment on the friction stir welded AA6061 Al-alloy plates with different temper conditions. *Metallurgical and Materials Transactions A*, 45(2): 864–877.

Jafari, M., Abbasi, M., Poursina, D., Gheysarian, A. and Bagheri, B. 2017. Microstructures and mechanical properties of friction stir welded dissimilar steel-copper joints. *Journal of Mechanical Science and Technology*, 31(3): 1135–1142.

Karakizis, P.N., Pantelis, D.I., Fourlaris, G. and Tsakiridis, P. 2018. Effect of SiC and TiC nanoparticle reinforcement on the microstructure, microhardness, and tensile performance of AA6082-T6 friction stir welds. *International Journal of Advanced Manufacturing Technology*, 95(9–12): 3823–3837.

Kumar, A., Mahapatra, M.M., Jha, P.K., Mandal, N.R. and Devuri, V. 2014. Influence of tool geometries and process variables on friction stir butt welding of Al-4.5%Cu/TiC in situ metal matrix composites. *Materials and Design*, 59(1): 406–414.

Kumar, P.V., Reddy, G.M. and Rao, K.S. 2015. Microstructure, mechanical and corrosion behavior of high strength AA7075 aluminium alloy friction stir welds—Effect of post weld heat treatment. *Defence Technology*, 11(4): 362–369.

Marzoli, L.M., Strombeck, A.V., Dos Santos, J.F., Gambaro, C. and Volpone, L.M. 2006. Friction stir welding of an AA6061/Al_2O_3/20p reinforced alloy. *Composites Science and Technology*, 66(2): 363–371.

Mastanaiah, P., Sharma, A. and Reddy, G.M. 2016. Dissimilar friction stir welds in AA2219-AA5083 aluminium alloys: Effect of process parameters on material intermixing, defect formation, and mechanical properties. *Transactions of the Indian Institute of Metals*, 69(7): 1397–1415.

Mehta, K.P. and Badheka, V.J. 2016. A review on dissimilar friction stir welding of copper to aluminum: Process, properties, and variants. *Materials and Manufacturing Processes*, 31(3) 233–254.

Mehta, K.P. and Badheka, V.J. 2017. Influence of tool pin design on properties of dissimilar copper to aluminum friction stir welding. *Transactions of Nonferrous Metals Society of China (English Edition)*, 27(1): 36–54.

Mohammadzadeh Jamalian, H., Farahani, M., Besharati Givi, M.K. and Aghaei Vafaei, M. 2016. Study on the effects of friction stir welding process parameters on the microstructure and mechanical properties of 5086-H34 aluminum welded joints. *International Journal of Advanced Manufacturing Technology*, 83(1–4): 611–621.

Murugan, B., Thirunavukarasu, G., Kundu, S., Kailas, S.V. and Chatterjee, S. 2018. Interfacial microstructure and mechanical properties of friction stir welded joints of commercially pure aluminum and 304 stainless steel. *Journal of Materials Engineering and Performance*, 27(6): 2921–2931.

Ouyang, J., Yarrapareddy, E. and Kovacevic, R. 2006. Microstructural evolution in the friction stir welded 6061 aluminum alloy (T6-temper condition) to copper. *Journal of Materials Processing Technology*, 172(1): 110–122.

Qin, Q.D., Huang, B.W., Wu, Y.J. and Su, X.D. 2017. Microstructure and mechanical properties of friction stir welds on unmodified and P-modified Al-Mg_2Si-Si alloys. *Journal of Materials Processing Technology*, 250(1): 320–329.

Rafiei, R., Ostovari Moghaddam, A., Hatami, M.R., Khodabakhshi, F., Abdolahzadeh, A. and Shokuhfar, A. 2017. Microstructural characteristics and mechanical properties of the dissimilar friction-stir butt welds between an Al–Mg alloy and A316L stainless steel. *International Journal of Advanced Manufacturing Technology*, 90(9–12): 2785–2801.

Sabari, S.S., Malarvizhi, S. and Balasubramanian, V. 2016. The effect of pin profiles on the microstructure and mechanical properties of underwater friction stir welded AA2519-T87 aluminium alloy. *International Journal of Mechanical and Materials Engineering*, 11(5): 1–10.

Salari, E., Jahazi, M., Khodabandeh, A. and Ghasemi-Nanesa, H. 2014. Influence of tool geometry and rotational speed on mechanical properties and defect formation in friction stir lap welded 5456 aluminum alloy sheets. *Materials and Design*, 58(1): 381–389.

Saravanan, V., Rajakumar, S. and Muruganandam, A. 2016. Effect of friction stir welding process parameters on microstructure and mechanical properties of dissimilar AA6061-T6 and AA7075-T6 aluminum alloy joints. *Metallography, Microstructure, and Analysis*, 5(6): 476–485.

Selvakumar, S., Dinaharan, I., Palanivel, R. and Ganesh Babu, B. 2017. Characterization of molybdenum particles reinforced Al6082 aluminum matrix composites with improved ductility produced using friction stir processing. *Materials Characterization*, 125: 13–22.

Shanavas, S., Edwin Raja Dhas, J. and Murugan, N. 2018. Weldability of marine grade AA 5052 aluminum alloy by underwater friction stir welding. *International Journal of Advanced Manufacturing Technology*, 95(9–12): 4535–4546.

Sharma, C., Dwivedi, D.K. and Kumar, P. 2013. Effect of post weld heat treatments on microstructure and mechanical properties of friction stir welded joints of Al-Zn-Mg alloy AA7039. *Materials and Design*, 43: 134–143.

Shen, Z., Chen, Y., Haghshenas, M. and Gerlich, A.P. 2015. Role of welding parameters on interfacial bonding in dissimilar steel/aluminum friction stir welds. *Engineering Science and Technology, an International Journal*, 18(2): 270–277.

Shi, H., Chen, K., Liang, Z., Dong, F., Yu, T., Dong, X. and Shan, A. 2017. Intermetallic compounds in the banded structure and their effect on mechanical properties of Al/Mg dissimilar friction stir welding joints. *Journal of Materials Science and Technology*, 33(4): 359–366.

Shirazi, H., Kheirandish, S. and Safarkhanian, M.A. 2015. Effect of process parameters on the macrostructure and defect formation in friction stir lap welding of AA5456 aluminum alloy. *Measurement: Journal of the International Measurement Confederation*, 76: 62–69.

Singarapu, U., Adepu, K. and Arumalle, S.R. 2015. Influence of tool material and rotational speed on mechanical properties of friction stir welded AZ31B magnesium alloy. *Journal of Magnesium and Alloys*, 3(4): 335–344.

Singh, R.K.R., Sharma, C., Dwivedi, D.K., Mehta, N.K. and Kumar, P. 2011. The microstructure and mechanical properties of friction stir welded Al-Zn-Mg alloy in as welded and heat treated conditions. *Materials and Design*, 32(2): 682–687.

Sivaraj, P., Kanagarajan, D. and Balasubramanian, V. 2014. Effect of post weld heat treatment on tensile properties and microstructure characteristics of friction stir welded armour grade AA7075-T651 aluminium alloy. *Defence Technology*, 10(1): 1–8.

Taban, E., Gould, J.E. and Lippold, J.C. 2010. Dissimilar friction welding of 6061-T6 aluminum and AISI 1018 steel: Properties and microstructural characterization. *Materials and Design*, 31(5): 2305–2311.

Tan, S., Zheng, F., Chen, J., Han, J., Wu, Y. and Peng, L. 2017b. Effects of process parameters on microstructure and mechanical properties of friction stir lap linear welded 6061 aluminum alloy to NZ30K magnesium alloy. *Journal of Magnesium and Alloys*, 5(1): 56–63.

Tan, Y.B., Wang, X.M., Ma, M., Zhang, J.X., Liu, W.C., Fu, R.D. and Xiang, S. 2017b. A study on microstructure and mechanical properties of AA 3003 aluminum alloy joints by underwater friction stir welding. *Materials Characterization*, 127(1): 41–52.

Uzun, H. 2007. Friction stir welding of SiC particulate reinforced AA2124 aluminium alloy matrix composite. *Materials and Design*, 28(5): 1440–1446.

Uzun, H., Dalle Donne, C., Argagnotto, A., Ghidini, T. and Gambaro, C. 2005. Friction stir welding of dissimilar Al 6013-T4 To X5CrNi18-10 stainless steel. *Materials and Design*, 26(1): 41–46.

Verma, J., Taiwade, R.V., Reddy, C. and Khatirkar, R.K. 2018. Effect of friction stir welding process parameters on Mg-AZ31B/Al-AA6061 joints. *Materials and Manufacturing Processes*, 33(3): 308–314.

Vijay, S.J. and Murugan, N. 2010. Influence of tool pin profile on the metallurgical and mechanical properties of friction stir welded Al-10wt.% TiB_2 metal matrix composite. *Materials and Design*, 31(7): 3585–3589.

Wang, Q., Zhao, Z., Zhao, Y., Yan, K. and Zhang, H. 2015. The adjustment strategy of welding parameters for spray formed 7055 aluminum alloy underwater friction stir welding joint. *Materials and Design*, 88: 1366–1376.

Wang, Q., Zhao, Z., Zhao, Y., Yan, K., Liu, C. and Zhang, H. 2016. The strengthening mechanism of spray forming Al-Zn-Mg-Cu alloy by underwater friction stir welding. *Materials and Design*, 102: 91–99.

Watanabe, T., Takayama, H. and Yanagisawa, A. 2006. Joining of aluminum alloy to steel by friction stir welding. *Journal of Materials Processing Technology*, 178(1–3): 342–349.

Yan, Y., Zhang, D.T., Qiu, C. and Zhang, W. 2010. Dissimilar friction stir welding between 5052 aluminum alloy and AZ31 magnesium alloy. *Transactions of Nonferrous Metals Society of China (English Edition)*, 20(1): 619–623.

Yazdipour, A. and Heidarzadeh, A. 2016. Effect of friction stir welding on microstructure and mechanical properties of dissimilar Al 5083-H321 and 316L stainless steel alloy joints. *Journal of Alloys and Compounds*, 680(1): 595–603.

Zhang, J., Shen, Y., Yao, X., Xu, H. and Li, B. 2014. Investigation on dissimilar underwater friction stir lap welding of 6061-T6 aluminum alloy to pure copper. *Materials and Design*, 64: 74–80.

Zhao, Y., Lu, Z., Yan, K. and Huang, L. 2015. Microstructural characterizations and mechanical properties in underwater friction stir welding of aluminum and magnesium dissimilar alloys. *Materials and Design*, 65: 675–681.

Zhao, Y., Wang, Q., Chen, H. and Yan, K. 2014. Microstructure and mechanical properties of spray formed 7055 aluminum alloy by underwater friction stir welding. *Materials and Design*, 56(1): 725–730.

18 Effect of Solution Treatment and Artificial Ageing on Strength Properties of Friction Stir Welded AA2014-T6 Aluminium Alloy

C. Rajendran

CONTENTS

18.1 INTRODUCTION TO EFFECT OF SOLUTION TREATMENT AND ARTIFICIAL AGEING ON FRICTION STIR WELDING JOINTS

High-strength aluminium alloys (2xxx and 7xxx series) are used in multiple fields, namely automobile, aircraft and military vehicles, because of their superior strength-to-weight ratio and good corrosion resistance (Zhou et al. 2006). However, the welding of copper (CU) containing aluminium alloy (AA2014) is an arduous task using the fusion welding process, as it results in hot cracking, alloy segregation, partially melted zones and porosity (Squillance et al. 2004). Moreover, because of the dendritic structure growth in the weld region caused by fusion welding, there is a significant reduction in strength (Genevosis et al. 2005). Friction Stir Welding (FSW) is a solid-state process in which metals are joined through hot shear. A nonconsumable tool pin profile is rotated and gently plunged into the area around the joint line of the overlapped sheet material. Aluminium alloys, namely 2xxx and 7xxx series, which were previously considered to possess poor weldability properties, are now being welded using FSW. AA2014 aluminium alloy is an age-hardened alloy, and it has qualities such as high strength due to the formation of Al_2CuMg and $CuAl_2$ (copper aluminide) phases as a result of heat treatment processes, such as quenching, tempering and artificial ageing treatments. During the FSW, all the phases of precipitates formed previously are dissolved; however, because of the applied thermal cycle in the thermomechanically affected zone (TMAZ), a fraction of the precipitates are coarsened.

Though the FSW joints yielded higher strength than the fusion welded joints, the disparity between the strengths of weld metal and base metal is very high as a result of the soft regions developed in the FSW joints, which causes the degradation of mechanical properties. To retain the lost strength, various alternate processes, namely underwater FSW, cold and hot FSW and post-weld heat treatment (PWHT) were used. The effect of traverse speed on the mechanical properties and microstructure of AA2219 aluminium alloy using underwater FSW was examined. It was found that by increasing the welding speed there was a considerable reduction in the distribution of precipitates in the TMAZ and heat-affected zone (HAZ) (Liu et al. 2011). Water cooling and its influence on mechanical properties and the microstructure of AA2014 was investigated, and it was noticed that the lowest hardness zone (LHZ) was observed on both the advancing side (AS) and the retreating side (RS) in air cooling as well as water cooling condition. However, the LHZ was located in different zones for air cooling and water cooling. The LHZ was located at the HAZ for joint weld in air and the TMAZ for the underwater FSW joint (Zang et al. 2014).

Wang et al. (2014) used two PWHT, such as low-temperature ageing (LTA) and LTA with deep cryogenic treatment, and found two soft zones located at the AS and RS of the HAZ in the as-welded (AW) joint. A single LTA process was reduced by the soft region of the HAZ. Moreover, after LTA with deep cryogenic treatment featured the soft region similar to that of LTA.

Rajendran et al. (2016) reported that the solution treatment followed by ageing () solution treatment and artificial ageing (STA) was beneficial for recovering the tensile properties of the joint compared to the AW and artificial ageing (AA) treatments. Aydin et al. (2009) investigated the effect of PWHT on microstructure and mechanical properties of FSW joints of AA2024 aluminium alloy. The PWHT procedure caused abnormal grain coarsening in the stir zone (SZ), which resulted in a drop in microhardness at the SZ compared to the base material. The ageing treatment was found to be more beneficial than other heat treatments. Hu et al. (2011) found that the fine equiaxed grains were stable and retained in the SZ of the weld, and grain in the TMAZ became coarse and equiaxed as annealing temperature increases. fu et al. (2013) used heat input conditions, and found that, at high-heat input conditions, the hardness of the SZ was lower than the base material. Under low-heat input conditions, the hardness of the SZ was high.

Sharma et al. (2013) investigated the effect of PWHT on microstructure and mechanical properties of FSW AA7039 aluminium alloy and found that the applied

PWHT increases the size of the alpha grain in all the regions of FSW joints. Abnormal grain growth was observed in the entire region modified by the FSW in the case of solution treatment with or without ageing. Rajendran et al. (2016) investigated the effect of PWHT on the tensile strength of FSW butt joints of AA2014-T6 aluminium alloy and reported that solution treatment, followed by artificial ageing, improved tensile strength of FSW butt joints of AA2014-T6 aluminium alloy.

18.2 METHOD OF MANUFACTURING ADVANCE MATERIAL

The parent metal (PM) used in this investigation was rolled sheets of 2 mm thick AA2014-T6 alloy. Table 18.1 and Table 18.2 represent the chemical composition and the mechanical properties of the PM, respectively.

The sheets were cut into 150 mm × 150 mm pieces and clamped firmly to achieve a lap joint configuration (Figure 18.1). The joints were performed perpendicular to the rolling direction of the sheet using CNC-FSW. A high-speed steel tool (dimensions shown in Figure 18.1b) was used to fabricate the joints. Optimum conditions or parameters used for the fabrication of the joints were tool rotational speed of 900 rpm, welding speed of 110 mm/min, shoulder diameter of 12 mm and tool tilt angle of 2°.

Figure 18.1c depicts a photograph of the FSW lap joints. To study the effect of solutionizing time on strength of joints of AA2014 aluminium, the welded joints were subjected to solution treatment followed by artificial ageing. The lap-shear specimens were prepared as per the ANSI/AWS/SAE/D8.9-97 specification (Babu et al. 2014), and the lap-shear test was carried out using 100 kN electromechanical controlled universal testing machine at a crosshead velocity of 1.5 mm/min. Figure 18.1d shows a photograph of lap-shear specimens before testing and after testing. The composite specimens were cut across the transverse cross-section of the joints.

TABLE 18.1
Chemical Composition (%wt.) of Parent Metal

Si	Fe	Cu	Mn	Mg	Zn	Cr	Ti	Al
0.81	0.21	4.2	0.65	0.51	0.02	0.02	0.02	Balance

TABLE 18.2
Mechanical Properties of Parent Metal

0.2% Yield Strength (MPa)	Ultimate Tensile Strength (MPa)	% Elongation in 50 mm Gauge Length	Vickers Hardness (HV) (0.5 N, 15 s)
432	460	7.5	152

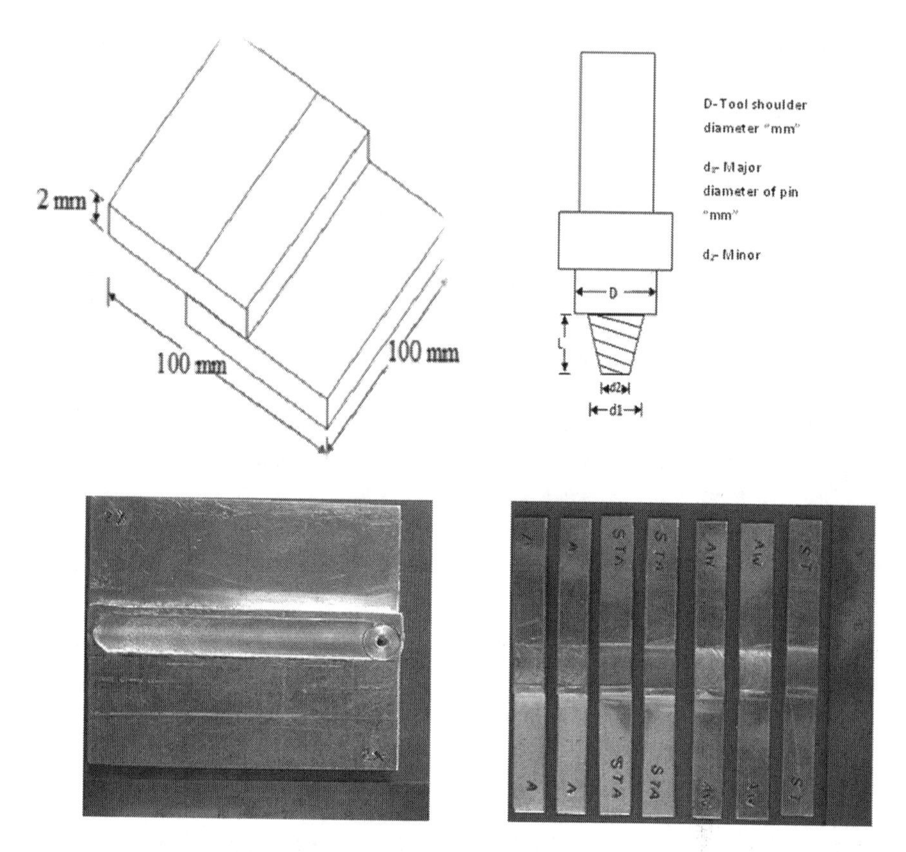

FIGURE 18.1 Schematic diagram of (a) joint configuration, (b) tool configuration, (c) photograph of fabricated joint and (d) tensile specimens (before testing) (Rajendran et al. 2016).

The standard metallographic technique was followed to prepare the specimens, and they were etched by Keller's reagent in order to reveal the grain size of the different weld regions. An optical microscope was used for the microstructural analysis. The fracture surfaces of the tensile tested specimens were analyzed by using a scanning electron microscope (SEM). The microhardness measurements were taken across the weld centreline on the top sheet using a Vickers microhardness tester with a load of 50 N and dwell time of 15 seconds. A transmission electron microscope (TEM) was used to investigate the distribution of precipitates and the dislocation cell structure evolved in the SZ of the joint.

18.3 STRENGTH AND MICROSTRUCTURAL PROPERTIES OF JOINT

The strength properties of lap joint in AW and different solutionizing temperature are presented in Table 18.3. Three specimens were tested in each condition, and the average value is presented in Table 18.3.

TABLE 18.3
Tensile Test and Hardness Test Results

Sl. No	Joint Type	Tensile Shear Fracture Load (TSFL) "kN"	Fracture Location	Microhardness VHN 0.5 N, 15 s		
				SZ	TMAZ	HAZ
1	AW Joint	12.76	HAZ	125	125	115
2	STA	13.37	SZ/TMAZ	125	112	124
3		18.34			155	

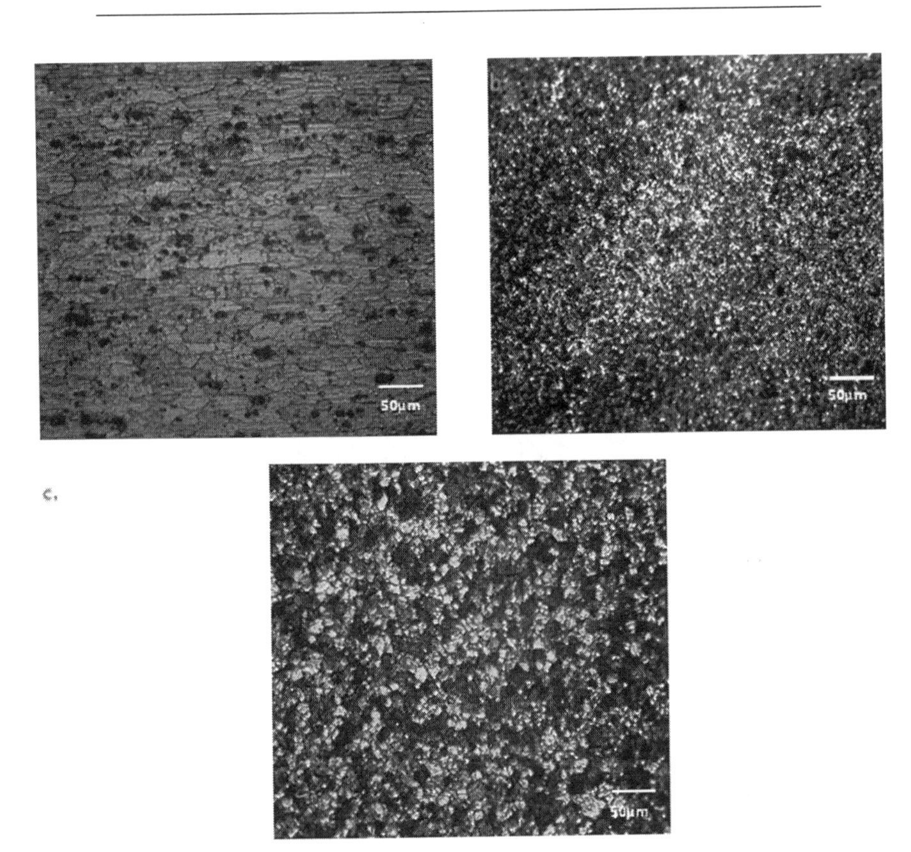

FIGURE 18.2 Optical micrograph of (a) base metal, stir zone of (b) AW and (c) STA joint.

The PM showed tension to the shear failure load of 18.34 kN. A joint exhibited a lower shear fracture load (SFL) of 12.76 kN compared to the PM. The STA joint yielded tensile shear fracture load (TSFL) of 15.27 kN, an increment of 20% compared to the AW joint. Figure 18.2a–c shows the optical micrograph of the PM and SZ of AW and STA joints. The PM is composed of coarse and elongated grains of an

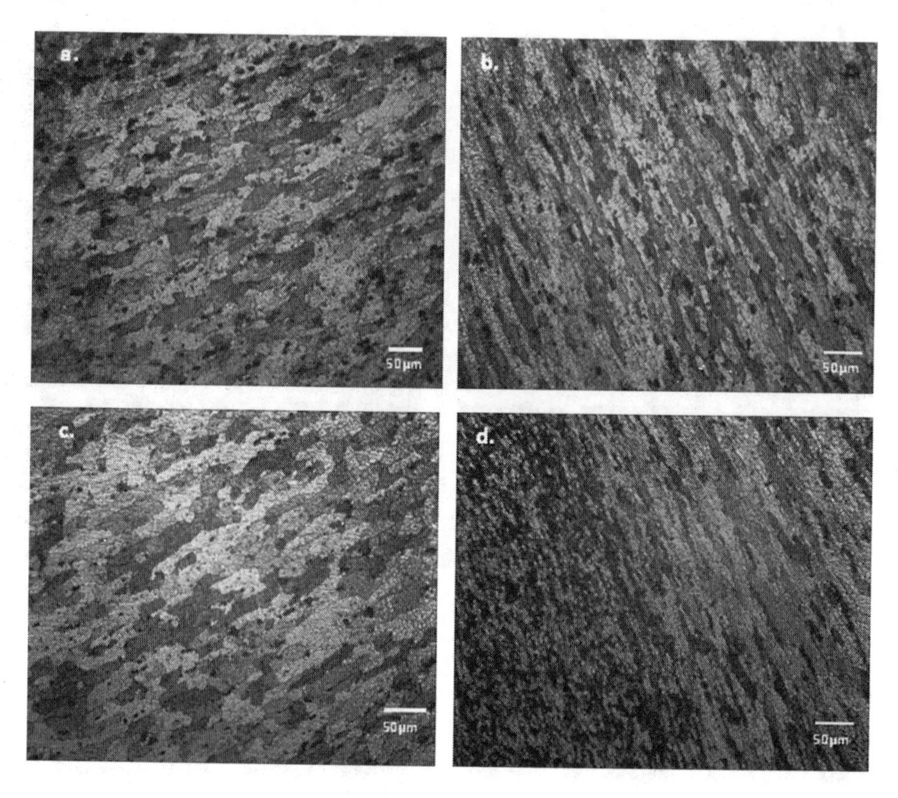

FIGURE 18.3 Optical micrograph of AW joint (a) RS-TMAZ, (b) AS-TMAZ, solution-treated joint (c) RS-TMAZ and (d) AS-TMAZ.

average grain size of 30 μm. The SZ of the AW joint (Figure 18.2b) consists of fine and equiaxed grain, which is due to severe plastic deformation followed by dynamic recrystallization (DRX) that occurred during the FSW process. The SZ of the STA joint (Figure 18.2c) reveals the marginal increase in grain size due to solutionizing and followed by ageing treatment.

The optical micrographs of RS-TMAZ and AS-TMAZ in AW condition are shown in Figure 18.3a and Figure 18.3b, respectively. The optical micrographs of RS-TMAZ and AS-TMAZ in the STA joint are shown in Figure 18.3c and Figure 18.3d, respectively. AS-TMAZ and RS-TMAZ show the partially strengthened conditions of the highly deformed matrix, which might be the fine precipitates precipitated during the STA treatment of the region. The TEM micrograph of PM (shown in Figure 18.4a) reveals two types of precipitates (coarse and fine). The coarse precipitates vary in size from 50 to 100 nm, while the fine precipitates vary from 10 to 50 nm in size.

The TEM of AW joint SZ (Figure 18.4b) reveals precipitates with fine, needle-like structure and spherical morphology. The finer Cu Al2 and Al2CuMg precipitate in the SZ completely dissolved in the matrix because of friction heat during the FSW cycle (Babu et al. 2016). SZ of the STA joint (Figure 18.4c) shows the

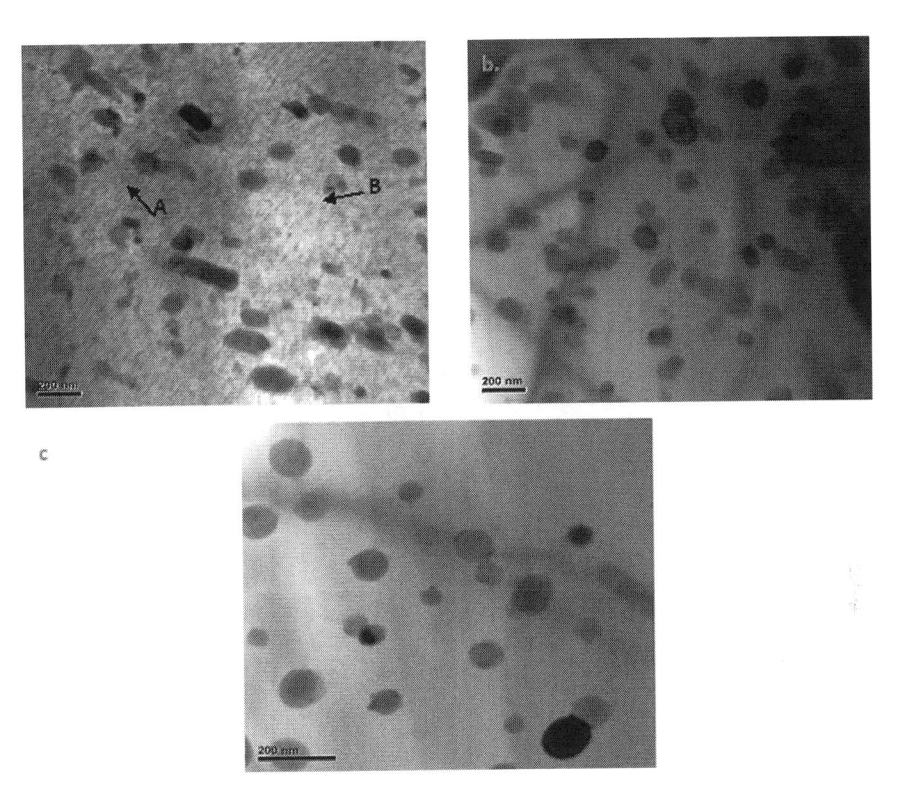

FIGURE 18.4 TEM micrograph of (a) PM, (b) AW and (c) STA joint.

dissolution of all the agglomerated coarse precipitates in the matrix except for few coarse precipitates.

It can be understood that the STA treatment coarsened the alpha aluminium grains in all the regions of joints, which resulted in large size alpha aluminium grains; the grain growth appears to be a natural consequence of heat treatment (Charit and Mishra 2008). The size and distribution of strengthening precipitates in the STA-treated joints were more uniform than other joints. This may be the reason for the higher hardness and superior tensile strength of STA joints. The solution temperature is another important factor to control the stability of the grains in the FSW joints. The increase in solution temperature resulted in increases in grain growth (Chen et al. 2007). In this study, solution treatment was carried out at 500°C and a 1-hour soaking period. A close examination of the fracture surface of the specimen, which was tensile tested, can provide useful information on the role and contribution of the inherent microstructural features on the strength and ductility of the joint (Srivatsan et al. 2007; Suvillian and Robson 2008). Figure 18.5(a–c) shows the fracture morphology of the tensile-tested samples treated by the STA joint. The fracture surfaces of the tensile-tested joints are characterized by the SEM, which was taken from the centre of the fracture region.

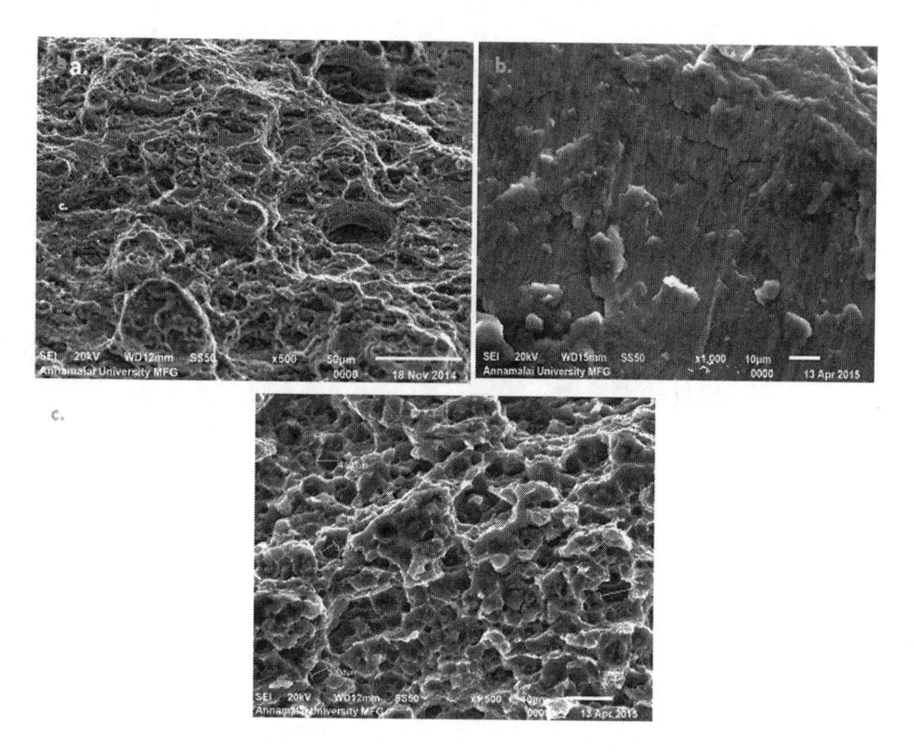

FIGURE 18.5 SEM-fractographs of (a) PM, (b) AW and (c) STA joint.

18.4 ANALYSIS OF PRECIPITATE MORPHOLOGY OF FSW JOINTS

The stir zone was solutionized owing to a high thermal cycle (420°–480°C) (Rhodes et al. 1997; Mahoney et al. 1998), which resulted in the dissolution of $CuAl_2$ precipitates. It is possible that this was due to low joint stiffness or hardness in AW condition. The mechanical properties of the precipitation hardening aluminium alloy FSW joints are dependent on precipitate distribution and volume rather than grain size (Attalh and Salem 2005). Ageing treatment is the most efficient method of precipitate precipitation, which leads to an increase in strength compared to the AW joint. The precipitation hardening aluminium alloy (2xxx) produces fine needle-like and spherical shaped $CuAl_2$, precipitating Al_2CuMg with an average size of 100 nm. During the FSW process, the precipitates become rough or dissolve. This result reduces the joint's strength. The STA treatment significantly improved the properties of the soft area in the AW joint. The joint strength was significantly increased by the two PWHT from the results, and the STA joint offered better strength (15.27kN) compared to the AW joint. Ultimately, the FSW joint fracture occurred from the weak region of the joints during the tensile test. The location of the soft region is found to vary from HAZ, SZ/TMAZ interface, depending upon the condition of joints.

The AW joint, fractured from HAZ (115 HV) on the AS and the fracture location of the AA and STA joint was observed at the SZ/TMAZ interface on the AS. The

AW joint fracture surface exhibited deeper and larger dimples (Figure 18.5b). Similar behaviour was also found in the STA joint (Figure 18.5c), which showed an elongated, dimpled surface. These elongated dimples indicated dimple shearing during tensile testing, and fracture mode was ductile. The width of the dimples in the STA joint is high, which might be suggestive that at the tip of the crack there is a large stretch area, resulting in a large plastic region ahead of the crack (Aydin et al. 2009).

18.5 SUMMARY OF EFFECT OF SOLUTION TREATMENT AND ARTIFICIAL AGEING ON FSW JOINTS

The solution treatment followed by ageing cycle (500° C, for 1 hr. and 170° C for 10 hrs.) is found to be more beneficial to increase the tensile shear fracture load-carrying capability of FSW lap joints of AA2014 aluminium alloy, and enhancement is approximately 20%. STA joint yielded higher load-carrying capability than other joints because of the dissolution of all the agglomerated coarse precipitates (during AA) in the matrix except for a few coarse precipitates. The artificial ageing processes in the STA treatment caused the reprecipitation of fine θ' precipitates.

REFERENCES

Attallah, M.M., Salem, H.G. 2005. Friction stir welding parameters: A tool for controlling abnormal grain growth during subsequent heat treatment. *Materials Science & Engineering. part A*, 39: 51–9.

Aydin, H., Bayram, A., Uguz, A., Akay, S.K. 2009. The tensile properties of friction stir welded joints of 2024 aluminum alloy in different heat treated state. *Materials and Design*, 30: 2211–21.

Babu. S., Janaki Ram, G.D. Venkitakrishnan, P.V., Madhusudhan Reddy, G., Prasad Rao, K. 2012. Microstructure and mechanical properties of friction stir lap welded aluminum alloy AA2014. *Materials Science and Technology*, 28: 414–26.

Charit, I., Mishra, R.S. 2008. Abnormal grain growth in friction stir processed alloys. *Scripta Materlia*, 58: 367–71.

Chen, Y.C., Feng, J., Liu, H.J. 2007. The stability of the grain structure in 2219-O aluminum alloy friction stir welds during solution treatment. *Materials Characterization*, 58: 174–8.

Fu, R., Zhang, J.F., Li, Y.J., Kang, J., Liu, H.J., Zhang, F.C. 2013. Effect of heat input and post welding natural aging on the hardness of stir zone for friction stir welded 2024-T3 aluminum alloy thin sheet. *Materials Science and Engineering. part A*, 599: 319–24.

Genevosis C, Deschamps A, Denquin A, Doisneau-coottignies B. 2005. Quantities investigation of precipitation a mechanical behavior for AA2024 friction stir weld. *Acta Materlia*, 53: 2447–58.

Hu, Z., Yuan, S., Wang, X., Liu, G., Huang, Y. 2011. Effect of post weld heat treatment on the microstructure and plastic deformation of friction stir welded 2024. *Materials and Design*, 32: 5055–5060.

Liu, H.J., Zhang, H.J., Yu, L. 2011. Effect of welding speed on microstructures and mechanical properties of underwater friction stir welded 2219 aluminum alloy. *Materials and Design*, 32: 1548–53.

Mahoney, M.W., Rhodes, C.G., Flint, J.G., Spurling, R.A., Bingel, W.H. 1998. Properties of friction stir welded 7075 T651Al. *Metallurgy and Materials Transaction A*, 29: 1955–64.

Rajendran, C., Srinivasan, K., Balasubramanian, V., Balaji, H., Selvaraj, P. 2016. Influence of post weld heat treatment on tensile strength and microstructural characteristics of friction stir welded butt joints of AA2014-T6 aluminium alloy. *Journal of Mechanical Behaviour of Materials*, 25(3–4): 89–98.

Rhodes, C.G., Mahoney, M.W., Bingel, W.H. 1997. Effect of FSW on the microstructure of 7075 Al. *Scripta Materlia*, 36: 69–75.

Sharma, C., Dwivedi, K.P. 2013. Effect of post weld heat treatment on microstructure and mechanical properties of friction stir welded joints or Al-Zn-Mg alloy by AA7039. *Materials and Design*, 43: 134–43.

Squillance, A., De Renzo, A., Giorleo, G., Bellucci, F. 2004. A comparison between FSW and TIG welding techniques: modification of microstructure and pitting corrosion resistance in AA2024-T3 butt joints. *Journal of Materials Processing Technology*, 152: 97–105.

Srivatsan, T.S., Vasudevan, S., Park, L. 2007. Tensile deformation and fracture behavior of friction stir welded aluminum alloy 2024. *Materials Science and Engineering. part A*, 456: 235–45.

Suvillian, A., Robson, J.D. 2008. Microstructural properties of friction stir welded and post welded heat treatment 7449 aluminum alloy thick plate. *Materials Science and Engineering. part A*, 478 (1–2): 351–60.

Wang, J., Fu, R., Li, Y., Zang, F. 2014. Effect of deep cryogenic treatment and low temperature aging on the mechanical properties of friction-stir-welded joints of 2024-T351. *Materials Engineering A*, 609: 147–53.

Zang, Z., Xiao, M.Z.Y. 2014. Influence of water cooling on microstructure and mechanical properties of friction stir welded 2014 aluminum alloy. *Materials Engineering A*, 614: 6–15.

Zhou, C., Yang, X., Luan, G. 2006. Effect of root flaws on the fatigue property of friction stir welds in 2024-T3 aluminum alloys. *Materials Science and Engineering. part A*, 418: 155–60.

19 Effect of Oscillation Frequency on Microstructure and Tensile Properties of Linear Friction Welded Ti-6Al-4V Alloy Joints

C. Mukundhan, P. Sivaraj, C. Rajarajan, Vijay Petley, Shweta Verma and V. Balasubramanian

CONTENTS

19.1 INTRODUCTION

Ti-6Al-4V (Ti-64) alloy is widely used in aerospace, automobile, nuclear and petrochemical industries because of its corrosion resistance, high-temperature mechanical properties, and low density (Gao et al. 2013). In particular, Ti-64 is used in bladed disk (blisk) assembly in aero engines (Bhamji et al. 2010, 2011). Generally, the gas tungsten arc welding (GTAW) process is employed to join sheet metals in superalloys and titanium alloys, but higher heat input of fusion welding resulted in distortion and inclusion defects (Karadge et al. 2007; Guo et al. 2017). The energy density processes, such as an electron and laser beam weldings, are preferred for joining titanium alloys and its alloys. The critical cooling rate possesses porosity and weld crack in weldment (Fall et al. 2017). To overcome these issues, researchers recommended solid-state welding processes, such as friction welding, friction stir welding and diffusion bonding to join similar and dissimilar joints of titanium alloys and superalloys in aerospace applications. Among the different solid-state welding processes recommended, friction welding offers many advantages, such as the elimination of consumables, reduced welding time, higher joint efficiency, etc. Linear friction welding (LFW) is a variant of the friction welding process in which the joint between two materials is made by the relative motion and the compressive forces. In LFW, one part is kept stationary and the other part oscillates linearly. During this process, frictional heat is generated between the surfaces, and the plasticized region forms at the interface. After this, the forging force is applied to produce a final joint with a limited thermomechanically affected zone (TMAZ) (Wanjara and Jahazi 2005).Abbasi et al.(2017) investigated the effect of filler metal on microstructure and mechanical properties of Ti-6Al-4V joints. The joint fabricated with matching filler exhibited higher tensile strength than others. The microstructure of the weld metal consisted of both acicular and basket weave morphologies. Babu and Raman (2006) studied the effect of current pulsing and postweld heat treatment on microstructure and mechanical properties of TIG weldments of Ti64. The current pulsing resulted in refinement of prior β grains, which improved both strength and ductility of the weldments. An increase in ductility and reduction in strength was observed for the postweld heat treated weldments due to the coarsening of α grains, reduction in defect density and decomposition of martensite. Balasubramanian et al. (2008) investigated the corrosion behaviour of pulsed gas tungsten arc welded Ti64 joints. The corrosion resistance increased with increasing pulsing frequency and peak current and then decreased. The finer grains developed in the fusion zone were responsible for the increased corrosion resistance. Cao and Jahazi (2009) studied the effect of welding speed on microstructure and mechanical properties of laser-welded Ti64 joints. The presence of α' in the fusion zone increased the hardness by 20% with respect to the base metal. The microstructure was inhomogeneous across the weld joint and the tensile strength of the joint is found to be increased with a reduction in ductility. Romero et al. (2009) studied the effect of forging pressure on microstructure and residual stress development of Ti64 linear friction welds. From his study, they concluded that the forging pressure has a strong influence on weld width and TMAZ. During welding, the temperature developed at the weld region and TMAZ exceeds β transition temperature. At low forging pressure, the amount of α-Ti is higher, and was decreased as the forging pressure increased. An increase in forging pressure decreased the residual stresses in both x- and y-direction. Li et al. (2008) studied the influence of friction time asymmetric flash formation and behavioural changes in LFW steel joints. The burn-off length increased with increasing friction time. They observed undulating-ribbon structure flashes in the vertical direction with curly edges. The numerical study was carried out to understand the influence of welding parameters in LFW discussed by Wen Ya Li et al. (2010). At higher oscillating frequency, the interface temperature increased quickly, and axial shortening occurred at a faster rate. Similar behaviour was also observed for amplitude and friction pressure. Liu and Dong (2014) elaborated the nugget behaviour through contour method. They found that within the weld zone of the TC17 LFW weld, the through-thickness stress was not uniform; the interior stress was higher than that near the top or bottom surface. Zhang et al. (2013) found that after PWHT in TC4 alloy, the martensite decomposed, and the acicular became more distinct and were coarsened, resulting in a decrease in hardness. Fine precipitates were found to form in the weld region in TC17 alloy, resulting in a sharp increase in hardness. Sivaraj et al. (2019) examined the mechanical and metallurgical behaviours of high strength aluminium alloy through LFW. They obtained a joint efficiency of 75%, and fracture occurred in the TMAZ region. Su et al. (2018) found that the ß phase reduced gradually from base metal (BM) to weld centre zone (WCZ), while there is none in WCZ because of the higher content of a phase stable elements, and the weld metal is more easily converted to a phase when it cools down after reaching the ß-transus temperature. Ballat-Durand et al. (2019) found that heat treating the Ti17 joint permitted recovery of an equilibrium a+ß Widmanstätten microstructure across the whole assembly, leading to both homogenized stiffness values and plastic behaviour. From the literature review, it is understood that many authors focused on studying the effect of friction time, amplitude and forging pressure. The published information on the effect of oscillating frequency of linear friction welded Ti-6Al-4V joints are limited in number. Hence, the present investigation focused on studying the effect of oscillating frequency of linear friction welded Ti-6Al-4V joints.

19.2 EXPERIMENTAL DETAILS

The effect of oscillation frequency on linear friction welding of titanium joints was studied on

$60 \times 30 \times 6$ mm (length × width × thickness) plate. The percentage of α stabilizers, β stabilizers and other alloying elements of the as-received Ti64V alloy and tensile strength of the alloy before welding were tested and presented in Tables 19.1 and 19.2.

The optical micrograph and scanning electron microscopy (SEM) image of Ti64 alloy before welding bimodal α and β structure is shown in Figures 19.1a and 19.1b. The optical and SEM micrographs revealed the presence of β phases as grain boundaries of the α phase. The blade and disk assembly section of the gas turbine engine is shown in Figure 19.2a. The indigenously designed LFW machine was used for this investigation and is shown in Figure 19.2c. The working principle for LFW progress with four stages (i) initial phase, (ii) transition phase (iii) equilibrium phase and (iv) deceleration or forging phase is shown in Figure 19.2b. Initial welding trials were performed to identify the visible range of welding parameters viz. oscillation frequency range (11–19 Hz), friction time (30s), friction pressure (22 MPa), forging pressure (11 MPa) and forging time (3s) were kept constant. Figure 19.3a represents

TABLE 19.1
Chemical Composition of Ti64 Base Metal (wt %)

Elements (wt %)	Al	V	Fe	O	C	N	H	Ti
Ti-6Al-4V	6	4	0.19	0.15	0.06	0.04	0.01	Bal.

TABLE 19.2
Mechanical Properties of Ti64 Base Metal

0.2 Yield Strength (MPa)	Ultimate Tensile Strength (MPa)	Elongation in 50 mm GL (%)	Reduction in Cross-Sectional Area (%)	Hardness in Hv
980	1030	12	24	439

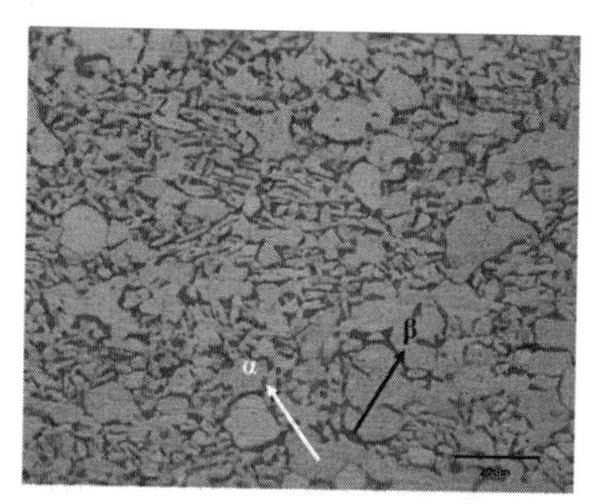

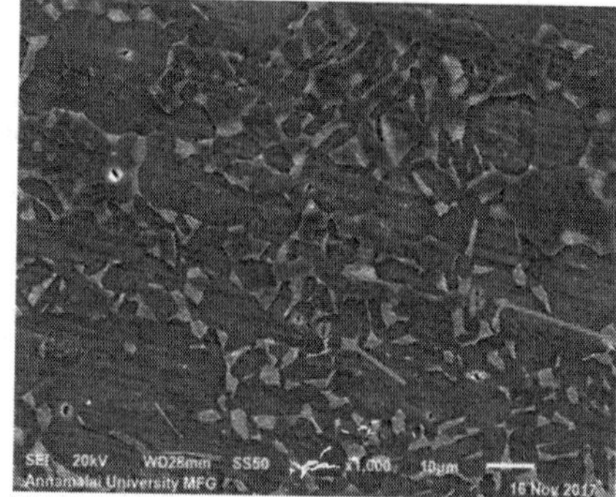

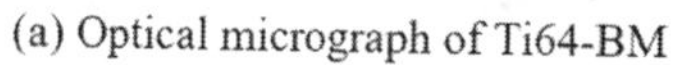
(a) Optical micrograph of Ti64-BM

(b) SEM image of Ti64-BM

FIGURE 19.1 Micrographs of Ti64 base metal.

the fabricated LFW titanium alloy joints. The effect of the oscillation frequency on the tensile strength of the welded specimen was tested using a servo-hydraulic controlled universal testing machine at a constant strain rate of 2.4×10^{-3} s^{-1}. The specimen configuration for the transverse tensile test was set as per Figure 19.3b. The load-displacement curves recorded using the data acquisition system during the process were then converted into stress-strain curves. The microhardness variations across the joints were studied using Vicker's microhardness tester in different zones of welded joints viz. WCZ, TMAZ, heat affected zone (HAZ) and BM. In a further investigation, the LFW joint was cut in sections by a power hacksaw. The sectioned metal cuts were mounted with phenolic resin in the wet cooling method for convenient handling of the material for polishing procedure. The mounted specimen was polished was using the following techniques in the following order: grinding, dish

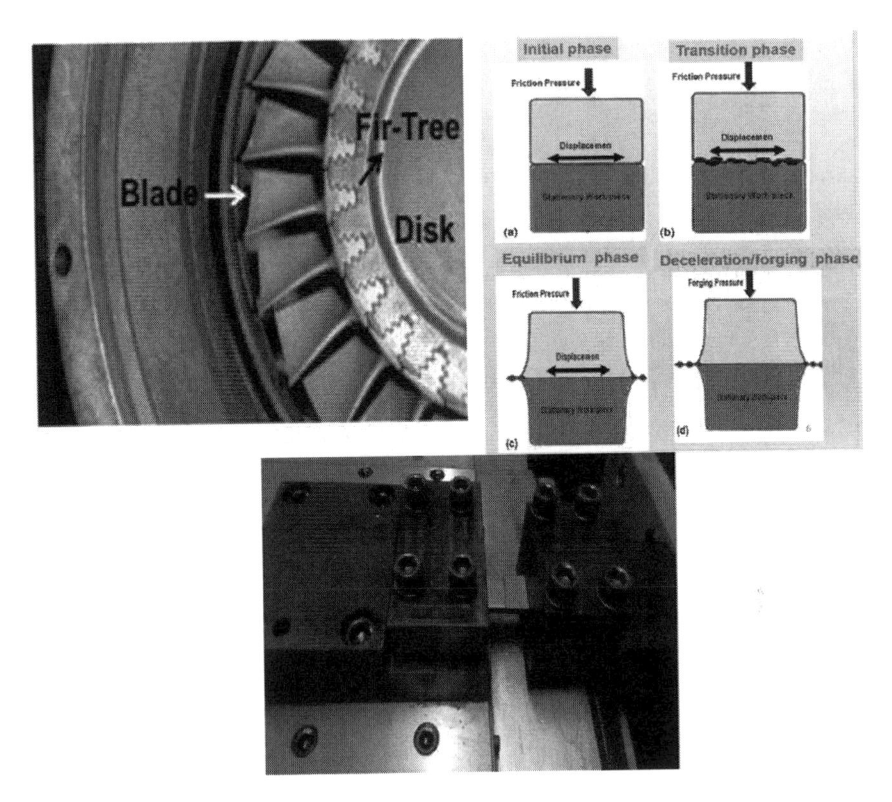

FIGURE 19.2 (a) Aero-engine blade assembly, (b) working principle of LFW and (c) LFW machine setup.

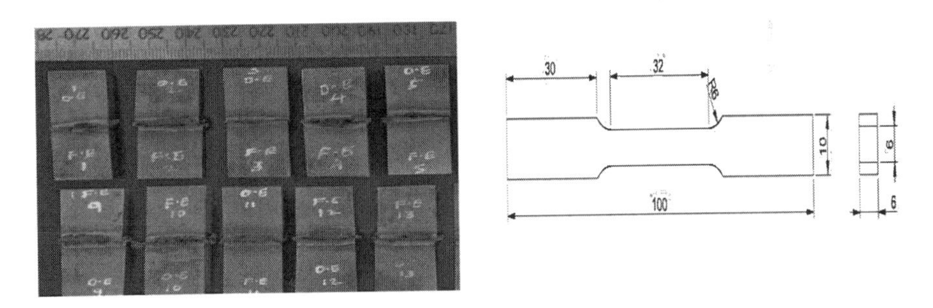

FIGURE 19.3 (a) Fabricated LFW joints and (b) specimen configuration for tensile test.

polishing and etching of the mounted specimen. Higher quality mirror polishing was accomplished with the aid of fine granular diamond paste. Moreover, the microstructural features of defect-free joints were revealed using Kroll's reagent (100 ml H_2O + 2 ml HF + 5 ml HNO_3) under optical microscopy (OM) and SEM. The fracture characteristics were analyzed to reveal the mode and behaviour growth of weld failure in the LFW joint.

19.3 RESULTS

19.3.1 Macrostructure Analysis

The macrostructure has shown evident quality of the weld joint in the lower magnification. The photograph and macrograph of the joint welded with 11–19 Hz oscillating frequency range are displayed in Table 19.3. At 11 Hz oscillating frequency, the rubbing action between the faces of the material was unable to plasticize at the mating surface, producing no joints. According to Sivaraj et al. (2019), low-range friction and heat generation do not tend to bond with the mating material. The 13 Hz oscillating frequency produced notable heat by the rubbing action at the interface, causing the material to bond on the edges and a marginal area in the middle surface. At the oscillating frequencies of 15 Hz and 17 Hz, the flash formation was appreciable at the oscillating end as the heat generated between the mating surfaces was sufficient enough to push out the plasticized material, forming a strong bond at the interface and produced sound welds. These joints were characterized as defect-free joints and were subjected to further characterizations. Further, the 17 Hz joint was higher, and at a maximum oscillation frequency of 19 Hz, excess flash formation was caused by the high rubbing action between the mating surfaces, causing the material to flow out excessively and produce joints with visible cracks (Mukundhan et al. 2020).

19.3.2 Tensile Properties

The tensile properties, such as yield strength, ultimate tensile strength and elongation of the defect-free joints (15 Hz and 17 Hz oscillatory frequencies) are shown in Table 19.4. The 15 Hz and 17 Hz oscillatory frequency joints exhibited 94% and 96% joint efficiencies, which resemble the characteristic feature of the LFW process. The tensile strength of the welded joint found close to the base material indicates the coalescence of the process. Figure 19.4 represents the strength-to-displacement graph for 15 Hz and 17 Hz oscillation frequency. The superior strength of the joint is also due to the nucleation of fine grains in the WCZ. The nucleation of these fine grains is due to the fine rubbing action at the interface of the process, which helps the material to plasticize (Wang et al. 2019; Hua et al. 2014). Moreover, both the joints failed in the TMAZ region because the coarser grains formed as a result of heat dissipation through the process. The fall in the percentage reduction of elongation also indicates the reduction in ductility of the LFW joints (Zhao and Fu 2020).

19.3.3 Microhardness Variations across the Joints

The microhardness survey reveals the welded joints (395Hv) exhibit higher hardness values than the base metal (340Hv), as shown in Figure 19.5. The WCZ of 15 Hz and 17 Hz frequency joints were recorded finer grain structure then the base metal. The hardness of the WCZ for both the joints was found to be higher than in other regions of the weld. This higher hardness is due to the fine grains in the

TABLE 19.3
Effect of Oscillation Frequency on Macrograph of LFW Joints

S.No	Oscillation Frequency (HZ)	Photograph	Macrograph	Observation
1	11 Hz	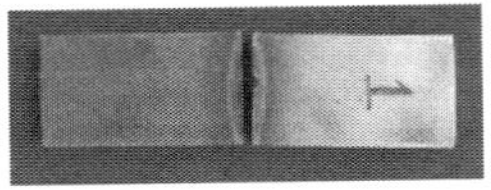	---	Not welded
2	13 Hz	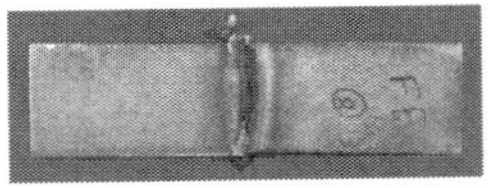	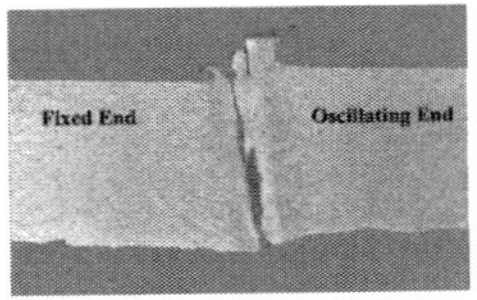	Welded with macro-level defect
3	15 Hz	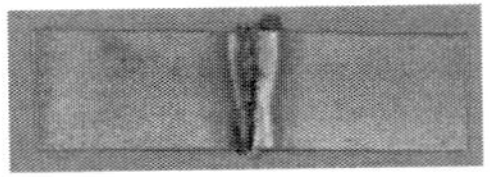	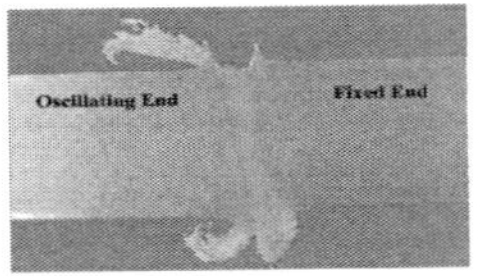	Welded with the nonuniform flash formation
4	17 Hz	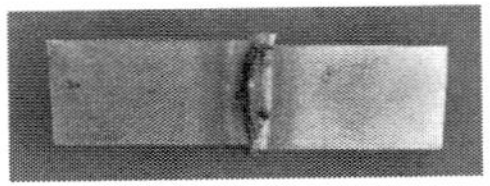	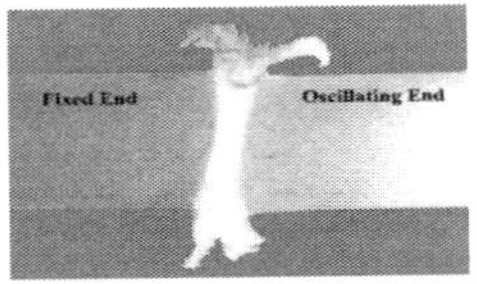	Welded with the good flash formation on both sides
5	19 Hz	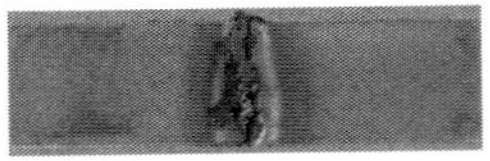	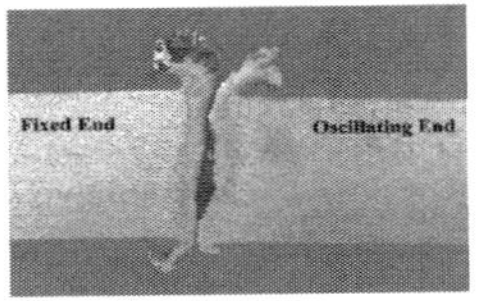	Welded with macro-level defect

TABLE 19.4
Tensile Properties of LFW Joints

Process Condition	0.2 Yield Strength (MPa)	Tensile Strength (MPa)	Elongation in 50 mm GL (%)	Joint Efficiency (%)	Location of Failure
BM	980	1030	12	–	–
15 Hz Joint	900	975	8.4	94	TMAZ
17 Hz Joint	926	1011	7.5	98	TMAZ

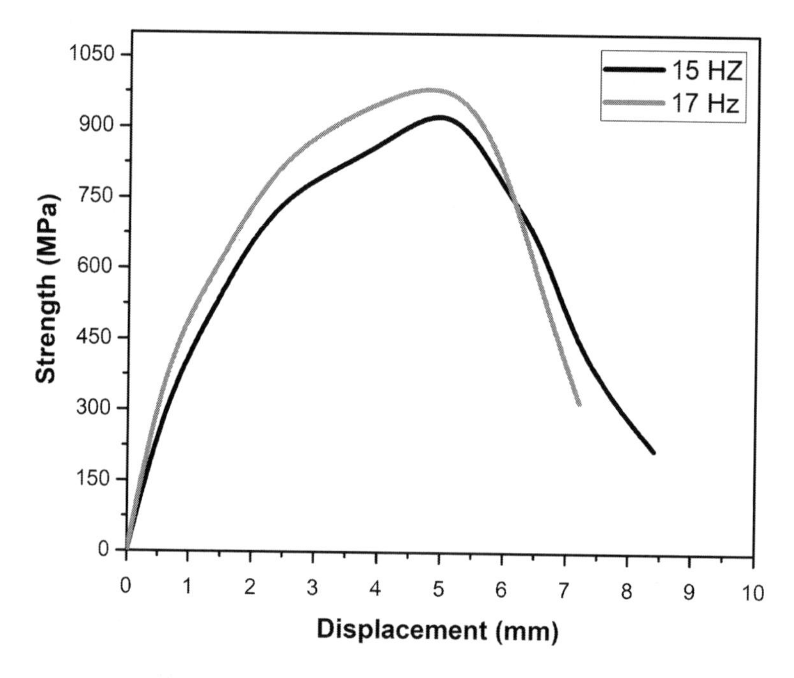

FIGURE 19.4 Strength-to-displacement graph for LFW joint.

WCZ that were formed as a result of the high rubbing action generated by the high heat at the interface during the process. This heat generation was the result of the combined rubbing action with the application of frictional pressure and forging pressure (Dalgaard 2011).

The TMAZ and HAZ recorded lower values than WCZ and BM, which was due to the coarser grains formed in these regions. Ji et.al (2014) recorded similar microhardness behaviour of LFW titanium alloy joints. they recorded lower microhardness in the TMAZ or Partially Deformed Zone (PDZ) region due to grain deformation. The heat dissipation from the WCZ to TMAZ through HAZ to BM is the reason behind the drop in the hardness values of these regions. Further, the decrease in hardness profile from WCZ to TMAZ and HAZ also caused failure during tensile testing (Jackson et al. 2009).

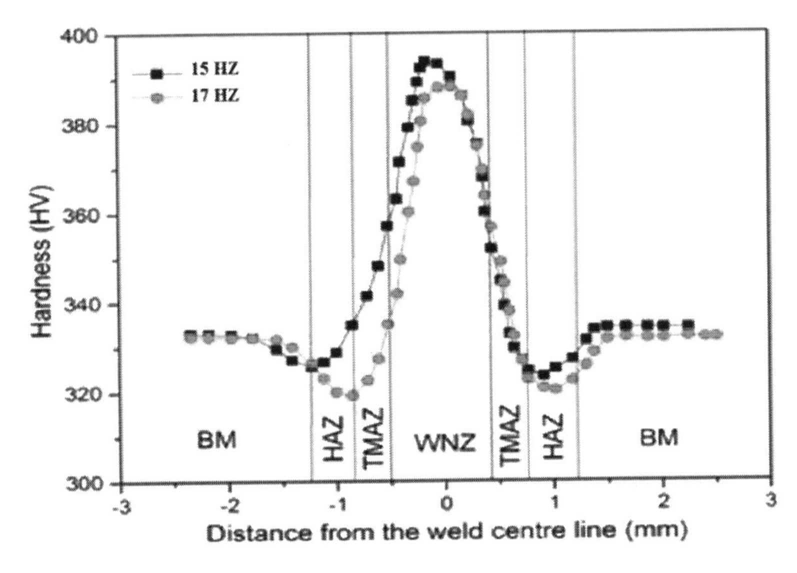

FIGURE 19.5 Microhardness variation across the mid-thickness in LFW joint.

19.3.4 Microstructural Features of the LFW Joint

The microstructures along the weld regions were found to be heterogeneous and classified as WCZ, TMAZ, HAZ and unaffected BM. The different microstructural regions of 15 Hz and 17 Hz oscillation frequencies are shown in Figure 19.6 (a) and (d).

The microstructure of WCZ metal is transformed into the Widmanstätten (basketweave) structure from the bimodal microstructure of alpha and beta grains exhibited in the base materials. During LFW, the interface temperature will exceed 995°C, causing the transformation of the bimodal alpha and beta grains into the single-phase beta field (Buffa and Fratini 2017). The higher cooling rate also resulted in the diffusionless transformation of beta grains into the martensitic (Widmanstätten) structure. This was the reason for the high microhardness recorded in this region of weld. The formation of TMAZ was due to the dissipation of heat generated by the combined action of friction pressure and forging pressure on the mating surfaces (Chamanfar et al. 2011). The microstructure in the HAZ is almost the same as the TMAZ region. In HAZ, the grains are softened because of the conventional heat transfer. The grain sizes were found to be almost the same for joints at both the frequencies. The temperature generated at these regions is well below the beta transition temperature, and

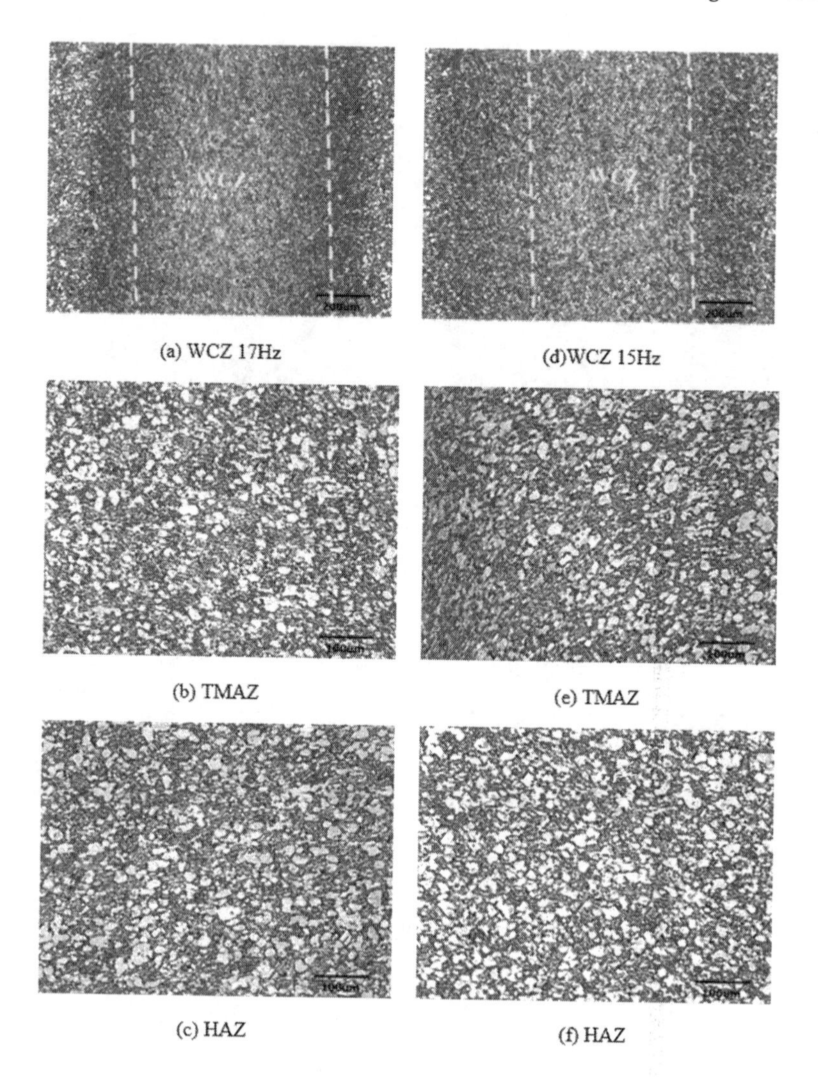

FIGURE 19.6 Microstructural features of LFW joints. (a), (b) and (c) – oscillation frequency = 15 Hz.; (d), (e) and (f) – oscilation frequency = 17 Hz.

this retards the phase transformation of these regions(McAndrew et al. 2018). This was also characterized by the decrease in hardness and tensile values of the joints. This phenomenon was the reason behind the failure of the joints in the TMAZ region, showing good joint integrity of the welds. The microstructure of the TMAZ is shown in Figure 19.6 (b–e). This region is very narrow compared to the WCZ, and it is highly deformed because of the combined action of frictional heat and forging pressure. The temperature in this region is well below the beta transition temperature (below 995°C); therefore, no phase transformation occurred in this region. The original bimodal alpha and beta grains are reoriented during LFW (Mironov et al. 2008). The microstructure in the HAZ is almost the same as the TMAZ region. In Figure 19.6 (c and f) the HAZ is shown. The grains are softened because of the conventional heat transfer. The grain sizes are almost the same for both of the frequencies. This inhomogeneous microstructure is consistent with the hardness profiles and tensile properties. Because of the failure of the tensile specimens at the TMAZ region due to the coarse alpha and beta grains.

19.3.5 Analysis of SEM Microstructure

For a better understanding of the morphology of the weld region, SEM was used to observe the microstructures. It was observed that the bimodal microstructure of alpha and beta grains in the base metal was totally transformed into the Widmanstätten (basketweave) structure, especially in the WCZ, which is shown in Figure 19.7a_1. The colour mapping of microstructures gives a better understanding of Fully Deformed Zone FDZ as represented in Figure 19.7a_2. The partially deformed region was composed of finer and coarser grain structures. Figure 19.7 b_1 illustrates the PDZ microstructure of titanium alloy, which consists of martensite and prior β transition grains. The colour map distinctly shows the soft and hard phases in the PDZ regions of martensite, alpha and prior β grain structures (Li et al. 2009). Figure 19.7 (c_1–c_2) displayed the HAZ of the LFW joint at 17 Hz. However, the fine secondary alpha existing in the parent material was no longer observable in the HAZ because of the heating during the welding process. Also, the HAZ had lower deformation areas by force and frictional heat by solid-state welding processes (Ma et al. 2011). The

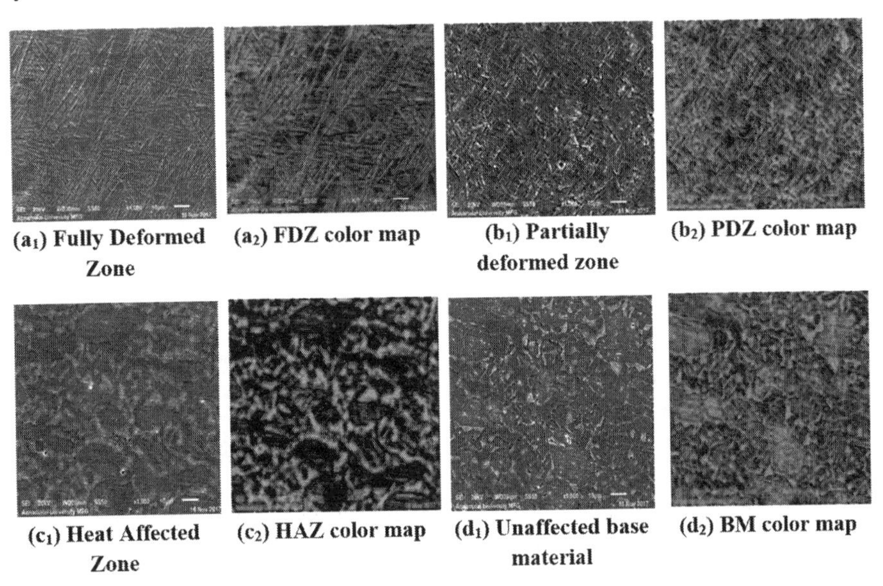

FIGURE 19.7 SEM microstructure of various zones in LFW joints.

interface between the beta phase and the primary alpha phase also became less clear presumably due to the composition change (Wang and Wu 2012).

19.3.6 Fracture Surface Analysis

The topography features revealed mode tensile failures, which caused weld failure. The fracture morphology was classified into three sections: (i)crack initiation, (ii) propagation region and (iii)disintegration (Sivaraj et al.2018). The fracture surfaces of tensile specimens are shown in Figure 19.6 (a and b). Both 15 Hz and 17 Hz frequency joints exhibited a ductile mode of failure that was characterized by a large number of finer dimples in TMAZ regions. The difference in ductility of the welded joints at 15 Hz and 17 Hz frequency was characterized by finer and elongated dimples. According to Sivaraj et al. (2019), the size of the dimple describes the joint efficiency and strength of the weld joint. Finer dimples obtained greater strength than coarser dimple features. From the above dimple features, the 17 Hz LFW joint recorded finer dimples pattern than the 15 Hz weld joint (Figure 19.8).

19.4 DISCUSSION

The LFW has more advantages over various fusion welding processes, as it yields maximum tensile strength properties over 96% to the parent metal. While welding, the rubbing action against the mating material produces the asymmetrical flash formation in the Ti64 alloys. The flash formation increases with increasing bonding tendency and strength upto maximum burn off length (Hynes and Velu 2018). The macrostructure analysis helps to understand the characteristic behaviour changing with oscillation frequency in the LFW joint. The lower range of frequency having an inferior tendency to join an alpha-beta alloy. From the transverse tensile test, the two ranges (15 Hz and 17 Hz) of the welded joints produced better weld strength; however, they possess lower ductility compared to the parent metal. Weld failure in the LFW joint was not observed in the weld nugget. Because the weld nugget obtains finer microstructure and superior hardness properties compared with other regions, similar

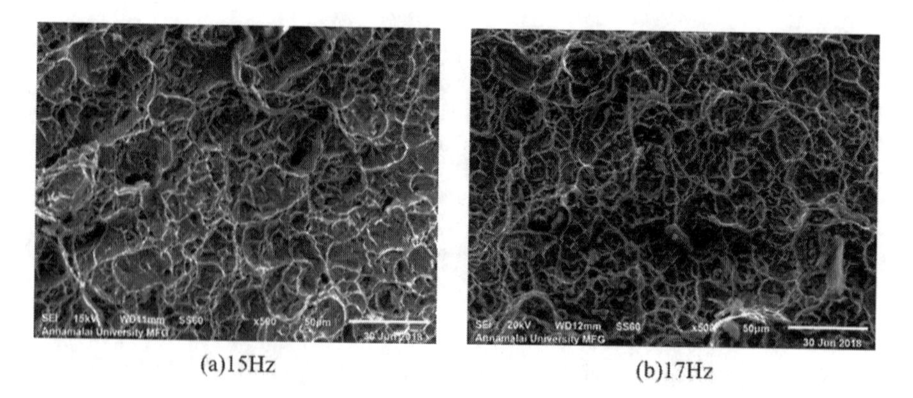

FIGURE 19.8 SEM fractograph of tensile tested specimens.

microhardness was recorded in LFW joints (Baeslack et al. 1994; Wang et al. 2017). The grain deformation and elastic strain are prime causes for LFW weld failures. The superior tensile and hardness properties may be attributed to the phase transformation and refinement of the grains. The failure initiates from the TMAZ region, as it is found to be the weakest region among the weld cross-section (Li et al. 2008). The weld nugget microstructure consists of needle-like and basketweave patterns. During weld deformation, the bimodal α-β grains changed into Widmanstätten (basketweave) and martensite (ά) phases. He observed phase transformation of refined prior-β grains at the weld centre (Li et al. 2016). The β transition temperature of Ti64 alloy is about 995°C; therefore, the nugget region was not expected to display phase changes in welded joints. Similarly, phase transition was observed in tempered and untempered regions (Fonda et al.2008; Wang et al. 2017). The material flow patterns of TMAZ or PDZ region were composed of reciprocating motion and extrusion of material in the weld thermal cycle. Further, the TMAZ region classified inner and outer regions.

The TMAZ region has relatively finer grain structure similar to martensite structure in weld nugget. But the outer region had comparably coarser grains of prior-β and alpha titanium. According to Dalgaard et al. (2010), partially deformed regions were categorized into coarser and finer sections. The coarser regions led to recorded inferior microhardness values in LFW titanium alloys. The heterogeneous grain deformation and softening affected the LFW titanium alloy joints. The finer dimple patterns confirmed the ductile mode of failure in the Ti64 alloy joint, also the size of the dimples indicated the joint strength of the weld cross-section (Uday et al. 2010).

19.5 CONCLUSIONS

The effect of oscillation frequency on linear friction welded titanium alloys joints were studied and the following conclusions were derived.

1. Oscillation frequencies <13 Hz was not sufficient enough to produce the rubbing action necessary to generate sufficient heat between the mating surfaces to join.
2. The oscillating frequency of 15 Hz and 17 Hz yielded defect-free joints with high structural integrity. A joint efficiency of 94% and 96% were obtained at these frequencies.
3. The joint efficiency of 15 Hz and 17 Hz frequency joints were calculated as 94% and 98% with 975 and 1011 MPa ultimate tensile strength.
4. The joint produced at 15 Hz showed maximum hardness 395Hv at WCZ and other areas.
5. The inhomogeneous hardness distribution was due to the transformation of the bimodal alpha and beta grains into the single-phase beta field in WCZ, and there was no phase transformation in TMAZ and HAZ.
6. The microstructural analysis revealed the formation of Widmanstätten structure (basketweave) at the weld nugget zone and reoriented and elongated bimodal alpha and beta grains in the TMAZ region.
7. The fracture surface of the specimen showed fine elongated dimple structures, confirming the ductile mode of failure.

REFERENCES

Abbasi, K., Beidokhti, B. and Sajjadi, S.A. 2017. Microstructure and mechanical properties of ti-6al-4v welds using α, near-α and α+β filler alloys. *Materials Science and Engineering. Part A*, 702: 272–278. doi:10.1016/j.msea.2017.07.027

Babu, N.K. and Sundara Raman, S.G. 2006. Influence of current pulsing on microstructure and mechanical properties of ti-6al-4v tig weldments. *Science and Technology of Welding and Joining*, 11(4): 442–447. doi:10.1179/174329306X120750

Baeslack, W.A., Broderick, T.F., Juhas, M. and Fraser, H.L. 1994. Characterization of solid-phase welds between Ti6A12Sn4Zr2Mo0.1Si and Ti13 5A121.5Nb titanium aluminide. *Materials Characterization*, 33(4): 357–367. doi:10.1016/1044-5803(94)90140-6

Balasubramanian, M., Jayabalan, V. and Balasubramanian, V. 2008. Effect of pulsed gas tungsten arc welding on corrosion behavior of Ti-6Al-4V titanium alloy. *Materials and Design*, 29(7): 1359–1363. doi:10.1016/j.matdes.2007.06.009

Ballat-Durand, D., Bouvier, S. and Risbet, M. 2019. Contributions of an innovative post-weld heat treatment to the micro-tensile behavior of two mono-material linear friction welded joints using: The β-metastable Ti–5Al–2Sn–2Zr–4Mo–4Cr (Ti17) and the near-α Ti–6Al–2Sn–4Zr–2Mo (Ti6242) ti-alloys. *Materials Science and Engineering. Part A*, 766: 138334. doi:10.1016/j.msea.2019.138334

Bhamji, I., Preuss, M., Threadgill, P.L. and Addison, A.C. 2011. Solid state joining of metals by linear friction welding: A literature review. *Materials Science and Technology*, 27(1): 2–12. doi:10.1179/026708310X520510

Bhamji, I., Preuss, M., Threadgill, P.L. and Addison, A.C. 2010. Solid state joining of metals by linear friction welding: A literature review. *Materials Science & Technology*. 2011, 27(1). Https://Www.Twi-Global.Com/Technical-Knowledge/Published-Papers/Solid-State-Joining-of-Metals-by-Linear-Friction-Welding-a-Literature-Review 1/33 SOLID STATE JOINING OF META

Buffa, G. and Fratini, L. 2017. Strategies for numerical simulation of linear friction welding of metals: a review. *Production Engineering*, 11(3): 221–235. doi:10.1007/s11740-017-0726-7

Cao, X. and Jahazi, M. 2009. Effect of welding speed on butt joint quality of Ti-6Al-4V alloy welded using a high-power Nd:YAGlaser. *Optics and Lasers in Engineering*, 47(11): 1231–1241. doi:10.1016/j.optlaseng.2009.05.010

Chamanfar, A., Jahazi, M., Gholipour, J., Wanjara, P. and Yue, S. 2011. Mechanical property and microstructure of linear friction welded waspaloy. *Metallurgical and Materials Transactions. part A, Physical Metallurgy and Materials Science*, 42(3): 729–744. doi:10.1007/s11661-010-0457-2

Dalgaard, E., Coghe, F., Rabet, L., Jahazi, M., Wanjara, P. and Jonas, J. J. 2010. Texture evolution in linear friction welded Ti-6Al-4V. *Advanced Materials Research*, 89: 124–129. doi:10.4028/www.scientific.net/AMR.89-91.124

Dalgaard, E., Wanjara, P., Gholipour, J. and Jonas, J. J. 2011. Evolution of microstructure, microtexture and mechanical properties of linear friction welded IMI 834. *The Canadian Journal of Metallurgy and Materials Science*, 51(3): 269–276. doi:https://doi.org/10.1179/1879139512Y.0000000014

Fall, A., Jahazi, M., Khdabandeh, A.R. and Fesharaki, M.H. 2017. Effect of process parameters on microstructure and mechanical properties of friction stir-welded Ti–6al–4v joints. *International Journal of Advanced Manufacturing Technology*, 91(5–8): 2919–2931. doi:10.1007/s00170-016-9527-y

Fonda, R.W., Knipling, K.E. and Bingert, J.F. 2008. Microstructural evolution ahead of the tool in aluminum friction stir welds. *Scripta Materialia*, 58(5): 343–348. doi:10.1016/j.scriptamat.2007.09.063

Gao, X.L., Zhang, L.J., Liu, J. and Zhang, J.X. 2013. A comparative study of pulsed Nd: YAG Laser welding and tig welding of thin Ti6Al4V titanium alloy plate. *Materials Science and Engineering. Part A*, 559: 14–21. doi:10.1016/j.msea.2012.06.016

Guo, Y., Attallah, M.M., Chiu, Y., Li, H., Bray, S. and Bowen, P. 2017. Spatial variation of microtexture in linear friction welded Ti-6Al-4V. *Materials Characterization*, 127: 342–347. doi:10.1016/j.matchar.2017.03.019

Hua, K., Xue, X., Kou, H., Fan, J., Tang, B. and Li, J. 2014. Characterization of hot deformation microstructure of a near beta titanium alloy Ti-5553. *Journal of Alloys and Compounds*, 615: 531–537. doi:10.1016/j.jallcom.2014.07.056

Rajesh Jesudoss Hynes, N. and Shenbaga Velu, P. 2018. Effect of rotational speed on Ti-6Al-4V-AA 6061 friction welded joints. *Journal of Manufacturing Processes*, 32: 288–297. doi:10.1016/j.jmapro.2018.02.014

Jackson, M., Jones, N.G., Dye, D. and Dashwood, R.J. 2009. Effect of initial microstructure on plastic flow behaviour during isothermal forging of Ti-10V-2Fe-3Al. *Materials Science and Engineering. Part A*, 501(1–2): 248–254. doi:10.1016/j.msea.2008.09.071

Ji, Y., Chai, Z., Zhao, D. and Wu, S. 2014. Linear friction welding of Ti-5Al-2Sn-2Zr-4Mo-4Cr alloy with dissimilar microstructure. *Journal of Materials Processing Technology*, 214(4): 979–987. doi:10.1016/j.jmatprotec.2013.11.006

Karadge, M., Preuss, M., Lovell, C., Withers, P.J. and Bray, S. 2007. Texture development in Ti-6Al-4V linear friction welds. *Materials Science and Engineering. Part A*, 459(1–2): 182–191. doi:10.1016/j.msea.2006.12.095

Li, W.Y., Ma, T.J., Yang, S.Q., Xu, Q.Z., Zhang, Y., Li, J.L. and Liao, H.L. 2008a. Effect of Friction Time on Flash Shape and Axial Shortening of Linear Friction Welded 45 Steel. *Materials Letters*, 62(2): 293–296. doi:10.1016/j.matlet.2007.05.037

Li, W.Y., Ma, T. and Li, J. 2010. Numerical simulation of linear friction welding of titanium alloy: Effects of processing parameters. *Materials and Design*, 31(3): 1497–1507. doi:10.1016/j.matdes.2009.08.023

Li, W.Y., Ma, T., Li, Jinglong and Yang, S. 2009. Numerical simulation of linear friction welding: effects of processing parameters. In *Welding in the World*. 53(Special Issue): 443–448.

Li, W.Y., Ma, T., Zhang, Y., Xu, Q., Li, J., Yang, S. and Liao, H. 2008b. Microstructure characterization and mechanical properties of linear friction welded Ti-6Al-4V alloy. *Advanced Engineering Materials*, 10(1–2): 89–92. doi:10.1002/adem.200700034

Li, W., Vairis, A., Preuss, M. and Ma, T. 2016. Linear and rotary friction welding review. *International Materials Reviews*, 61(2): 71–100. doi:10.1080/09506608.2015.1109214

Liu, C. and Dong, C.L. 2014. Internal residual stress measurement on linear friction welding of titanium alloy plates with contour method. *Transactions of Nonferrous Metals Society of China (English Edition)*, 24(5): 1387–1392. doi:10.1016/S1003-6326(14)63203-9

Ma, T., Chen, T., Li, W.Y., Wang, S. and Yang, S. 2011. Formation mechanism of linear friction welded Ti-6Al-4V alloy joint based on microstructure observation. *Materials Characterization*, 62(1): 130–135. doi:10.1016/j.matchar.2010.11.009

McAndrew, A.R., Colegrove, P.A., Bühr, C., Flipo, B.C.D. and Vairis, A. 2018. A literature review of Ti-6Al-4V linear friction welding. *Progress in Materials Science*, 92: 225–257. doi:10.1016/j.pmatsci.2017.10.003

Mironov, S., Zhang, Y., Sato, Y.S. and Kokawa, H. 2008. Development of grain structure in β-phase field during friction stir welding of Ti-6Al-4V alloy. *ScriptaMaterialia*, 59(1): 27–30. doi:10.1016/j.scriptamat.2008.02.014

Mukundhan, C., Sivaraj, P., Petley, V., Verma, S. and Balasubramanian, V. 2020. Effect of forging pressure on microstructural characteristics and tensile properties of linear friction welded Ti-6Al-4V alloy joints. *Materials Today: Proceedings*. doi:10.1016/j.matpr.2020.03.485 (In Press)

Romero, J., Attallah, M.M., Preuss, M., Karadge, M. and Bray, S.E. 2009. Effect of the forging pressure on the microstructure and residual stress development in Ti-6Al-4V linear friction welds. *ActaMaterialia*, 57(18): 5582–5592. doi:10.1016/j.actamat.2009.07.055

Sivaraj, P., AhamedBahavudeen, J. and Balasubramanian, V. 2018. Optimizing linear friction welding parameters to attain maximum tensile strength in aluminum alloy joints. *Journal of Advanced Microscopy Research*, 13(2): 204–210. doi:10.1166/jamr.2018.1380

Sivaraj, P., Hariprasath, P., Rajarajan, C. and Balasubramanian, V. 2019. Analysis of grain refining and subsequent coarsening along on adjacent zone of friction stir welded armour grade aluminium alloy joints. *Materials Research Express*, 6(6): 066566. doi:10.1088/2053-1591/ab0e37

Sivaraj, P., Vinoth Kumar, M. and Balasubramanian, V. 2019. Microstructural characteristics and tensile properties of linear friction-welded AA7075 aluminum alloy joints. In *Lecture Notes in Mechanical Engineering*. 1: 467–476. doi:10.1007/978-981-13-1780-4_45

Su, Y., Li, W., Wang, X., Ma, T., Yang, X. and Vairis, A. 2018. On microstructure and property differences in a linear friction welded near-alpha titanium alloy joint. *Journal of Manufacturing Processes*, 36: 255–263. doi:10.1016/j.jmapro.2018.10.017

Uday, M.B., Ahmad Fauzi, M.N., Zuhailawati, H. and Ismail, A.B. 2010. Advances in friction welding process: A review. *Science and Technology of Welding and Joining*, 15(7): 534–558. doi:10.1179/136217110X12785889550064

Wang, S.Q., Ma, T.J., Li, W.Y., Wen, G.D. and Chen, D.L. 2017a. Microstructure and fatigue properties of linear friction welded TC4 titanium alloy joints. *Science and Technology of Welding and Joining*, 22(3): 177–181. doi:10.1080/13621718.2016.1212971

Wang, S. and Wu, X. 2012. Investigation on the microstructure and mechanical properties of Ti-6Al-4V alloy joints with electron beam welding. *Materials and Design*, 36: 663–670. doi:10.1016/j.matdes.2011.11.068

Wang, X.Y., Li, W.Y., Ma, T.J. and Vairis, A. 2017b. Characterisation studies of linear friction welded titanium joints. *Materials and Design*, 116: 115–126. doi:10.1016/j.matdes.2016.12.005

Wang, X., Li, W., Ma, T., Yang, X. and Vairis, A. 2019. Effect of welding parameters on the microstructure and mechanical properties of linear friction welded Ti-6.5Al-3.5Mo-1.5Zr-0.3Si Joints. *Journal of Manufacturing Processes,* 46: 100–108. doi:10.1016/j.jmapro.2019.08.031

Wanjara, P. and Jahazi, M. 2005. Linear friction welding of Ti-6Al-4V: Processing, microstructure, and mechanical-property inter-relationships. *Metallurgical and Materials Transactions. Part A, Physical Metallurgy and Materials Science*, 36(8): 2149–2164. doi:10.1007/s11661-005-0335-5

Zhang, C.C., Zhang, T.C., Ji, Y.J. and Huang, J.H. 2013. Effects of heat treatment on microstructure and microhardness of linear friction welded dissimilar Ti alloys. *Transactions of Nonferrous Metals Society of China (English Edition)*, 23(12): 3540–3544. doi:10.1016/S1003-6326(13)62898-8

Zhao, P. and Fu, L. 2020. Numerical and experimental investigation on power input during linear friction welding between TC11 and TC17 alloys. *Journal of Materials Engineering and Performance*, 29: 2061–2072.. doi:10.1007/s11665-020-04745-6

20 Linear Friction Welding of Ni-Co-Cr Superalloy for Blisk Assembly

P. Sivaraj, D. Manikandan, Vijay Petley, Shweta Verma and V. Balasubramanian

CONTENTS

20.1 INTRODUCTION

Superalloys were, and continue to be, developed for elevated temperature service. They are utilized at a higher proportion of their actual melting point than any other class of broadly commercial metallic material. They are divided into three classes, namely nickel-based superalloys, cobalt-base superalloys and iron-base superalloys. Superalloys have found applications in aircraft, marine and industrial gas turbines, as well as in rocket engines, nuclear reactors and petrochemical equipment (Ma et al. 2016).

Aircraft engines are high technology products, the manufacture of which involves exotic techniques such as investment casting and high-end precision milling. Likewise, aero-engines face the need for unceasing improvements of its technical competences in terms of achieving higher efficiencies concerning lower fuel consumption, enhanced reliability and safety, meeting the restrictive environmental regulations.

Airplane motors and mechanical gas turbines generally utilize bladed blower plates with singular airfoils tied down by nuts and bolts. An improvement of the part plate in addition to cutting edges is the blisk, in which circular discs and blade edges are manufactured as a single solid piece. The term 'blisk' is composed of the words 'blade' and 'disk'. Blisks are also called integrated bladed rotors (IBR), meaning that blade roots and blade locating slots are no longer required (Mateo 2011).

Welding has proven to be an economical way of fabricating components and repairing service-damaged turbine parts. Unfortunately, unlike other alloys, Ni-superalloys that contain a considerable amount of Ti and Al are very tough to weld because of their high vulnerability to heat-affected zone (HAZ) cracking during conventional fusion welding processes and strain age cracking during postweld heat treatment (PWHT) (Li et al. 2016). The source of this cracking, which is typically intergranular, has been attributed to the liquation of several phases in the alloy, subsequent wetting of the grain boundaries by the liquid and decohesion along with one of the solid-liquid interfaces due to on-cooling tensile stresses. This mechanism of cracking is usually referred to as the HAZ grain boundary liquation cracking. This study aims to establish process parameters for linear friction welding (LFW) of Inconel-100, a nickel superalloy, as LFW is being progressively utilized for near-net-shape assembly of materials in aviation and power generating gas turbines because it provides a better quality joint and metallurgical features.

20.1.1 Friction Welding

Friction welding technologies generate heat from the applied mechanical energy so the parts can be joined. Amalgamation of metals occurs under compressive contact between the parts involved in making the joint, in which one material oscillates with respect to other. The frictional heat at the interface sufficiently heats the material. The friction welding process does melt the material, instead the process parameters will plasticize the material at the interface of the base materials (Messler 2004).

Friction welding techniques have significant advantages:

- No external fillers are used.
- Neither fluxes nor shielding gases are essential.
- Efficient utilization of thermal energy is generated.
- Joint preparation is minimal.
- Consistent and repetitive process.
- Suitable for single piece manufacturing to mass production.
- Eco-friendly process.
- Being a solid-state process, characteristic fusion welding defects are eliminated.
- Creates narrow HAZ.

20.1.1.1 Linear Friction Welding Process

LFW is a relatively new friction welding process. It is the extension of existing applications of friction welding on nonaxisymmetric components, which could be of similar or dissimilar materials. The key modification between this process and the variants of rotary friction welding is that, while inertia friction and continuous-drive friction welding processes involve heat generation under rotational-type motion, LFW makes use of heat generated under a reciprocating linear translational motion. In LFW, one of the workpieces is held in a fixed position while the other workpiece reciprocates. The interfaces become softened by the frictional heat. The motion is subsequently stopped by aligning the moving workpiece with the stationary workpiece and a forging force is applied to fuse the workpieces. Important parameters of this process include frequency and amplitude of oscillation, burn-off (or upset or axial shortening) and friction and forging pressures (or forces) (Mateo 2011).

Phase I – Initial Phase: The two workpieces are brought into contact with each other under pressure. The surfaces begin to wear and the surface area of real contact increases. Heat is generated from solid friction as the parts rub against each other. There is no weld penetration or any noticeable axial shortening at this stage. At lower rubbing speeds under the applied compressive force, the frictional heat generated will be insufficient to overcome the conduction and radiation losses. This will eventually lead to insufficient thermal softening, which will prevent the next phase from occurring; however, if the parameters are chosen such that sufficient heat is generated at the interface, the real contact area approaches 100%, and a plasticized layer begins to develop at the interface.

Phase II – Transition Phase: As the sufficient heat softens the interface material and a sufficiently plasticized layer is being established, wear particles begin to get expelled from the interface. The real contact area is 100% at Stage II, and the HAZ expands. The soft plasticized layer formed between the two materials is no longer able to support the axial load, leading to Phase III.

Phase III – Equilibrium Phase: Axial shortening begins noticeably as a result of materials being expelled from the interface. The axial shortening varies approximately linearly with time. Materials continue to be extruded from the interface into the flash under the influence of high local stresses in the plasticized layer, which is assisted by the oscillatory movement of the workpieces. If the workpieces are aligned properly at the beginning of the process, the temperature distribution at the interface will be reasonably uniform; however, if the temperature increases excessively in one part of the interface, away from the centreline of oscillation, the plasticized layer becomes thicker in that section, causing more plastic material to be extruded. This can result in the rotation of the original plane and has been attributed to an original misalignment of the workpieces.

Phase IV – Deceleration Phase: As soon as the desired axial shortening (or upset) is achieved, the two workpieces are brought to rest swiftly, as the weld is consolidated by the applied forging/upsetting force.

Ma et al. (2016) reported the influence of process parameters on microstructure and mechanical properties of IN-718 joints. The results indicate the fine recrystallized grains in the weld zone could be obtained under optimum friction pressure, oscillation of amplitude and friction time processing conditions.

Ola et al. (2011) reported the weldability of Inconel-738 by the fusion welding process and LFW process with precipitation-hardened nickel superalloy that contains a substantial amount of Al and Ti. The fusion welding of Inconel-738 revealed its high vulnerability to HAZ cracking. Moreover, crack-free welding of the alloy was obtained by LFW.

Wang et al. (2017) investigated the microstructure and texture evolution during LFW of pure titanium joints with a scanning electron microscope, transmission electron microscope and electron back-scattered diffraction. The exclusive combination of high temperatures and strain rates during LFW causes limited continuous dynamic recrystallization in the weld centre zone (WCZ), leading to a mixed microstructure of refined grains with severe elongated grains. Under the combined effect of axial pressure and shear stress, the texture across the weld line changes significantly. The characterization results revealed the c-axis of grains of the parent metal is parallel to the rolling direction (RD) plane and has an angle of 45° to transverse direction (TD). In the TMAZ, the c-axis turns parallel to the welding interface and along with the transverse direction (TD). And in WCZ the c-axis turns to the normal direction (ND) and the P1($\{1\ 0\ \bar{1}\ 0\}$ $\langle 1\ 1\ \bar{2}\ 0\rangle$) texture forms. Compared with the ideal hexagonal cubic packing (HCP) () shear textures, the current results show that material flow during LFW of titanium indeed arises from the simple-shear deformation and is governed by prismatic slip rather than twinning. The strong texture of the LFW specimens is the reason for anisotropic mechanical properties of the joints, and the fatigue failure is characterized by fatigue striations with secondary cracks.

Wang et al. (2017) studied the microstructure and fatigue properties of LFW TC4 titanium alloy joints. From the study, they found that the strain ratio had a strong effect on the cyclic deformation characteristics of the joint, with hysteresis loops being different at different strain ratios. However, the difference of fatigue life of the joint was small with varying strain ratios. The stress amplitude of LFW TC4 joint showed essentially cyclic softening until failure at all strain ratios. Fatigue cracks initiated from the near-surface of base metal and welded dissimilar joints of titanium and stainless steel. Two different types of joints were studied: AISI 304–Ti6Al4V and AISI 316–Ti6Al4V. Particular attention was paid to characterizing the intermetallic compounds using scanning electron microscopy, electron probe microanalysis and X-ray diffractometry. Zones with different microstructures were observed. Due to the diffusive phenomena occurring during the welding, the Kirkendall effect and the occurrence of several intermetallics were observed. Moreover, it was found that the joint with AISI 316 formed brittle intermetallic compounds, which led to crack formation close to the weld line.

Ji et al. (2014) studied the microstructure evolution and temperature distribution of LFW Ti–5Al–2Sn–2Zr–4Mo–4Cr joint. From the study, they found that the typical microstructures of the weld centre were recrystallized β grains with some acicular α'' martensite. In the thermo-mechanically affected zone, the partial re-crystallized grains are formed found as severely deformed microstructures under a mass of dislocations However, dislocations were rarely found in the recrystallized β grains of the weld centre, and the calculated temperature field of the weld joints was consistent with the microstructural evolution.

Li et al. (2012) studied the effect of as-welded and postwelded heat treatment (PWHT) on mechanical properties and microstructural characteristics of titanium alloy Ti-6Al-4V. They found that tensile strength of the joint was improved to about 71% of the parent Ti-6Al-4V through PWHT at 950°C for 1 hour, and the failure of specimens still took place across the bond line with their presentative cleavage fracture due to the formation of basket-weave structure at the weld centre. Postwelded condition led to the deformation of weld zone and became insufficient with the reduction of friction and forging pressures, resulting in a much thicker weld and presence of spherical grains unexpectedly formed near the bond line, causing the sudden drop in the joint tensile strength, which is about 44% of the parent Ti-64. But in as-welded condition, results demonstrate the presence of fine microstructure in the weld centre zone was formed, which was attributed to higher tensile strength than the parent-6Al-4V. All the tensile testing specimens exhibited failure in the base metal with a typical ductile rupture.

Micari et al. (2014) studied the tool material selection in welding of Ti–6Al–4V sheets of 100mm x 200mm and 3mm by friction welding process In particular, rotating speeds of 300, 700 and 1000 rpm were selected, and the tool was made of tungsten and rhenium alloy. Fixed advancing speed equal to 35 mm/min, nuting angle equal to 2° and tool shoulder sinking of 0.2 mm were considered for all the welds. The tool sinking speed was kept constant and equal to 0.6 mm/min for all the welds. All the tools failures were observed between the end of the sinking stage and the first few mm of welding. The best results, both interims of tool life and weld quality, were obtained with the W25Re tool. Fulfiling instrument life was acquired and no obvious indication of debasement was seen during the existence cycle. Both the tool materials resulted for greater defect-free lengths of the weld due to the proper stirring action and process parameters of the process.

Sorina-Muller et al. (2010) studied the FEM simulation of the linear friction welding of titanium alloys Ti-6Al-2Sn-4Cr-6Mo in two conditions (ß and α+ß forged) was simulated using the finite element method. A full structural–thermal couple transient 3D-analysis was conducted in two different set of combinations to investigate on prismatic and blade-like structures. In the present calculations, temperature differences at the contact interface of the specimens with blade-like geometry up to 500 K were observed, but in general, the temperature differences (from the edges to the centre of the body) reached approx. 740–800 K. This is attributed to the very short time of overlap and to the losses via convection. At last, the results conclude that simulation is a very useful qualitative tool to define and optimize the processing parameters and other factors that influence the temperature distribution. It is expected that modelling will considerably decrease the number of experimental trials necessary to describe the processing parameters, thus saving considerable time and funds.

Vairis and Frost (1998) studied about extrusion phase of LFW of Ti-6Al-4V and found that frictional heat input at the interface depends on the amplitude of oscillation. As there was no movement of the specimens in that direction, it may be concluded that axial shortening proceeds in a stepwise fashion. The effect of this pumping action is that the material yields and extrudes impulses, therefore confining the HAZ close to the interface. Additionally, to the four phases mentioned previously, a fifth phase – the standing phase – can be introduced. According to the calculations, the temperatures in the contact zones after the complete stop of the oscillation are still about 800–950 K. Recrystallization and creep can still proceed. During cooling, the material in the HAZ shrinks. The forging (compressive load) pressure is held for an adequate amount of time (5–9 s) to fuse the materials to form a weld joint.

Chamanfar et al. (2011) reported the linear friction of waspaloy and the thermal effects on the process. Temperature across the experiments was recorded by thermocouples in different weld interfaces. The interface temperature was recorded as 1280°, which is close to the melting point of the bulk alloy. Different microscopy images reveal the LFW waspaloy joints were defect-free. Moreover, the ϒ' liquation, a usual problem reported by researchers, was not found. The feasibility of LFW of waspaloy revealed considerable results.

Milligan et al. (2004) investigated the microstructure on high-temperature behaviour of LFW Inconel-100. The results revealed fine γ matrix grains in size of 3–5 μm. The fine microstructures were due to the heat developed at the interface of the material, causing the grains to undergo dynamic recrystallization at under high pressures.

Damodaram et al. (2013) reported the properties on preweld and postweld treatment on the microstructures and mechanical properties of IN 718 joints. The preweld treatments were carried out at solutionizing condition (ST) and solution treatment followed by ageing (STA). In the as-welded condition, samples welded with prior ST condition exhibited higher hardness in the weld zone compared to the base material in ST condition. This was attributed to grain refinement due to dynamic recrystallization. They concluded that they expected the dissolution of ϒ' (Ni3Nb) phase at the weld zone, as the temperature reached in the weld zone was found to be 1118°C. The direct ageing treatment after friction welding improved microhardness and ultimate tensile strength of friction welds. This was attributed to the combined effect of grain refinement due to dynamic recrystallization during friction welding and formation of precipitates during the direct ageing. Literature supports that LSW has been successfully tested in both ferrous, nonferrous and superalloys. Hence, this study focusses on the use of linear friction welding in nickel superalloy, Inconel-100.

20.2 EXPERIMENTAL CONDITIONS

For this study, Inconel-100 with dimensions on the fixed end included thickness at 6 mm, width at 50 mm and length at 60 mm. Dimensions on the stationary end included thickness at 4.7 mm, width at 30 mm and length at 50 mm as shown in Figure 20.1. The chemical composition, mechanical properties and microstructures of the parent material is presented in Table 20.1, Table 20.2 and Figure 20.2

The welding parameters used for fabricating the weld joints are presented in Table 20.3.

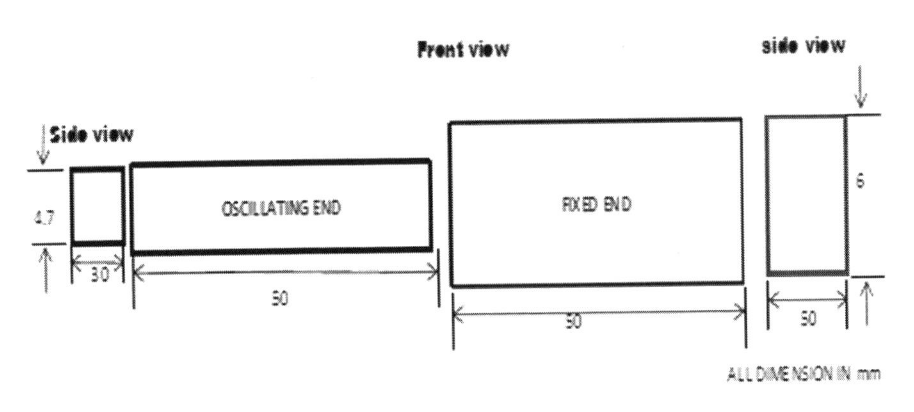

FIGURE 20.1 Schematic diagram of specimen preparation for linear friction welding.

TABLE 20.1
Chemical Composition (%wt) of Inconel-100

Elements	Ni	Co	Cr	AL	Ti	MO	V	Zr	B
Wt%	60	15	10	5.5	4.7	3	1	0.06	0.014

TABLE 20.2
Mechanical Properties of Parent Material

Sl.No.	0.2% Yield Strength (MPa)	Ultimate Tensile Strength (Mpa)	Elongation in 32 mm Gauge Length (%)	Hardness (Hv)
IN-100 Smooth	610	688	8.50	269
Notch	–	982	–	269

The fabricated joints made by using the optimized parameters are shown in Figure 20.3.

The welded joints and their specimen extraction for tensile testing according to ASTM standards (ASTM E08-13) are presented in Figure 20.4.

The transverse test was carried out with using a 500 kN electromechanical hydraulic-controlled universal testing machine (Make: MTS). The 0.2% offset yield strength was derived from the stress–strain curve from which other factors including strain rate, ultimate tensile strength and percentage of elongation were calculated. The strain rate is measured using an 8 mm extensometer (Make: MTS). The microhardness survey was taken along the welding regions of the Inconel joints (weld zone, thermomechanical affected zone [TMAZ], HAZ, BM) by the Vickers

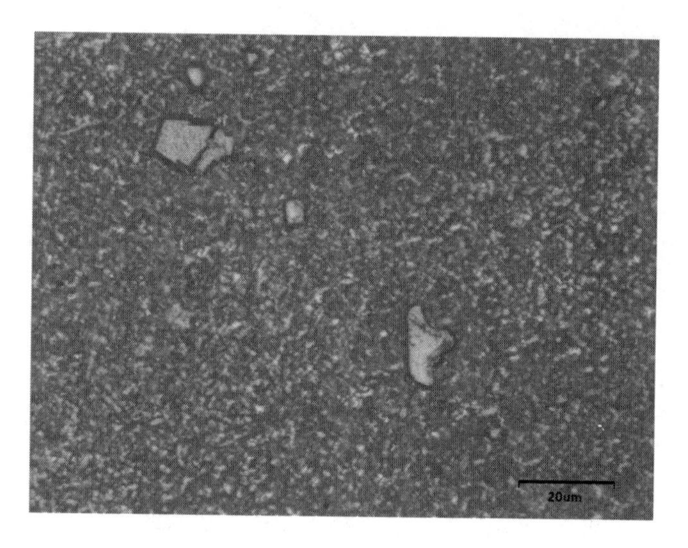

FIGURE 20.2 Base material microstructure.

TABLE 20.3
Linear Friction Welding Parameters

Parameters	
Friction Pressure	21.78 MPa
Friction Time	45 sec
Forging Pressure	8.16 MPa
Forging Time	03 sec
Frequency	20 Hz

microhardness testing machine through diamond indenter with variables of 0.3 MPa (500 grams) and dwell time of about 10 sec. The joints after tensile testing are presented in Figure 20.5.

The metallurgical features of LFW Inconel-100 were investigated through stereomicroscopy and optical microscopy (OM). The metallographic specimen of the weld joint was obtained through wire electrical discharge machining (WEDM) with a double side formed flash. The sliced specimen was fixed onto the holder by conductive strips The hot mount cycle time has a heating time of about 6 mins, and soaking and cooling time were 5 mins. The mounted specimen was polished with silicon abrasive paper of various grades (1000μ–2500μ), followed by disc polishing with alumina power (0.1μ) for superior polishing for the joint (Wusatowska-Sarnek et al. 2003). The polished coupon was etched using a suitable reagent to reveal its microstructural features under light and electron microscopes.

FIGURE 20.3 Linear friction welded joints.

FIGURE 20.4 Notched and unnotched tensile testing specimens.

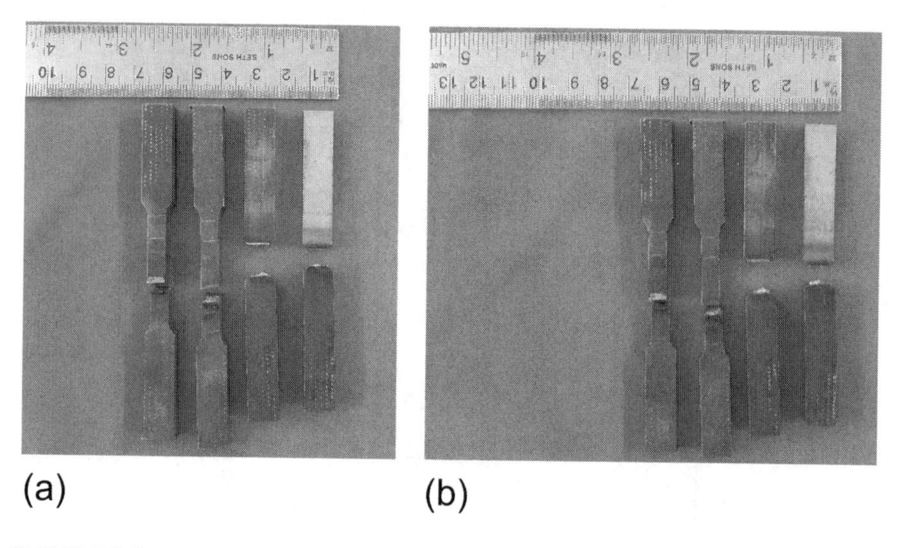

FIGURE 20.5 Fractured joints (a) smooth tensile specimen and (b) notch tensile specimen.

TABLE 20.4
Tensile Test Results of Linear Friction Welded Joints

Weld Joint	0.2% Yield Strength (MPa)	Ultimate Tensile Strength (MPa)	Elongation in 25 mm Gauge Length (%)	Strain Rate (%)	Joint Efficiency (%)	Failure Location
In-100 Smooth	385	465	12	88	67	TMAZ
Notch	–	827	–	–	84	Weld

20.3 RESULTS AND DISCUSSION

20.3.1 Transverse Tensile Properties

The tensile specimens were prepared according to ASTM E08-13 standards to test the mechanical properties, such as 0.2% yield strength, ultimate tensile strength (UTS), strain rate and the percentage of elongation, which are presented in Table 20.4.

The average values were obtained as the mean value of three tests. The unwelded parent material recorded an ultimate tensile strength in notched and unnotched conditions as 688 and 982 MPa, respectively. The LFW joints recorded an ultimate tensile strength in notched and unnotched conditions as 465 and 827 MPa, respectively, which is a decrease of 32.41% and 15.4% of ultimate tensile strengths in smooth and notched conditions respectively. On the other hand, the percentage of elongation for the friction weld was found to be a 4% higher. Moreover, the weld failure occurred at the nearer regions of the nugget zone, which also describes the fine grains of the weld centre zone and coarser grains structures in the TMAZ. This accounts for the deformation occurring at the oscillating side, where higher heat is generated than the stationary side. Similarly, for notched tensile specimens failed in the welded region because of artificial strain induced in the weld region by notches. The joints in notched and unnotched conditions exhibited a joint efficiency of 84% and 67%, respectively (Damodaram et al. 2013).

From the load vs. displacement of Figure 20.6, it can be seen that the joint showed displacement up to 2.65 mm for an applied load of around 13 kN, after which the joints start to dissipate the energy and start to fail.

20.3.2 Microhardness Measurement

The hardness of the welded joint was evaluated across the weld regions using the Vickers microhardness testing machine. From the Figure 20.7 of microhardness survey, it can be seen that the stationary side of the weld records higher hardness than the oscillating side. The maximum hardness recorded in the weld region was 420 Hv The interface nugget region recorded the low hardness value of about 378 Hv, which was a softened region near the weld, as the heat conduction stops at the region. Further, the microstructures of the specified region will be coarser when compared to the base metal. Further, this also accounts for the dissolution of γ^{I} and γ^{II} precipitates in the interface zone. Even though high hardness was recorded in the stationary side, failure occurs nearer the weld nugget region. This is because of the brittle intermetallic formed in the TMAZ region (Lee and Jung 2004).

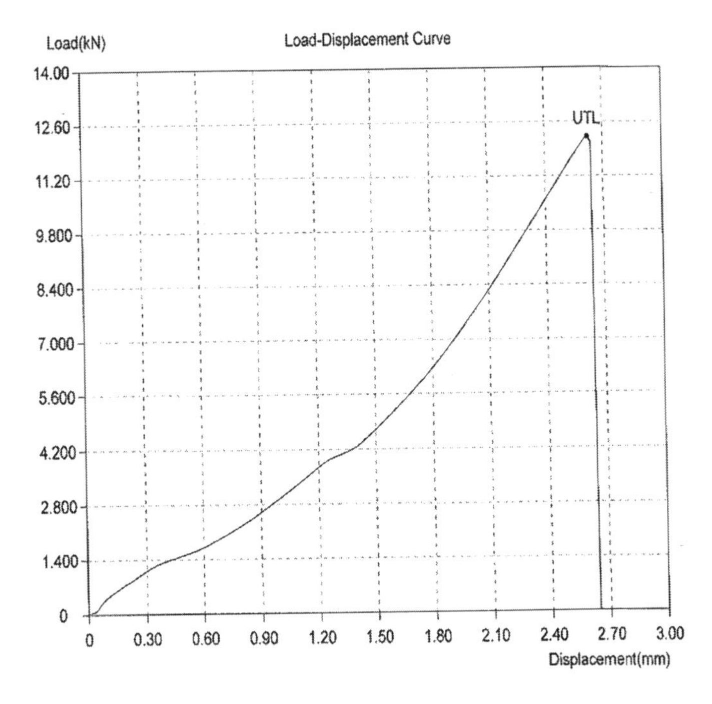

FIGURE 20.6 Load vs. displacement.

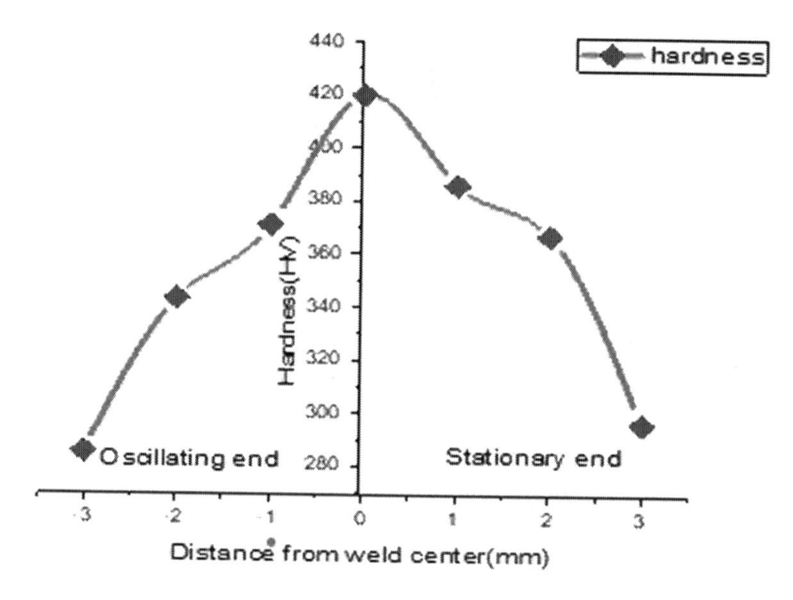

FIGURE 20.7 Microhardness distribution.

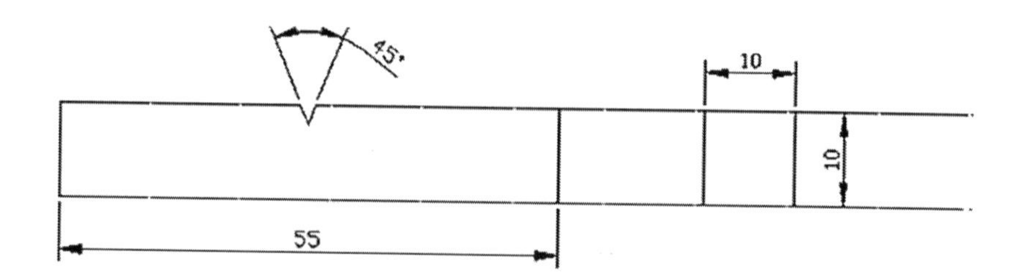

FIGURE 20.8 Impact specimen extraction scheme.

20.3.3 Impact Testing

The impact testing of the welded specimens was performed according to AWS SFA A5.17 standards as shown in Figure 20.8. The specimen extraction details are represented in Table 20.5. The impact tested specimens are presented in Figure 20.9. The impact testing of welded specimens showed an absorption of 14 energy joules, and the failure also occurred near the weld zone. The impact energy of the tested joint and test temperature is presented in Table 20.6.

20.3.4 Macrostructural and Microstructural Features of Inconel-100 Weld Joint

The macrograph of LFW Inconel-100 joints from Figure 20.10 reveal double-edge flash formation, proving the joint quality and integrity, which is characteristic of the process. Moreover, the macrograph clearly shows that more expulsion of

TABLE 20.5
Impact Specimen Extracted Dimensions

Length	55 mm
Width	10 mm
Thickness	10 mm
Angle	45°
Depth	2 mm

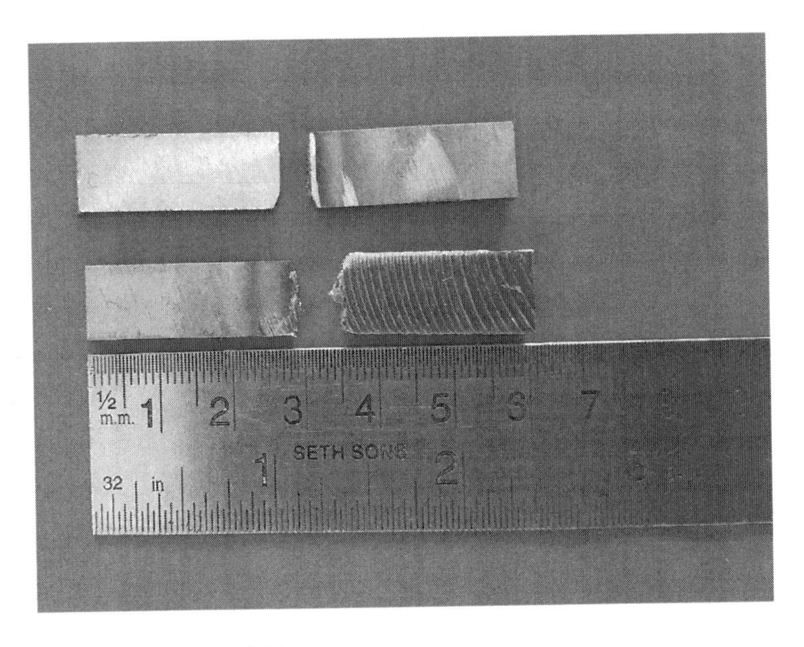

FIGURE 20.9 Impact tested joints.

material from the interface occurred at the fixed end, whereas the oscillating end had double-edge flash formation at the interface. This further shows the material can withstand high-temperature properties with good mechanical and metallographically sound structures. The flash was then removed after the macroexamination of the joints, and the surface was prepared for light and electron microscopy evaluation.

The optical micrographs from Figure 20.11 show distinct areas can be differentiated from one another, such as unaffected base material, weld zone, plastically deformed zones on oscillatory and stationary side, HAZ on oscillating and stationary sides (Sivaraj et al. 2018; Sivaraj et al. 2019). All zones can be characterized and differentiated by different grain sizes and several inclusions, which will be determined by chemical composition analysis X-ray diffraction.

TABLE 20.6
Impact Results of the Welded Joints

Test Position	Test Temperature	Energy Absorbed in Joules
Weld Zone	24°C	14

FIGURE 20.10 Macrostructure of linear friction welded Inconel-100.

From Figure 20.12, the distinct zones of microstructures in various zones viz. base material, weld centre zone, TMAZ and HAZ interface and TMAZ are clearly visible. Further, the SEM micrographs also reveal the different zones by different grain sizes, and the interfaces along the weld region and TMAZ-HAZ interfaces are distinctly visible from the micrographs under electron microscopes. The weld zone, which formed at the interface between the two LFW workpieces, extended to about 20 μm into the material on either side of the joints. The microstructure of these regions can be characterized by γ' precipitates and inclusions, such as γ-γ' eutectic constituents. Another microstructural characteristic was the occurrence of recrystallization within the weld zone of the LFW IN-100. A recrystallized microstructure along the weld line has been obtained in different Ni-based superalloys when welded by frictional processes (Ma et al. 2016; Li et al. 2016; Ola et al. 2011). It can be correlated with reports that the compressive loads applied during the forging phase of the process induce the formation of TMAZ where the heat conducts from the interface of the materials to a certain distance, which is the area where the dynamic recrystallization was observed. Moreover, this area exhibits a high level of strain at elevated temperatures with high strain rates. The extreme volume of plastic deformation happened at the weld line, which resulted in the formation of fine, fully

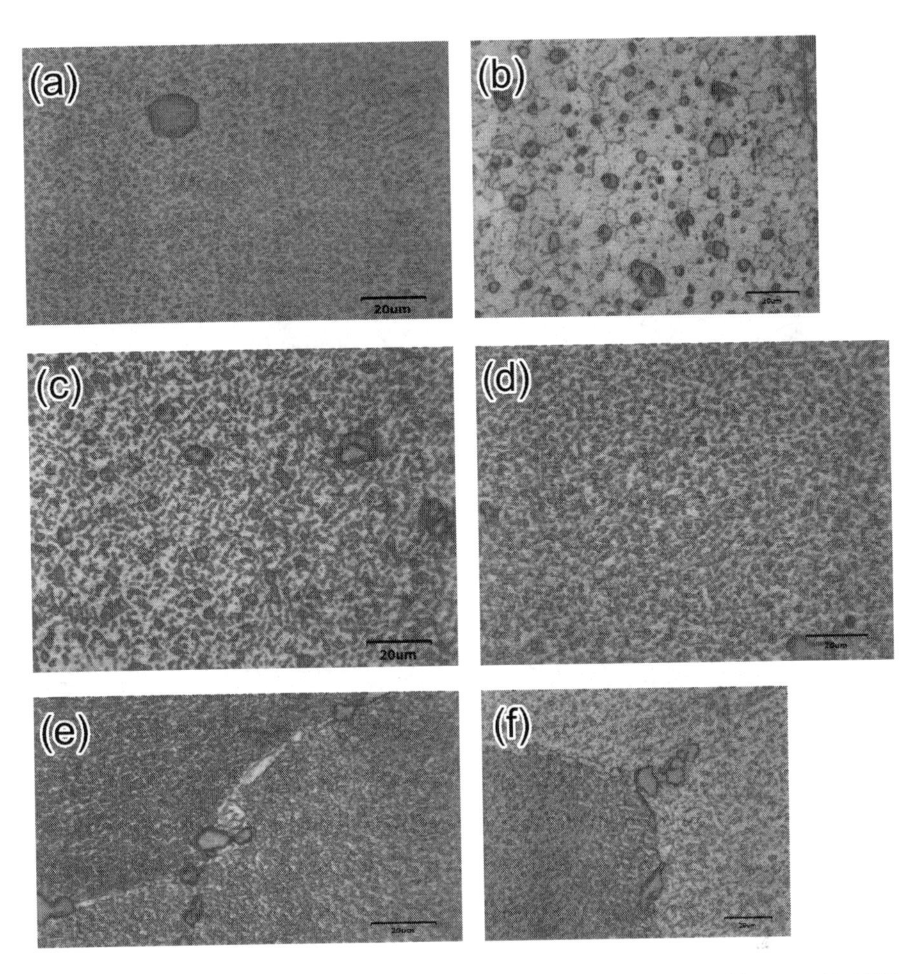

FIGURE 20.11 Optical micrographs of (a) unaffected zone, (b) weld zone, (c) PDZ-oscillating side, (d) PDZ-stationary side, (e) HAZ-oscillating side and (f) HAZ-stationary side.

recrystallized grains. Traversing away from the weld line, the microstructural evolution increasingly exhibited regions of deformed or partially recrystallized grains, such that recrystallization was localized along prior grain boundaries. On the whole, the observed changes in the microstructural characteristics (precipitates, carbides and grain structure) are related to the steep gradients in the temperature and deformation conditions within the weld zone.

The TMAZ, Region 2, exhibited a complete dissolution of the γ' precipitates and γ-γ' eutectic constituents, while the MC carbides remained (Sauer and Lütjering 2001). Hence, the coarse and secondary γ' precipitates that were present in the IN-100 dissolved completely in both the weld zone and TMAZ, and the main microstructural difference is related to the recrystallization characteristics as mentioned above. A higher magnification SEM study showed the occurrence of much finer γ'

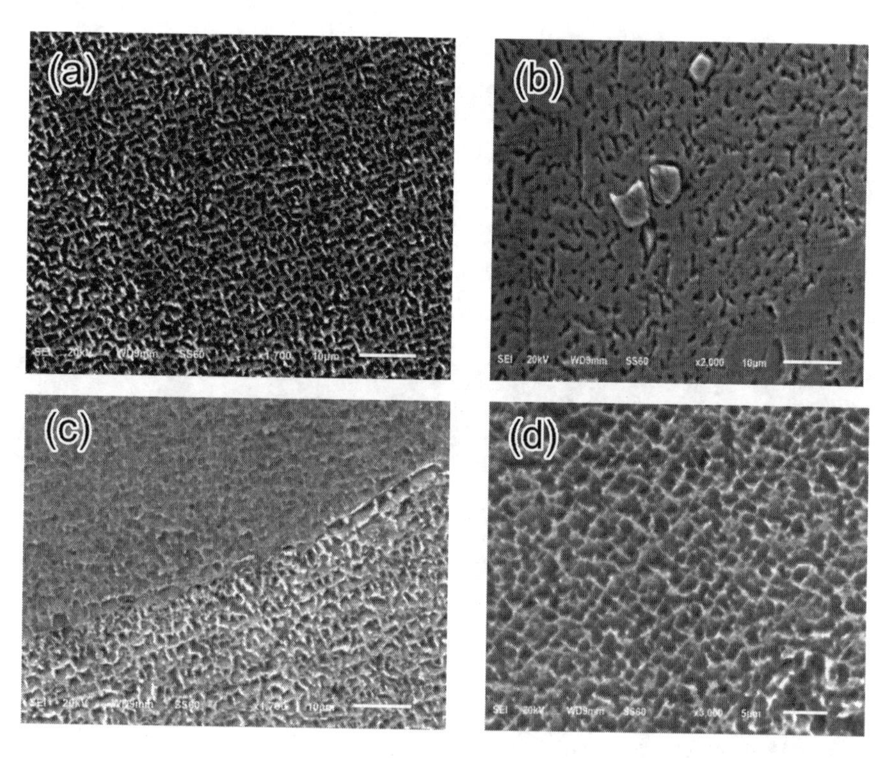

FIGURE 20.12 SEM microstructures of (a) base material, (b) weld zone, (c) interface between TMAZ and HAZ and (d) TMAZ.

precipitates in both the weld zone and TMAZ. The occurrence of very fine γ' particles in the weld zone and TMAZ can be related to the fast cooling rate after LFW. This rapid reprecipitation of very fine γ' particles during cooling of the LFW IN-100 contributed significantly to an increase in the microhardness of the material close to the weld line. Overall, the measurements show in increase in microhardness measurements from that of the base material at distances of up to about 1 mm from either side of the weld line. The region within the weld line (weld zone and TMAZ), where reprecipitated very fine γ' particles were observed by SEM examination, correlates with the region where the highest hardness values were recorded. The hardness then decreased gradually as the distance from this region into the base material increased. In the HAZ, Region 3, secondary γ' precipitates that were produced by the pre-weld heat treatment dissolved completely, while the primary γ' precipitates dissolved partially. A significant microstructural observation that is generally neither expected nor has been reported in LFW materials is grain boundary liquation.

20.3.5 Elemental Analysis (EDS) of Welded Joints

The percentage of elements in the weld zone is identified by the elemental analysis of the SEM. The Energy Dispersive Spectroscopy (EDS) spectrum of the joint

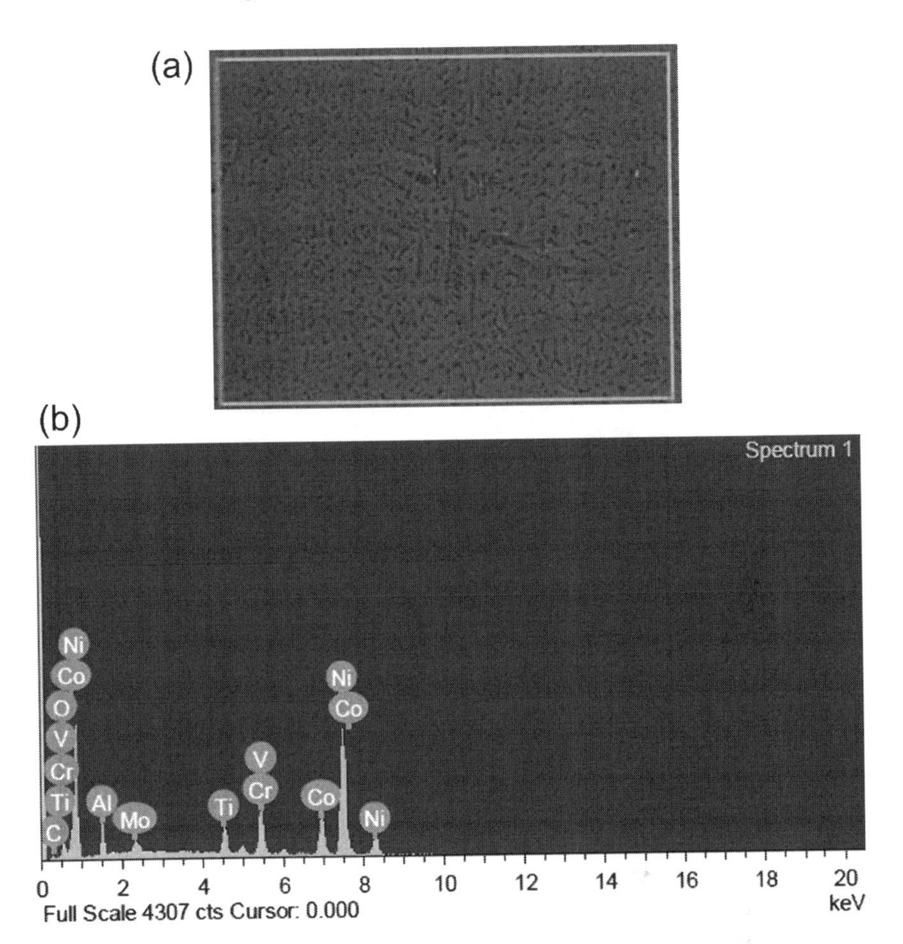

FIGURE 20.13 (a) EDS spectrum of the welded joint and (b) composition of elements from EDS.

analyzed for weld metal percentage composition is presented in Figure 20.13a, and the composition of elements in the weld is presented in Figure 20.13b. The compositional elements were analyzed along the mid-section of the weldments. From the EDS analysis, the majority of the elements in the weldments are identified as Ni, Co, Cr, AL and Ti, which are the major composition of the alloy as predicted earlier.

20.4 CONCLUSION

From the microstructural features of the weld joints, the following conclusions are drawn:

1. Weld interface appears as a dark line at the extremities of the joints, which approaching the central region, cannot be observed at the centre. The dark line appearing on the weld contains residues of carbides and oxides

2. Superfine grains are dispersed in the weld zone of LFW Inconel-100 joints. Deformed grains cannot be found in the TMAZ. The microstructures of the joints suggest that the welding parameters had significant effect on the properties and integrity of the joints. Moreover, the weld parameters, such as friction pressure and forging pressure, would determine the percentage of recrystallized grains in the weld centre zone. The extent of this dynamic recrystallization determines the mechanical properties of the weld joints.
3. Because of the formation of carbides and oxides, segregation on grain boundaries and reversion of γ'' and γ', mechanical properties of Inconel-100 joints are lower than those of the parent metal. The microhardness of joints all show an increasing from the parent metal to the weld zone. And the presence of a fine grain zone around ±1.5 mm from the weld line was attributed to the fluctuation of microhardness in the recrystallized zone. The average tensile strength of the joint achieves only 67% of the parent metal, respectively and tensile specimens all failed in the TMAZ or the weld zone.

REFERENCES

ANTEC and Society of Plastics Engineers, eds. 2003. *Conference Proceedings, Annual Technical Conference/ANTEC* 2003, May 4–8, 2003, *Nashville, Tennessee.* Technical Papers/Society of Plastics Engineers, vol. 61. Brookfield, CT: Society of Plastics Engineers.

Antonio, M. Mateo García (November 4th 2011). BLISK Fabrication by Linear Friction Welding, Advances in Gas Turbine Technology, Ernesto Benini, IntechOpen, DOI: 10.5772/21278. Available from: https://www.intechopen.com/books/advances-in-gas-turbine-technology/blisk-fabrication-by-linear-friction-welding.

Bhadeshia, H.K.D.H. and DebRoy, T. 2009. critical assessment: friction stir welding of steels. *Science and Technology of Welding and Joining*, 14 (3): 193–196.

Bhamji, I., Preuss, M., Threadgill, P.L. and Addison, A.C. 2011. Solid state joining of metals by linear friction welding: A literature review. *Materials Science and Technology*, 27 (1): 2–12. doi:10.1179/026708310X520510

Bhamji, I., Preuss, M., Threadgill, P.L., Moat, R.J., Addison, A.C. and Peel, M.J. 2010. Linear friction welding of AISI 316L stainless steel. *Materials Science and Engineering: A*, 528 (2): 680–690.

Bußmann, M., Kraus, J. and Bayer, E. 2005. An integrated cost-effective approach to blisk manufacturing. In *17th Symp. on Airbreathing Engines.*

Ceschini, L., Morri, A., Rotundo, F., Jun, T.S. and Korsunsky, A.M. 2010. A study on similar and dissimilar linear friction welds of 2024 Al alloy and 2124Al/SiCP composite. *Advances in Materials Research*, 89: 461–466. Trans Tech Publ.

Chamanfar, A., Jahazi, M., Gholipour, J., Wanjara, P. and Yue, S. 2011. Mechanical property and microstructure of linear friction welded waspaloy. *Metallurgical and Materials Transactions A*, 42 (3): 729–744. doi:10.1007/s11661-010-0457-2

Corzo, M., Mendez, J., Villechaise, P., Rebours, C., Ferte, J.P., Gach, E., Roder, O., Llanes, L., Anglada, M. and Mateo, A. 2006. High-cycle fatigue performance of dissimilar linear friction welds of titanium alloys. In *Proceedings of the 9th Integr. Fatigue Congress.*

Corzo, V., Casals, O., Alcala, J., Mateo, A. and Anglada, M.. 2007. Mechanical evaluation of linear friction welds in titanium alloys through indentation experiments. *Welding International*, 21 (2): 125–129.

Damodaram, R., Sundara Raman, S.G. and Rao, K.P. 2013. Microstructure and mechanical properties of friction welded alloy 718. *Materials Science and Engineering: A*, 560: 781–786.

Donachie, M.J. 2000. *Titanium: A Technical Guide*. Materials Park, OH: ASM international.

Harvey, R.J., Strangwood, M. and Ellis, M.B.D. 1995. Bond line structures in friction welded Al_2O_3 particulate reinforced aluminium alloy metal matrix composites. In *Proceedings 4th Int. Conference on Trends in Welding Research*, 803–808.

Ji, Y., Chai, ., Zhao, D. and Wu, S. 2014. Linear friction welding of Ti–5Al–2Sn–2Zr–4Mo–4Cr alloy with dissimilar microstructure. *Journal of Materials Processing Technology*, 214 (4): 979–987. doi:10.1016/j.jmatprotec.2013.11.006

Karadge, M., Preuss, M., Withers, P.J. and Bray, S. 2008. Importance of crystal orientation in linear friction joining of single crystal to polycrystalline nickel-based superalloys. *Materials Science and Engineering: A*, 491 (1–2): 446–453.

Kauzlarich, J.J. and Maurya Ramamurat, R. 1969. Reciprocating friction bonding apparatus. *USA, US3420428*.

Lee, W.-B. and Jung, S.-B. 2004. The joint properties of copper by friction stir welding. *Materials Letters*, 58 (6): 1041–1046. doi:10.1016/j.matlet.2003.08.014

Li, W., Vairis, A., Preuss, M. and Ma, T. 2016. Linear and rotary friction welding review. *International Materials Reviews*, 61 (2): 71–100.

Li, W., Wu, H., Ma, T., Yang, C. and Chen, Z. 2012. Influence of parent metal microstructure and post-weld heat treatment on microstructure and mechanical properties of linear friction welded Ti-6Al-4V joint. *Advanced Engineering Materials*, 14 (5): 312–318. doi:10.1002/adem.201100203

Ma, T.J., Chen, X., Li, W.Y., Yang, X.W., Zhang, Y. and Yang, S.Q. 2016. Microstructure and mechanical property of linear friction welded nickel-based superalloy joint. *Materials and Design*, 89: 85–93. doi:10.1016/j.matdes.2015.09.143

Magudeeswaran, G., Rajapandiyan, P., Balasubramanian, V. and Sivaraj, P. 2019. Establishing the process parameters of linear friction welding process for dissimilar joints. *International Journal of Mechanical Engineering and Technology*, 10 (6): 1–10.

Mary, C. and Jahazi, M.. 2007. Linear friction welding of IN-718 process optimization and microstructure evolution. *Advances in Materials Research*, 15: 357–362. Trans Tech Publ.

Mateo, A. 2011. BLISK fabrication by linear friction welding. doi:10.5772/21278

Messler, R.W. 2004. *Joining of Materials and Structures: From Pragmatic Process to Enabling Technology*. Butterworth-Heinemann, Elseiver Publications.

Micari, F. , Buffa, G., Pellegrino, S. and Fratini, L. 2014. Friction stir welding as an effective alternative technique for light structural alloys mixed joints. *Procedia Engineering*, 81 (2014)74–83. doi:10.1016/j.proeng.2014.09.130

Milligan, W.W., Orth, E.L., Schirra, J.J. and Savage, M.F. 2004. Effects of microstructure on the high temperature constitutive behavior of IN100. *Superalloys*, 2004, 331–340.

Mishra, R.S. and Mahoney, M.W. 2007. *Friction Stir Welding and Processing*. Materials Park, OH: ASM International.

Nunn, M. E. 2005. Aero Engine Improvements through linear friction welding. In *Proceedings of 1st International Conference on Innovation and Integration in Aerospace Sciences*, Belfast, UK.

Ola, O.T., Ojo, O.A., Wanjara, P. and Chaturvedi, M.C. 2011. Enhanced resistance to weld cracking by strain-induced rapid solidification during linear friction welding. *Philosophical Magazine Letters*, 91 (2): 140–149.

Ola, O.T., Ojo, O.A., Wanjara, P. and Chaturvedi, M.C. 2011. Crack-free welding of IN 738 by linear friction welding. *Advanced Materials Research*, 278: 446–453. Trans Tech Publ.

Roder, O., Ferte, J., Gach, E., Mendez, J. and Anglada, M. 2008. Development and validation of a dual titanium alloy dual-microstructure blisk. *VDIBERICHT*, 2028: 309.

Sauer, C. and Lütjering, G. 2001. Influence of α layers at β grain boundaries on mechanical properties of Ti-alloys. *Materials Science and Engineering: A*, 319: 393–397.

Sivaraj, P., Ahamed Bahavudeen, J. and Balasubramanian, V. 2018. Optimizing linear friction welding parameters to attain maximum tensile strength in aluminum alloy joints. *Journal of Advanced Microscopy Research*, 13 (2): 204–210. doi:10.1166/jamr.2018.1380

Sivaraj, P., Vinoth Kumar, M. and Balasubramanian, V. 2019. Microstructural characteristics and tensile properties of linear friction-welded AA7075 aluminum alloy joints. In *Advances in Materials and Metallurgy*, edited by A. K. Lakshminarayanan, S. Idapalapati and M. Vasudevan, pp. 467–476. Singapore: Springer. doi:10.1007/978-981-13-1780-4_45

Sorina-Müller, J., Rettenmayr, M., Schneefeld, D., Order, O. and Fried, W. 2010. FEM simulation of the linear friction welding of titanium alloys. *Computational Materials Science*, 48 (4): 749–758. doi:10.1016/j.commatsci.2010.03.026

Threadgill, P.L., Leonard, A.J., Shercliff, H.R. and Withers, P.J. 2009. Friction stir welding of aluminium alloys. *International Materials Reviews*, 54 (2): 49–93. doi: 10.1179/174328009X411136

Vairis, A. and Frost, M. 1998. High frequency linear friction welding of a titanium alloy. *Wear*, 217 (1): 117–131. doi:10.1016/S0043-1648(98)00145-8

Wang, S.Q., Ma, T.J., Li, W.Y., Wen, G.D. and Chen, D.L. 2017. Microstructure and fatigue properties of linear friction welded TC4 titanium alloy joints. *Science and Technology of Welding and Joining*, 22 (3): 177–181. doi:10.1080/13621718.2016.1212971

Wang, X.Y., Li, W.Y., Ma, T.J. and Vairis, A. 2017. Characterisation studies of linear friction welded titanium joints. *Materials and Design*, 116: 115–126. doi:10.1016/j.matdes.2016.12.005

Wilhelm, H., Furlan, R. and Moloney, K. 1995. Linear friction bonding of titanium alloys for aero-engines applications, RR-PNR--92225, http://hdl.handle.net/10068/644776, UK.

Wusatowska-Sarnek, A.M., Blackburn, M.J. and Aindow, M. 2003. Techniques for microstructural characterization of powder-processed nickel-based superalloys. *Materials Science and Engineering: A*, 360 (1–2): 390–395.

Index